Springer Series in Statistics

Advisors:
P. Bickel, P. Diggle, S. Fienberg, K. Krickeberg,
I. Olkin, N. Wermuth, S. Zeger

Springer
New York
Berlin
Heidelberg
Hong Kong
London
Milan
Paris
Tokyo

Springer Series in Statistics

Andersen/Borgan/Gill/Keiding: Statistical Models Based on Counting Processes.
Atkinson/Riani: Robust Diagnostic Regression Analysis.
Atkinson/Riani/Cerioli: Exploring Multivariate Data with the Forward Search.
Berger: Statistical Decision Theory and Bayesian Analysis, 2nd edition.
Borg/Groenen: Modern Multidimensional Scaling: Theory and Applications.
Brockwell/Davis: Time Series: Theory and Methods, 2nd edition.
Bucklew: Introduction to Rare Event Simulation.
Chan/Tong: Chaos: A Statistical Perspective.
Chen/Shao/Ibrahim: Monte Carlo Methods in Bayesian Computation.
Coles: An Introduction to Statistical Modeling of Extreme Values.
David/Edwards: Annotated Readings in the History of Statistics.
Devroye/Lugosi: Combinatorial Methods in Density Estimation.
Efromovich: Nonparametric Curve Estimation: Methods, Theory, and Applications.
Eggermont/LaRiccia: Maximum Penalized Likelihood Estimation, Volume I: Density Estimation.
Fahrmeir/Tutz: Multivariate Statistical Modelling Based on Generalized Linear Models, 2nd edition.
Fan/Yao: Nonlinear Time Series: Nonparametric and Parametric Methods.
Farebrother: Fitting Linear Relationships: A History of the Calculus of Observations 1750-1900.
Federer: Statistical Design and Analysis for Intercropping Experiments, Volume I: Two Crops.
Federer: Statistical Design and Analysis for Intercropping Experiments, Volume II: Three or More Crops.
Ghosh/Ramamoorthi: Bayesian Nonparametrics.
Glaz/Naus/Wallenstein: Scan Statistics.
Good: Permutation Tests: A Practical Guide to Resampling Methods for Testing Hypotheses, 2nd edition.
Gouriéroux: ARCH Models and Financial Applications.
Gu: Smoothing Spline ANOVA Models.
Györfi/Kohler/Krzyżak/ Walk: A Distribution-Free Theory of Nonparametric Regression.
Haberman: Advanced Statistics, Volume I: Description of Populations.
Hall: The Bootstrap and Edgeworth Expansion.
Härdle: Smoothing Techniques: With Implementation in S.
Harrell: Regression Modeling Strategies: With Applications to Linear Models, Logistic Regression, and Survival Analysis.
Hart: Nonparametric Smoothing and Lack-of-Fit Tests.
Hastie/Tibshirani/Friedman: The Elements of Statistical Learning: Data Mining, Inference, and Prediction.
Hedayat/Sloane/Stufken: Orthogonal Arrays: Theory and Applications.
Heyde: Quasi-Likelihood and its Application: A General Approach to Optimal Parameter Estimation.

(continued after index)

James Antonio Bucklew

Introduction to Rare Event Simulation

With 50 Illustrations

Springer

James Antonio Bucklew
Department of Electrical and
Computer Engineering
University of Wisconsin-Madison
Madison, WI 53706
USA
bucklew@orisha.ece.wisc.edu

Library of Congress Cataloging-in-Publication Data
Bucklew, James A.
Introduction to rare event simulation
James Antonio Bucklew.
p. cm. — (Springer series in statistics)
Includes bibliographical references and index.
ISBN 0-387-20078-9 (alk. paper)
1. Large deviations. I. Title: Introduction to rare event simulation. II Series.
QA273.67.B84 2003
519.5′34—dc22 2003062208

ISBN 0-387-20078-9 Printed on acid-free paper.

Printed in the United States of America. (EB)

9 8 7 6 5 4 3 2 1 SPIN 10951101

Springer-Verlag is a part of *Springer Science+Business Media*

springeronline.com

To my best friend, Victor.

Preface

This book is an attempt to present a unified theory of rare event simulation and the variance reduction technique known as importance sampling from the point of view of the probabilistic theory of large deviations. This framework allows us to view a vast assortment of simulation problems from a single unified perspective. It gives a great deal of insight into the fundamental nature of rare event simulation.

Unfortunately, this area has a reputation among simulation practitioners of requiring a great deal of technical and probabilistic expertise. In this text, I have tried to keep the mathematical preliminaries to a minimum; the only prerequisite is a single large deviation theorem dealing with sequences of $\mathcal{R}^d$-valued random variables. (This theorem and a proof are given in the text.) Large deviation theory is a burgeoning area of probability theory and many of the results in it can be applied to simulation problems. Rather than try to be as complete as possible in the exposition of all possible aspects of the available theory, I have tried to concentrate on demonstrating the methodology and the principal ideas in a fairly simple setting.

Madison, Wisconsin 2003 *James Antonio Bucklew*

Contents

1. Random Number Generation

Anyone who considers arithmetical methods of producing random numbers is, of course, in a state of sin. *John Von Neumann*
Nothing is random, only uncertain. *Gail Gasram*

In this chapter, we give a quick overview and discussion of some of the principal methods for generating good random numbers with a specified distribution on a computer.

1.1 Uniform Generators

All computer systems and computational languages have random number generators. For the most part they work remarkably well. Indeed it has been said of random number generators that when they're good they're wonderful, and when they're bad, they're still pretty good.

The problem that electrical engineers (and others) face in simulation is that we typically need to model very large, complex, and highly reliable systems. To do this we need large quantities of good random numbers. Depending on the computer system you're using you may or may not have access to a good generator. Hence, you might need to know a little about them so that you can either program one or choose among existing programs. In this section, we talk a little about a few types of random number generators and some pros and cons associated with them. We also recommend highly the excellent survey article [39] dealing with the problems of random number generation.

We should note that from a practitioner's point of view, a random number generator is "good" if it yields the correct result in as many applications as possible. We should also note that there is no (and in fact there can't be) universally "good" random number generator. Every generator has to fail in those situations where correlations in the random numbers and in the simulated system interfere constructively. This unpleasant situation is a fact of life and can't be circumvented. A random number generator is nothing more than a deterministic algorithm operating on a finite state machine (the computer) that produces numbers with certain distributional properties. These sequences of "random" numbers are always periodic. Their task really is not to simulate "randomness" but rather to give the correct results in a simulation. We should mention that there have in recent years been several companies that have put out physical devices that take time samples from a "hot resistor" or make use of "Johnson noise" or a variety of other "real random" sources. What has been found is that all of these that have been tested to date fail spectacularly various stringent tests of randomness [57]. Hence, unless rigorous testing is done beforehand, the practitioner is advised to stick with computer random number generators whose failings are at least well documented. In the face of such uncertainty, how are we to proceed if we can't even trust our random number generators? Perhaps the simplest answer is that we shouldn't trust any ONE random number generator. Always check your results with several widely different generators, before taking them seriously.

The most popular generator is the so-called congruential generator:

$$c_k = [ac_{k-1} + C]\text{mod}(M) \tag{1.1}$$

where $M > 0$ is a large prime number called the modulus ; $0 < a < M$ is the multiplier; C is the increment; and $0 < c_0 < M$ is the initial value (or seed). This recursion produces a sequence of integers in the range $[0, M - 1]$ which we convert into a sequence of numbers on the interval $[0, 1]$ by

$$u_k = c_k/M.$$

Of course (1.1) describes a finite state machine with M states. Thus the maximum period of the output sequence is M. A rule of thumb for linear generators in general is that the usable sample size is close to $\sqrt{P}$, where P is the period of the generator. Note that this implies that for a 32-bit machine, the period length must be less than 2^{32} or a maximum usable sample of $2^{15} \approx 33,000$, which is much too small for demanding simulations.

Due to the finite word length of the computer, we are always constrained to choose our generator parameters as rational numbers (after all an irrational number would require an infinite number of bits for its binary representation). Hence, one can show that the congruential generator will generate pairs (c_{k-1}, c_k) that lie on parallel lines. The trick is to choose a so that there are as many parallel lines as possible in the unit square $[0, 1] \times [0, 1]$.

Example 1.1.1. The most common choices for this type of generator are:

$$c_k = [16807c_{k-1}]\text{mod}(2^{31} - 1) \tag{1.2}$$

which will give an almost maximal period of $2^{31} - 2$ (this was the random number generator used in MATLAB before version 5) and

$$c_k = [69069c_{k-1}]\text{mod}(2^{32}) \tag{1.3}$$

which actually attains the maximal period of 2^{32}. This last one was (as of 1993) the system generator for Vax computers. The following generator (and example initial condition) with the same (popular) multiplier is also used heavily

$$c_k^* = [69069c_{k-1}^* + 1234567]\text{mod}(2^{32}) \quad c_0^* = 380116160. \tag{1.4}$$

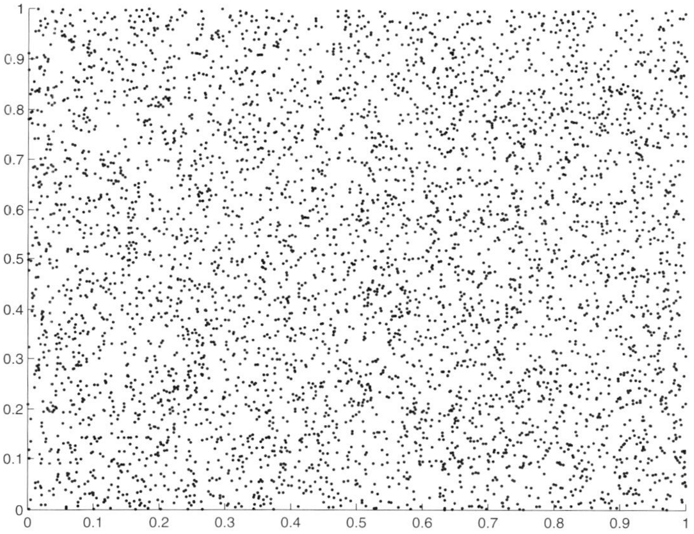

Fig. 1.1. 5000 pairs of (u_n, u_{n-1}) generated from c^*.

A generator that has a long period is not necessarily a good generator. The number of "good" numbers generated is typically much less than the period. But, what do we mean by "good" (uniformly distributed) random numbers? Obviously the relative frequency of the numbers should be uniform but we also would like to demand that they be independent. There is a large literature on developing tests for independence. Prof. G. Marsaglia (and others) have developed tests [56] based upon the relative frequency of n-tuples of numbers produced from the generators. He maintains a website where one can download a large variety of statistical tests of randomness (famously called the DIEHARD battery of tests). The congruential generators with prime moduli (as in (1.2)) perform well on all these tests. Unfortunately the generators with prime moduli are somewhat slower than those with power of 2 moduli. The generators with power of 2 moduli evidently don't perform very well with respect to the trailing bits generated. For example, the Vax generator of (1.3) fails these relative frequency of n-tuple tests if we try to utilize more than 13 bits or so.

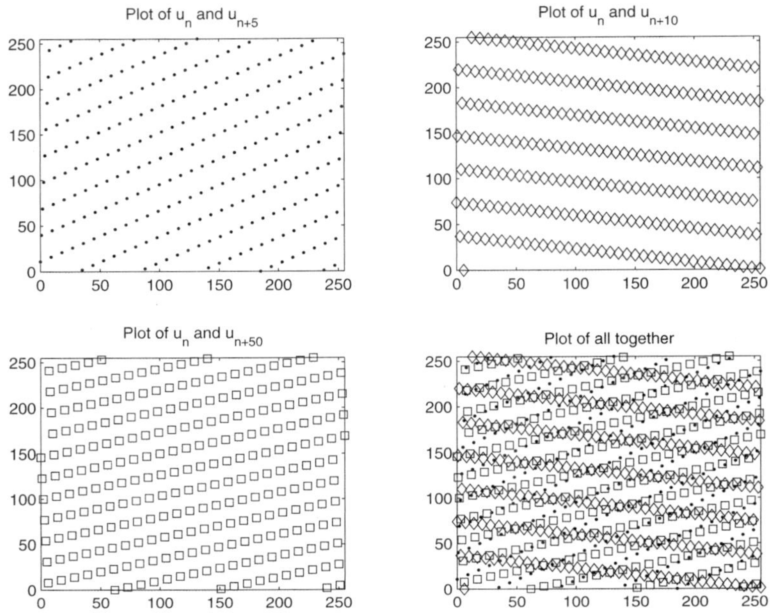

Fig. 1.2. Plot of the last eight bits (hence the numbers 0 to 255 of the pairs of (u_n, u_{n-k})) for $k = 5, 10, 50$ generated from c^*.

This issue of the goodness of the trailing bits is of more than a little importance. In many situations, one doesn't need all of the bits generated by the random number at once. For example, we may be simulating the toss of a coin. Clearly it would be very wasteful to generate the whole b bits (b is usually 32 or 16) to decide heads or tails, an event that could be decided on the outcome of just the first bit. In fact, in principle, if all of the bits were "good," we could perform b such fair coin flips, before we generated another number. In such usages, clearly the trailing bits are of the same importance as the leading ones.

Another class of simple generators is known as the Lagged–Fibonacci generators. We start off with an initial set of $f_1, f_2, \ldots, f_r$ elements and two "lags" r and s ($r > s$). We then implement the recursion

$$f_n = f_{n-r} \diamond f_{n-s}, \tag{1.5}$$

where the f_n are in a finite set over which there is a binary operation $\diamond$. Some examples of $\diamond$ are $+, -$ on integers $\mathrm{mod}(2^k)$; $*$ on odd such integers; or exclusive-or (xor) $\oplus$ on binary vectors.

G. Marsaglia [56] claims that these last (using odd integers and multiplication modulo 2^{32}) are the best of this class. Their periods are about 2^{32+r}. (To get around the problems of the other $\diamond$ operations, he has proposed the

4-lag generator $f_n = (f_{n-256} + f_{n-179} + f_{n-119} + f_{n-55})\mathrm{mod}(2^{32})$, which seems to pass all the usual tests and has a period of about 2^{287}.)

Example 1.1.2. The classical Fibonacci sequence (with example initial condition) taken modulo 2^{32} is

$$f_{n+1} = (f_n + f_{n-1})\mathrm{mod}(2^{32}) \quad f_0 = 224466889 \quad f_1 = 7584631.$$

This generator is evidently not that good but can be used in conjunction with other generators.

A very important class of random number generators is the so-called *shift register* generators (these are also called Tauseworthe generators). For a given $k \times k$ matrix T whose entries are all 0 or 1, the associated shift register generator is given by

$$sr_{n+1} = T sr_n,$$

where sr_n is represented as a vector of binary coordinates $e_{ni} \in \{0, 1\}$; that is, $sr_n = \sum_{i=0}^{k-1} e_{ni} 2^i$. Multiplication by T corresponds to shifting (and adding modulo-2) the contents of coordinates. A certain type of T is of particular interest, that is, the right and left shifts defined as $R(e_1 e_2 \cdots e_k)^t = (0 e_1 e_2 \cdots e_{k-1})^t$ and $L(e_1 e_2 \cdots e_k)^t = (e_2 e_3 \cdots e_k 0)^t$. Note that $R^{16} L^{16}$ corresponds to the operator

$$A = \begin{bmatrix} \mathbf{0} & \mathbf{0} \\ \mathbf{0} & \mathbf{I} \end{bmatrix},$$

where $\mathbf{0}$ corresponds to a 16×16 matrix of all zeros and $\mathbf{I}$ is a 16×16 matrix of all ones.

Example 1.1.3 (Shift Register Generator).

$$sr^*_{n+1} = (I \oplus L^5)(I \oplus R^{13})(I \oplus L^{17}) sr^*_n \qquad sr^*_0 = 123456789.$$

A class of excellent RNGs that at first glance appear very close to the congruential generators are the so-called multiply–with–carry generators that were developed to exploit the modulo 2^{32} arithmetic inherent in modern computers. They satisfy the recursion

$$mc_{n+1} = [a\ mc_n + carry]\mathrm{mod}(M). \tag{1.6}$$

Example 1.1.4 (Multiply–with–Carry). This algorithm combines two regular 16-bit multiply–with–carry algorithms (even though the modulo arithmetic is 32-bit) to create an excellent 32-bit fast generator that passes all the standard tests:

$$\begin{aligned} z_{n+1} &= [36969 R^{16} L^{16} z_n + R^{16} z_n]\mathrm{mod}(2^{32}) \quad z_0 = 362436069 \\ w_{n+1} &= [18000 R^{16} L^{16} w_n + R^{16} w_n]\mathrm{mod}(2^{32}) \quad w_0 = 521288629 \\ mc^*_n &= L^{16} z_n + w_n. \end{aligned}$$

Example 1.1.5. For a $2n$-bit sequence, x, define $top(x)$ = first n bits of x and $bot(x)$ = last n bits of x. Consider the sequence of 16-bit integers produced by the following recursion,

$$k_n = bot\Big(30903\ \big(bot(k_{n-1})\big) + top(k_{n-1})\Big)$$

Notice that this algorithm is merely multiplying the bottom half of the previous word by 30903, adds the top half, and then returns the new bottom. It turns out that this will produce a sequence of 16-bit integers with period greater than 2^{29}.

Example 1.1.6. The following segment in a (properly initialized) procedure will generate more than 2^{118} 32-bit random integers from six initial random seed values, $i_0, j_0, k_0, l_0, m_0, o_0$,

$$\begin{aligned}
i_n &= bot\Big(29013\ \big(bot(i_{n-1})\big) + \big(top(i_{n-1})\big)\Big)\\
j_n &= bot\Big(18000\ \big(bot(j_{n-1})\big) + \big(top(j_{n-1})\big)\Big)\\
k_n &= bot\Big(30903\ \big(bot(k_{n-1})\big) + \big(top(k_{n-1})\big)\Big)\\
l_n &= bot\Big(30345\ \big(bot(l_{n-1})\big) + \big(top(l_{n-1})\big)\Big)\\
m_n &= bot\Big(30903\ \big(bot(m_{n-1})\big) + \big(top(m_{n-1})\big)\Big)\\
o_n &= bot\Big(31083\ \big(bot(o_{n-1})\big) + \big(top(o_{n-1})\big)\Big)\\
z_n &= bot(k_n + i_n + m_n) + j_n + l_n + o_n.
\end{aligned}$$

Example 1.1.7 (KISS). The KISS generator of Marsaglia and Zaman [58] is a combination of two multiply–with–carry generators, a 3-shift shift register generator, and a congruential generator using addition and exclusive-or. The period is about 2^{123}. This generator is very highly recommended. It uses many of our previous examples.

$$k_{n+1} = mc^*_{n+1} \oplus c^*_{n+1} + sr^*_{n+1}.$$

The multiply–with–carry generator can be generalized to recur–with–carry as in

$$rc_{n+1} = [a_0 rc_n + \cdots a_{r-1} rc_{n-r+1}] \mathrm{mod}(2^{32}).$$

Example 1.1.8 (Mother of all RNGs). First we consider a 16-bit example that is easily implemented. It does such a great job of mixing the bits of the previous eight values that it would be difficult to imagine a test of randomness that it could not pass:

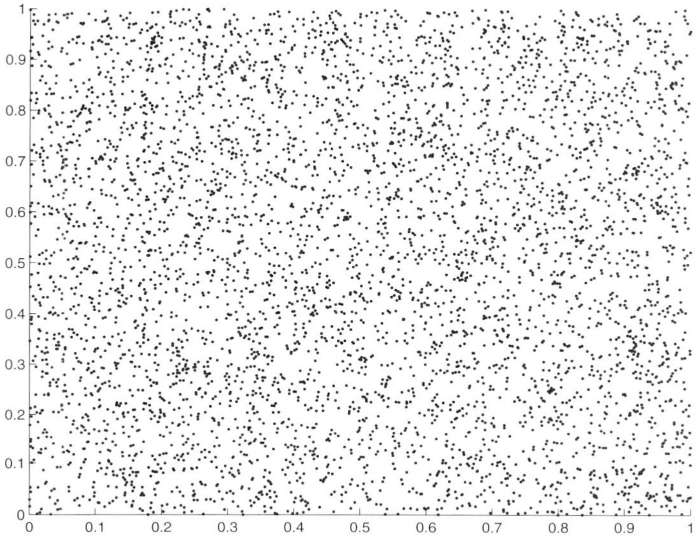

Fig. 1.3. 5000 pairs of (u_n, u_{n-1}) generated from the KISS algorithm.

$$\begin{aligned} y_n &= 12013x_{n-8} + 1066x_{n-7} + 1215x_{n-6} + 1492x_{n-5} + 1776x_{n-4} \\ &\quad +1812x_{n-3} + 1860x_{n-2} + 1941x_{n-1} + top(y_{n-1}) \\ x_n &= bot(y_n). \end{aligned}$$

The linear combination occupies at most 31 bits of a 32-bit integer. The bottom 16 is the output, the top 15 the next carry. It just provides 16-bit random integers, but very good ones. For 32 bits you would have to combine it with another, such as

$$\begin{aligned} y_n &= 9272x_{n-8} + 7777x_{n-7} + 6666x_{n-6} + 5555x_{n-5} + 4444x_{n-4} \\ &\quad +3333x_{n-3} + 2222x_{n-2} + 1111x_{n-1} + top(y_{n-1}) \\ x_n &= bot(y_n). \end{aligned}$$

Concatenating those two gives a sequence of 32-bit random integers (from 16 random 16-bit seeds) with period about 2^{250}. G. Marsaglia named this generator (semi-jokingly, since he considered it so awesome) the Mother of All Random Number Generators.

Example 1.1.9. Here is an interesting simple recur–with–carry generator for 32-bit arithmetic:

$$\begin{aligned} y_n &= 1111111464(x_{n-1} + x_{n-2}) + top(y_{n-1}) \\ x_n &= bot(y_n). \end{aligned}$$

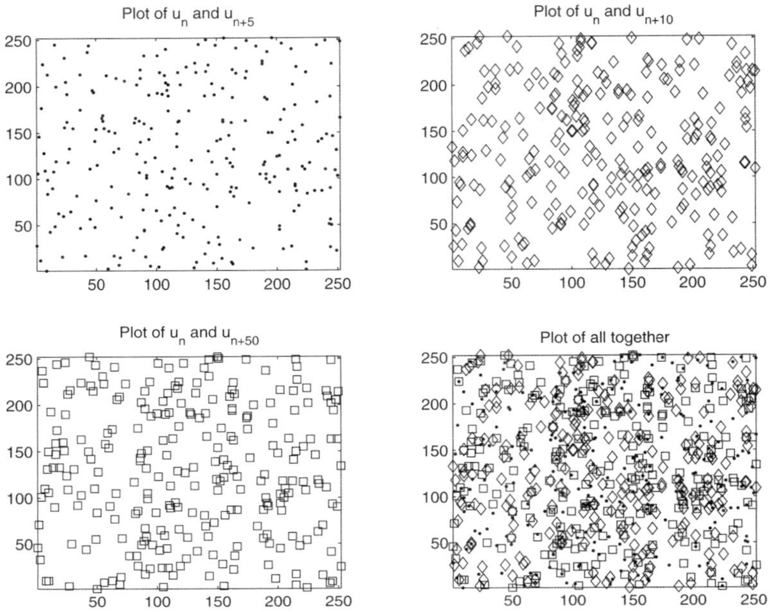

Fig. 1.4. Plot of the last eight bits of the pairs of (u_n, u_{n-k}) for $k = 5, 10, 50$ generated from the KISS algorithm.

With initial 32-bit x_1, x_0 and carry y_1, the above algorithm will give over 2^{92} random 32-bit numbers.

1.2 Nonuniform Generation

In this section we discuss some of the basic principles of nonuniform random number generation. We take as a starting point the supposition that we have ready access to a supply of uniform variates.

1.2.1 The Inversion Method

Let F be the distribution function for an $\mathcal{R}$-valued random variable. For $0 < u < 1$, let $F^{-1}(u) = \inf\{x : F(x) \geq u\}$.

Lemma 1.2.1. *For* $0 < u < 1$, $u \leq F(x)$ *if and only if* $F^{-1}(u) \leq x$.

Proof. Suppose $F(x) \geq u$. Then by definition, $F^{-1}(u) \leq x$, since $F^{-1}(\cdot)$ is the infimum of all such x.

Suppose $F^{-1}(u) \leq x$. Let $B_u = \{x : F(x) \geq u\}$. Now $F^{-1}(u)$ is the infimum of all the $x \in B_u$. Hence there exists a sequence $x_n \in B_u$ such that

x_n is converging monotonically downward to $F^{-1}(u)$. By the right continuity of the distribution function F, $F(x_n)$ is converging monotonically downward to $F\big(F^{-1}(u)\big)$. Now $x_n \in B_u$ implies $F(x_n) \geq u$ and hence

$$u \leq \lim_{n\to\infty} F(x_n) = F\big(F^{-1}(u)\big).$$

Also $F^{-1}(u) \leq x$ implies $F\big(F^{-1}(u)\big) \leq F(x)$. This then implies that $u \leq F(x)$, which completes the proof of the lemma. □

If U is a uniform $[0,1]$ random variable, then we claim that $F^{-1}(U)$ has distribution function F. This follows directly from our lemma since $\forall x \in \mathcal{R}$, $P\big(F^{-1}(U) \leq x\big) = P\big(U \leq F(x)\big) = F(x)$. Generating nonuniform random samples in this fashion from uniform random samples is known as the inverse method or the inversion method.

We can describe this method via the following algorithm:

The Inversion Method:

Generate a uniform $[0,1]$ random sample U.
RETURN $X \leftarrow F^{-1}(U)$

Note that our function $F^{-1}(\cdot)$ when F is a strictly monotone continuous distribution is the usual inverse function for F, which explains our notation.

We can then generate arbitrary random variables with continuous or discontinuous distribution functions given a supply of uniform random variables (provided that F^{-1} is explicitly known). The faster we can compute the inverse function, the faster we can generate the nonuniform variates. For the remaining part of this section we concentrate on continuous distribution functions and consider distribution functions with jumps in the next section.

Example 1.2.1. Suppose we wish to generate exponential λ random variables (with density $f(x) = \lambda \exp(-\lambda x)$, $x > 0$). $F(x) = 1 - \exp(-\lambda x)$. $F^{-1}(u) = -\frac{1}{\lambda}\log(1-u)$. To simplify even further, we can compute (noting that $1 - U$ has the same distribution as U),

$$X = -\frac{1}{\lambda}\log(U).$$

Example 1.2.2. Consider the Cauchy density: $f(x) = \sigma/[\pi(x^2+\sigma^2)]$.

$$F(x) = \frac{1}{2} + \frac{1}{\pi}\arctan\big(\frac{x}{\sigma}\big)$$
$$F^{-1}(u) = \sigma\tan\big(\pi(U - \frac{1}{2})\big).$$

Since $\tan(\pi x)$ is periodic 1, we can simplify and compute

$$X = \sigma\tan(\pi U).$$

Example 1.2.3. Consider the triangular density : $f(x) = \frac{2}{a}\left(1-\frac{x}{a}\right)$, $0 \le x \le a$. $F(x) = \frac{2}{a}\left(x - \frac{x^2}{2a}\right)$. $F^{-1}(u) = a(1 - \sqrt{1-u})$. Thus we need only generate

$$X = a(1 - \sqrt{U}).$$

Example 1.2.4. Consider the Pareto (a, b) density $f(x) = ab^a/x^{a+1}$, $0 \le b \le x$. $F(x) = 1 - \left(\frac{b}{x}\right)^a$. $F^{-1}(u) = b/(1-u)^{1/a}$. Thus we generate

$$X = \frac{b}{U^{1/a}}.$$

Example 1.2.5. The most famous exact method for generating Gaussian variants is the Box–Muller method. If X, Y are mean zero, variance one independent Gaussian random variables, then it is known that when we compute the polar coordinates of (X, Y) we have that the magnitude $R = \sqrt{X^2 + Y^2}$ has a *Rayleigh* density given by

$$f_R(r) = r\exp(-r^2/2), \qquad r > 0,$$

and distribution

$$F_R(r) = 1 - \exp(-r^2/2), \qquad r > 0.$$

The inverse of the distribution is

$$F^{-1}(u) = \sqrt{-2\log(1-u)}.$$

The phase angle Θ (by the circular symmetry of the bivariate Gaussian) is obviously uniform $[0, 2\pi]$. Let U_1, U_2 be two independent uniform $[0, 1]$ random variates. We can then easily generate the two polar variates by $(R, \Theta) = (\sqrt{-2\log(U_1)}, 2\pi U_2)$. We then transform them back to the rectangular variates by $(X, Y) = \big(R\cos(\Theta), R\sin(\Theta)\big)$. Thus two uniform variates give us two Gaussian variates.

When an exact form for F^{-1} is known, the inversion method is exact. When it is not known in closed form we can only try to solve the equation $F(X) = U$ numerically. That is given a U from our uniform generator, return the value of X that solves the equation. This brings about a whole other series of problems involving the numerical approximation to the equation (stopping criteria, methods, roundoff errors, numerical precision, etc.) We do not go further into this topic here.

1.2.2 The Acceptance–Rejection Method

Let us define an *acceptance function* $h : \mathcal{R}^d \to [0, 1]$. Suppose we generate independent and identically distributed (i.i.d.) $\mathcal{R}^d$-valued random variables

$X_1, X_2, \ldots$ from a probability density g. Suppose that for each X_i, we independently (of the other values of the X sequence) randomly accept it or reject it. We define this acceptance–rejection process through the $\{0, 1\}$ Bernoulli random variables $\{T_i\}$ by the relationship $T_i = 1$ if and only if X_i is accepted (and correspondingly $T_i = 0$ if and only if X_i is rejected. The probability structure is defined by $P(T = 1|X = x) = h(x)$. In other words if $\{X_i = x\}$, we reject it with probability $1 - h(x)$ (or accept it with probability $h(x)$). Let Y be an accepted sample. Note that

$$\begin{aligned} P(X \leq y \text{ and } T = 1) &= \mathbb{E}_X[P(X \leq y, T = 1|X)] \\ &= \int P(X \leq y, T = 1|X = x)g(x)dx \\ &= \int P(X \leq y|X = x)P(T = 1|X = x)g(x)dx \\ &= \int 1_{\{x \leq y\}} h(x)g(x)dx \\ &= \int_{-\infty}^{y} h(x)g(x)dx \end{aligned}$$

and also

$$\begin{aligned} P(T = 1) &= \mathbb{E}_X[P(T = 1|X)] \\ &= \int P(T = 1|X = x)g(x)dx \\ &= \int h(x)g(x)dx. \end{aligned}$$

Hence,

$$P(Y \leq y) = P(X \leq y \; |T = 1) = \frac{\int_{-\infty}^{y} h(x)g(x)dx}{\int_{-\infty}^{\infty} h(x)g(x)dx}.$$

Taking the derivative with respect to y shows that the accepted values have probability density function $gh/\int gh$.

The usual version of the rejection algorithm assumes that we have a dominating density g such that for some constant $c \geq 1$,

$$f(x) \leq cg(x).$$

We generate variables from the g density (which we hope is easy) and accept or reject them according to some criterion to have the accepted variables have the density f. We see that if we take our acceptance function h to be

$$h(x) = \frac{f(x)}{cg(x)},$$

the accepted values will have density $gh/\int gh = g(f/cg)/\int g(f/cg) = f$. Note also that the probability that we accept a sample from the g distribution is $\int gh = \int g(f/cg) = 1/c$.

The usual way to implement this procedure is to generate a random sample of X with density g and an independent sample from U uniform on $[0,1]$. Let $R = cg(X)/f(X)$. If $UR > 1$ we reject X. If $UR \leq 1$ we accept X. We easily see that if $\{X = x\}$, then we accept it if $\{Ucg(x)/f(x) \leq 1\} = \{U \leq f(x)/cg(x)\}$. This event has probability $f(x)/cg(x) = h(x)$ (because $P(U \leq z) = z\ 0 \leq z \leq 1$). The number of trials before a Y is accepted has a geometric distribution with mean c, so the algorithm works best if c is small.

The Acceptance–Rejection Method:

REPEAT
  Generate a random sample X with density g on $\mathcal{R}^d$.
  Generate a uniform $[0,1]$ sample U independent of X.
  Set $T \leftarrow c\frac{g(X)}{f(X)}$.
UNTIL $UT \leq 1$
RETURN X

The acceptance–rejection algorithm requires several things: (i) a dominating density g, (ii) a simple method for generating random variates with density g, and (iii) knowledge of c. (i) and (iii) are often possible by prior knowledge of the functional form of f. Note that the density g must have heavier tails and sharper infinite peaks than f does.

Example 1.2.6. Let f be the standard normal density. Obtaining an upper bound for f requires essentially that we find a lower bound for $x^2/2$. Note that

$$\begin{aligned}
\frac{1}{2}(|x|-1)^2 = \frac{x^2}{2} + \frac{1}{2} - |x| &\geq 0 \\
\frac{x^2}{2} &\geq |x| - \frac{1}{2} \\
\frac{1}{\sqrt{2\pi}} \exp\left(-\frac{x^2}{2}\right) &\leq \frac{1}{\sqrt{2\pi}} \exp\left(\frac{1}{2} - |x|\right) \\
&= \left(\sqrt{\frac{2e}{\pi}}\right)\frac{1}{2}\exp(-|x|) \\
&= cg(x),
\end{aligned}$$

where $g(x) = \exp(-|x|)/2$ is the Laplacian density. We, of course, can sample from the Laplacian using the inversion method.

Example 1.2.7. Let f be the density equal to

$$\frac{\exp(2\cos(x))}{I_0(2)}, \qquad x \in [0, \pi],$$

where $I_0(\cdot)$ is a modified Bessel function ($I_0(2) \approx 7.1619$). This would be a very hard density to generate via the inversion method. However, we can simply choose g to be uniform on $[0, \pi]$ and have

$$f(x) \leq 3.2412 g(x),$$

which gives a very easy way to generate these random variables.

The art of using the acceptance–rejection method is to find a suitable dominating density g from which it is easy to sample. There are a very large number of variations, twists, and tricks to implementing the acceptance–rejection method in various situations. It is a very general method that allows a great deal of creativity on the part of the random number generator designer.

1.3 Discrete Distributions

Suppose that we wish to generate samples of a discrete random variable X. By relabeling the state space, we can always assume that we are working with a nonnegative integer-valued random variable (a $\mathcal{N}^+$-valued random variable). Thus we may define

$$P(X = i) = p_i, \quad i = 0, 1, 2, \ldots .$$

We can use the inversion method to generate samples from this distribution. Generate a sample from a uniform $[0, 1]$ random variable U. Then define X as the integer such that

$$F(X-1) = \sum_{i<X} p_i < U \leq \sum_{i \leq X} p_i = F(X).$$

Then $P(X = i) = F(i) - F(i-1) = p_i$. To solve this system, the most straightforward solution is to do a sequential search: that is, initially set $X = 0$ and check the inequality; if it is satisfied stop; if it is not, set $X = X+1$ and continue. Let N be the number of trials we need to compute in the loop. $P(N = i) = P(X = i - 1) = p_{i-1}$ $i \geq 1$. Thus $\mathbb{E}[N] = \sum_{i=1}^{\infty} i p_{i-1}$ $= \sum_{i=1}^{\infty} (i-1) p_{i-1} + \sum_{i-1}^{\infty} p_{i-1} = \mathbb{E}[X] + 1$. This method can be very slow if $\mathbb{E}[X]$ is large. The inversion method with sequential search is a universal method but in many specialized situations there are better alternatives.

Example 1.3.1 (Poisson). One can use the inversion method to generate Poisson variates with parameter (mean value) λ. For the Poisson distribution we have

$$p_{i+1} = \frac{\lambda}{i+1} p_i, \quad p_0 = \exp(-\lambda).$$

Thus we can simplify the sequential search somewhat by recursively computing the $\{p_i\}$ values during the search as follows

Sequential Search Poisson Generator:

Generate a uniform $[0,1]$ random variate U.
Set $X \leftarrow 0, P \leftarrow \exp(-\lambda), S \leftarrow P$.
WHILE $U > S$ DO
    $X \leftarrow X + 1, P \leftarrow \frac{\lambda P}{X}, S \leftarrow S + P$.
RETURN X

The expected number of comparisons is equal to $\mathbb{E}[X + 1] = \lambda + 1$.

1.3.1 Inversion by Truncation of a Continuous Analog

Again, we suppose that we are trying to generate samples from the discrete distribution function F of a nonnegative integer-valued random variable. Suppose that we have a continuous distribution function G on $[0, \infty)$ that agrees with F on the nonnegative integers; that is,

$$G(i+1) = F(i) \quad i = 0, 1, \dots, \qquad G(0) = 0.$$

Generate a uniform [0,1] random variable U. Then define

$$X = \lfloor G^{-1}(U) \rfloor.$$

Note that for all $i \geq 0$, $P\big(X \leq i\big) = P\big(G^{-1}(U) < i + 1\big) = P\big(U < G(i+1)\big) = G(i+1) = F(i)$. This method can be extremely fast if G^{-1} is explicitly known. The task of finding a G satisfying the criteria is often quite simple.

Example 1.3.2. We can generate geometric random variables quite easily using this technique. Let $G(x) = 1 - \exp(-\lambda x)$, $x \geq 0$. Then

$$\begin{aligned} G(i+1) - G(i) &= \exp(-\lambda i) - \exp(-\lambda(i+1)) \\ &= \exp(-\lambda i)[1 - \exp(-\lambda)] \\ &= p^i(1-p), \end{aligned}$$

where $p = \exp(-\lambda)$. $G^{-1}(u) = [\log(1-u)]/(-\lambda)$. Thus $X = -\lfloor \frac{\log(U)}{\lambda} \rfloor$ has distribution $P\big(X = i\big) = p^i(1-p)$, $i = 0, 1, \dots$.

Example 1.3.3. Suppose $G(x) = 1 - x^{-b}$, $x \geq 1, G(1) = 0, b > 0$. $G(i+1) - G(i) = i^{-b} - (i+1)^{-b} = p_i$ $i \geq 1$. $G^{-1}(u) = (1-u)^{-1/b}$. Thus

$$X = \lfloor U^{-1/b} \rfloor$$

has probability vector p_i. In particular

$$X = \left\lfloor \frac{1}{U} \right\rfloor$$

has probability vector

$$p_i = \frac{1}{i(i+1)}, \qquad i \geq 1.$$

1.3.2 Acceptance–Rejection

The acceptance–rejection method is another universal method that remains valid for discrete distributions. We suppose that there exists a "dominating" probability vector g_i, $i \geq 0$ such that

$$p_i \leq c g_i,$$

where c is the rejection constant and g_i is an easy probability vector. The methodology to follow is now identical to that of the continuous case. We note that the family of heavy-tailed distributions of the previous example are one excellent possible source for easily generated dominating probability vectors.

Sometimes we can bound our probabilities by an easily generated probability density as

$$p_i \leq c g(x), \qquad x \in [i, i+1), \qquad i \geq 0.$$

We can create an "associated" probability density to our target probability mass function $\{p_i\}$ by defining

$$f(x) = p_i, \qquad x \in [i, i+1), \qquad i \geq 0.$$

The idea is that first we note $f(x) \leq c g(x)$. Then by an acceptance–rejection procedure, we generate samples from the $f(\cdot)$ density. We then round those values to the greatest integer less than or equal to the value. The rounding operation converts the variate with density f into a variate with probability mass function $\{p_i\}$. The algorithm appears as

Hybrid Rejection Algorithm:

REPEAT
    Generate a random variate Y with density g. Set $X \leftarrow \lfloor Y \rfloor$.
    Generate a uniform $[0,1]$ random variate U.
UNTIL $Ucg(Y) \leq p_X$
RETURN X

Example 1.3.4. Suppose our target probability mass function is given by

$$p_i = \frac{6}{\pi^2 i^2} \qquad i = 1, 2, 3, \ldots$$

For generating samples from a desired probability mass function using the hybrid algorithm, we need to select a density g from which we can generate the samples easily and such that g should have heavier tails than p. Since the Cauchy density has tails going to zero on the order of $o(x^2)$, we should be able to use this density for g. Also, since we are trying to generate a positive random variable, we can just use the absolute value of a Cauchy. Thus, let

$$g(x) = \frac{2}{\pi(1+x^2)}, \qquad x \geq 0.$$

There is a variety of ways to generate Cauchy random variates. One easy way to generate samples from the above density is to take the ratio of two standard normal random variables.

One of the main things is to make sure that $f(x) \leq cg(x)$ by proper selection of parameter c. We can derive an analytical expression in this Cauchy distribution setting.

Letting $f(x) = 6/(\pi^2 i^2)$, $x \in [i, i+1]$, $i = 1, 2, \ldots$, we find that our dominating criterion for $x \in [N, N+1]$ becomes

$$\begin{aligned}
\frac{2c}{\pi(1+x^2)} &\geq f(x) \qquad \forall\, x \in [N, N+1] \\
\frac{2c}{\pi\left(1+(N+1)^2\right)} &\geq \frac{6}{\pi^2 N^2} \\
2c &\geq \frac{6}{\pi}\frac{1+(N+1)^2}{N^2} \qquad \text{for } N = 1, 2, \ldots \\
c &\geq \frac{6}{\pi}\frac{5}{2} \approx 4.7746.
\end{aligned}$$

2. Stochastic Models

All models are wrong, some models are useful. *George Box*
Everything should be made as simple as possible, but not simpler. *Albert Einstein*

In the previous chapter, we were concerned almost exclusively with the problem of how to generate independent random variables with a specified distribution. In the simplest settings, this is the underlying statistical model and we need go no further. In many other situations, we have to simulate some sort of dependent data or noise process to act as inputs to our simulation model. Many dependent stochastic models can be simulated in an obvious way from their definitions. Nevertheless, some tricks sometimes are useful and we present a few of the more common ones below.

2.1 Gaussian Processes

We suppose that we have available a reliable source of i.i.d. Gaussian random numbers $\{X_i\}$, of mean zero and variance σ_x^2. We wish to simulate a Gaussian random process with a specified autocorrelation function (or equivalently a specified power spectral density). The usual idea is based on passing the i.i.d. Gaussian sequence through a specified digital filter. The most common (and useful) model is that of the following autoregressive moving average (ARMA(p, q)) structure,

$$Y_n = \sum_{m=1}^{p} a_m Y_{n-m} + \sum_{m=1}^{q} b_m X_{n-m} + X_n.$$

This is the equation of a discrete-time, linear, time-invariant system with transfer function given by

$$H(f) = \frac{1 + \sum_{m=1}^{q} b_m \exp(-i2\pi f m)}{1 - \sum_{m=1}^{p} a_m \exp(-i2\pi f m)},$$

where $i = \sqrt{-1}$. Thus the power spectrum of the Y process is given by the well-known relationship for wide sense stationary random processes and linear time-invariant systems as

$$S_Y(f) = S_X(f)|H(f)|^2 = \sigma_x^2 \left| \frac{1 + \sum_{m=1}^{q} b_m \exp(-i2\pi f m)}{1 - \sum_{m=1}^{p} a_m \exp(-i2\pi f m)} \right|^2 .$$

The problem that we now face is, given a specific power spectrum, how do we choose the coefficients $\{a_i\}$ and $\{b_j\}$ so that S_Y given above is "close" to the desired spectrum? There is a large body of literature on this subject that would take us far afield to try to cover. We point out that the simpler autoregressive version of the model has a much simpler solution. The autoregressive model (AR(p)) is given by

$$Y_n = \sum_{i=1}^{p} a_i Y_{n-i} + X_n.$$

It is simple to show that the autocorrelation function satisfies

$$R_Y(0) = \sum_{j=1}^{p} a_j R_Y(j) + \sigma_x^2$$

$$R_Y(k) = \sum_{j=1}^{p} a_j R_Y(k-j) \qquad k \geq 1.$$

This of course can be written as a matrix equation known as the Yule–Walker equation

$$\mathbf{r} = \mathbf{R}\mathbf{a},$$

where $\mathbf{r} = \big(R_Y(1)R_Y(2)\dots R_Y(p)\big)^T$, $\mathbf{R} = \big(R_Y(i-j)\big)$, $\mathbf{a} = (a_1 a_2 \dots a_p)$. Hence given a certain autocorrelation function, we can invert the Yule–Walker equation and find the required coefficients of the AR(p) model to emulate it.

Example 2.1.1. Suppose we wish to simulate a Gaussian random process with autocorrelation function, $R(\tau) = a^{|\tau|} + b^{|\tau|}$, $\tau = 0, \pm 1, \pm 2, \dots$, $|a| < 1$, $|b| < 1$. We then have the following representation for the power spectral density,

$$\begin{aligned} S(\omega) &= \sum_{\tau=-\infty}^{\infty} R(\tau)\exp(-i\omega\tau) \\ &= \frac{1-a^2}{\big(1-a\exp(i\omega)\big)\big(1-a\exp(-i\omega)\big)} \\ &\quad + \frac{1-b^2}{\big(1-b\exp(i\omega)\big)\big(1-b\exp(-i\omega)\big)} \\ &= \frac{|B(\omega)|^2}{|A(\omega)|^2}, \end{aligned}$$

where

$$B(\omega) = (1-a^2)\big(1-b\exp(-i\omega)\big)\big(1-b\exp(i\omega)\big) \\ +(1-b^2)\big(1-a\exp(-i\omega)\big)\big(1-a\exp(i\omega)\big)$$

and

$$A(\omega) = \big(1-a\exp(-i\omega)\big)\big(1-b\exp(-i\omega)\big) \\ \times\big(1-a\exp(i\omega)\big)\big(1-b\exp(i\omega)\big).$$

To generate a random process with this spectral density, we desire to factor $|B(\omega)|^2 = B(\omega)B^*(\omega)$ and $|A(\omega)|^2 = A(\omega)A^*(\omega)$ so that $B(\omega), A(\omega)$ are stable causal filters. The factorization for $A(\omega)$ is obvious; we take $A(\omega) = \big(1-a\exp(-i\omega)\big)\big(1-b\exp(-i\omega)\big)$. Now the denominator is a little more complicated,

$$\begin{aligned}
&|B(\omega)|^2 \\
&= -[(1-a^2)b + (1-b^2)a]\exp(-i\omega) + 2(1-a^2b^2) \\
&\quad -[(1-a^2)b + (1-b^2)a]\exp(i\omega) \\
&= -(1-ab)(b+a)\exp(-i\omega)\left[1 + \frac{2(1+ab)}{a+b}\exp(i\omega) + \exp(2i\omega)\right].
\end{aligned}$$

The term in brackets is a quadratic equation in the variable $\exp(i\omega)$, with roots

$$r_{1,2} = \frac{(1+ab) \pm \sqrt{(1-a^2)(1-b^2)}}{a+b}.$$

It is easy to see that $r_1 = 1/r_2$. Let r denote the root that is smaller in absolute value. We then put that root in with $B(\omega)$. Hence, we take $B(\omega) = \sqrt{(1-ab)(a+b)/r}(1-r\exp(-i\omega))$. Hence,

$$\frac{B(\omega)}{A(\omega)} = \frac{\sqrt{(1-ab)(a+b)/r}(1-r\exp(-i\omega))}{1-(a+b)\exp(-i\omega)+ab\exp(-2i\omega)}.$$

Therefore we can exactly fit an ARMA model with

$$\begin{aligned}
\sigma_x^2 &= \frac{(1-ab)(a+b)}{r} \\
a_1 &= r \\
b_1 &= a+b \\
b_2 &= ab.
\end{aligned}$$

Below is the power spectral density estimate computed from 5000 samples of a random process generated with the above ARMA model with $a = .4$, $b = .6$.

We now present another very useful technique for generating a stationary Gaussian time series with given autocorrelation coefficients $c_0, c_1, \ldots, c_n$. The method works best for small n and is a "batch" versus a "continuous"

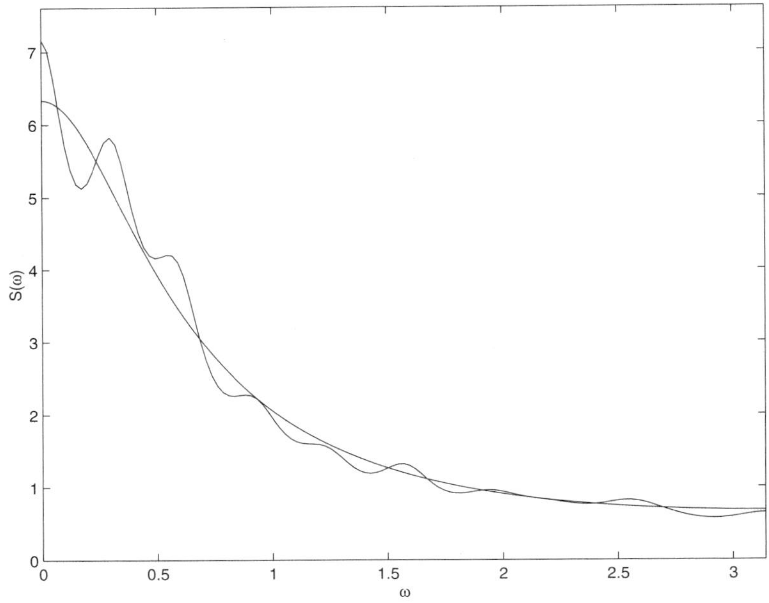

Fig. 2.1. Power spectral density estimate (this is the "wavy" line using a modified covariance method) and the actual density generated from an ARMA model.

method as given above. Let $\{g_k\}$ denote the discrete Fourier Transform of $c_0, c_1, \ldots, c_{n-1}, c_n, c_{n-1}, \ldots, c_1$; that is

$$g_k = \sum_{j=0}^{n-1} c_j \exp(\frac{2\pi ijk}{2n}) + \sum_{j=n}^{2n-1} c_{2n-j} \exp(\frac{2\pi ijk}{2n}),$$

which of course is implemented most efficiently by a Fast Fourier Transform algorithm. If $\{g_k\}$ were the discrete Fourier Transform of the whole autocorrelation sequence, then the whole sequence would be automatically nonnegative (in this case, the $\{g_k\}$ are samples of the power spectrum and hence must be nonnegative). If the $\{c_k\}$ are just the coefficients of (say) the first n lags of the autocorrelation, then we are effectively truncating the autocorrelation function after n lags and then computing its discrete Fourier Transform. In that case, the sequence $\{g_k\}$ can indeed take on negative values. We assume that the truncation (if there is any) is not so severe as to cause this phenomenon. Henceforth we just assume that all elements of the $\{g_k\}$ sequence are nonnegative.

Now let $\{Z_0, Z_1, \ldots, Z_{2n-1}\}$ be a sequence of complex, zero mean, Gaussian random variables, where Z_0 and Z_n are real with variance 2. We suppose $\{Z_1, \ldots, Z_{n-1}\}$ are independent and each has independent real and imaginary parts each of which in turn has variance one. Finally, we require $Z_k = \bar{Z}_{2n-k}$ for $n < k < 2n$. Then

$$X_j = \frac{1}{2\sqrt{n}} \sum_{k=0}^{2n-1} Z_k \sqrt{g_k} \exp\left(\frac{2\pi ijk}{2n}\right) \qquad j = 0, 1, \ldots, n$$

is the desired sequence. First of all note that the sequence is real, since the imaginary terms cancel themselves out. Then

$$\begin{aligned}\mathbb{E}[X_p X_q] &= \mathbb{E}[X_p \bar{X}_q] \\ &= \frac{1}{4n} \sum_{k=0}^{2n-1} \sum_{l=0}^{2n-1} \mathbb{E}[Z_k \bar{Z}_l] \sqrt{g_k g_l} \exp\left(\frac{2\pi i(pk - ql)}{2n}\right).\end{aligned}$$

Now note that $\mathbb{E}[Z_k \bar{Z}_l] = 0$ if $k \neq l$ by independence if $l \neq 2n - k$; and because $\mathbb{E}[Z_k \bar{Z}_{2n-k}] = \mathbb{E}[Z_k^2] = \mathbb{E}[(Re(Z_k))^2 - (Im(Z_k))^2 + 2Re(Z_k)Im(Z_k)] = 0$. Also note that $\mathbb{E}[Z_k \bar{Z}_k] = 2$. Hence

$$\mathbb{E}[X_p X_q] = \frac{1}{2n} \sum_{k=0}^{2n-1} g_k \exp\left(\frac{2\pi ik(p-q)}{2n}\right) = c_{p-q}$$

by the definition of the inverse discrete Fourier Transform.

2.2 Markov Processes

Given a discrete time Markov chain $\{X_n\}$ with state space S and transition matrix P, we can simulate it in the obvious way: If $X_n = s$, then select X_{n+1} as a sample from the discrete distribution $\{p_{si}\}$. Depending on the specific model we may or may not be able to do better than this straightforward emulation.

Continuous time chains can be simulated in the same way. Let $E_0, E_1, \ldots$ be the sequence of times between jumps and $J_0, J_1, \ldots$ be the sequence of states visited. Conditional on $J_0, \ldots, J_n$ and $E_0, \ldots, E_{n-1}$, E_n is exponential with parameter λ_{J_n}, and $\{J_i\}$ is a Markov chain (known as the embedded chain). In many cases the embedded chain contains all of the information of interest in the simulation and there may be no need to simulate the whole continuous time process.

2.3 Markov Chain Monte Carlo

Suppose that we have a very large discrete state space S and we wish to draw samples from some distribution π over this state space. In many situations in engineering and statistical physics, the size $|S|$ of the state space is so large that our conventional techniques from the previous chapter of drawing samples from a discrete distribution are not feasible. In fact, in many situations, the exact value of $|S|$ may not be known. The idea of Markov chain Monte

Carlo (MCMC) is that we construct an (easy) Markov chain on the state space of interest that will have stationary distribution π. One then runs this chain and from time to time, we take samples from it. If the chain has reached equilibrium, we may think of these samples as being from the distribution π. If we wait a sufficient amount of time between the samples, we can even view these samples as independent.

Let's consider the problem of designing the MCMC algorithm. One must find an ergodic transition matrix P on S, with stationary distribution π. There are infinitely many such transition matrices and among them are infinitely many that correspond to a reversible chain, that is, a Markov chain that satisfies the so-called detailed balance equation

$$\pi_i p_{ij} = \pi_j p_{ji}. \tag{2.1}$$

We try for solutions of the form

$$p_{ij} = q_{ij}\alpha_{ij} \qquad \text{for } j \neq i,$$

where $Q = (q_{ij})$ is an arbitrary irreducible transition matrix in S called the candidate–generating matrix. When the present state is i, the next *tentative* state j is chosen with probability q_{ij}. When $j \neq i$, this new state is accepted with probability α_{ij}. Hence, the probability of moving from i to $j \neq i$ is given by the displayed equation above. We now need a methodical way to choose the α_{ij} acceptance probabilities. A formulation due to Hastings gives a general framework that includes most well-known methods. He chooses

$$\alpha_{ij} = \frac{s_{ij}}{1 + t_{ij}},$$

where $\Sigma = (s_{ij})$ is a symmetric matrix and

$$t_{ij} = \frac{\pi_i q_{ij}}{\pi_j q_{ji}}.$$

Σ must be selected so that α_{ij} remains a valid probability, that is, lies in the range $[0, 1]$.

To verify that the resulting chain is reversible is easy. For $j \neq i$:

$$\begin{aligned}
\pi_i p_{ij} &= \pi_i q_{ij}\alpha_{ij} \\
&= \pi_i q_{ij}\frac{s_{ij}}{1 + t_{ij}} \\
&= \pi_i q_{ij}\frac{s_{ij}}{1 + \frac{\pi_i q_{ij}}{\pi_j q_{ji}}} \\
&= \pi_j q_{ji}\frac{s_{ij}\pi_i q_{ij}}{\pi_i q_{ij} + \pi_j q_{ji}} \\
&= \pi_j q_{ji}\frac{s_{ij}}{1 + \frac{\pi_j q_{ji}}{\pi_i q_{ij}}}
\end{aligned}$$

$$
\begin{aligned}
&= \pi_j q_{ji} \frac{s_{ji}}{1 + \frac{\pi_j q_{ji}}{\pi_i q_{ij}}} \\
&= \pi_j q_{ji} \alpha_{ji} \\
&= \pi_j p_{ji}.
\end{aligned}
$$

Thus the chain is time reversible and satisfies the detailed balance equation (2.1). (Note that it is trivially satisfied when $j = i$.) It is known that any chain that satisfies such a balance equation must have stationary distribution π [36].

To satisfy the constraint $\alpha_{ij} \in [0, 1]$, one must have $s_{ji} = s_{ij} \leq 1 + \min(t_{ij}, t_{ji})$. Suppose that we choose equality. Then (noting that $t_{ij} = 1/t_{ji}$) we have

$$
\begin{aligned}
s_{ij} &= 1 + \min(t_{ij}, t_{ji}) \\
\alpha_{ij} &= \frac{1 + \min(t_{ij}, t_{ji})}{1 + t_{ij}}.
\end{aligned}
$$

Note that if $\min(1, t_{ji}) = 1$, then $\min(t_{ij}, t_{ji}) = t_{ij}$ and $\alpha_{ij} = 1$. If $\min(1, t_{ji}) = t_{ji}$, then $\min(t_{ij}, t_{ji}) = t_{ji}$ and $\alpha_{ij} = (1 + t_{ji})/(1 + t_{ij}) = (1 + t_{ji})/(1 - 1/t_{ji}) = t_{ji}$. Thus

$$
\alpha_{ij} = \min(1, t_{ji}) = \min(1, \frac{\pi_j q_{ji}}{\pi_i q_{ij}}),
$$

which is called the *Metropolis algorithm.*

Example 2.3.1. Let $G = (V, E)$ be a connected graph. Let $\mathcal{C}$ be the set of all spanning subgraphs of G. (A spanning subgraph is a connected subgraph whose vertex set is the original V of G.) We note that for a large graph G, the size of $\mathcal{C}$, $|\mathcal{C}|$ can be astronomical. In general it is very difficult to even compute what the size is.

Consider a Markov chain with state space $\mathcal{C}$ as follows. Suppose we are at some $H \in \mathcal{C}$. Select an edge $e \in E$ uniformly and at random. If $e \notin H$, add e to H. If $e \in H$, remove e provided the resulting graph remains connected. If the removal of an edge disconnects the graph, then we stop and select another edge at random. The next step of the Markov chain is not decided until we select an edge that allows us to move to another connected graph. Hence the chain never remains where it is but at each step moves to a different connected subgraph of G.

Therefore, the algorithm moves from one subgraph to another subgraph that differs by only one edge. We can think of this algorithm as a MCMC with a symmetric candidate generating transition matrix; that is, $q_{ij} = q_{ji}$, for all i, j. Also since we always move from our current state to another state, we must have that $\alpha_{ij} = 1$. We now note that in a Metropolis algorithm with a uniform stationary distribution and symmetric Q matrix, we have $\alpha_{ij} = 1$ also. Hence our algorithm is a Metropolis algorithm with a uniform stationary distribution. What is so interesting and useful here is that this algorithm gives us a means to uniformly sample from this huge state space without even obtaining information about its size.

In the Hastings formulation, if we make the choice $s_{ij} = 1$, we have another algorithm, *Barker's algorithm*, for which

$$\alpha_{ij} = \frac{\pi_j q_{ji}}{\pi_j q_{ji} + \pi_i q_{ij}},$$

which we make use of in the following section.

2.3.1 Simulation of Markov Random Fields

Consider a Λ–valued process (elements of Λ are usually colors, gray levels, spin states, or whatever) on a collection of sites S. These sites are related by a graph structure so that each site has a number of neighbors. (In an image-processing example, it would be common to call each pixel a site, and each pixel has an obvious eight neighbors.) Call the set of neighbors of a site r, $\mathcal{N}_r$. A Markov random field has the special property that

$$P\big(Z_r = z | Z_s = z_s,\ \ s \in S, s \neq r\big) = P\big(Z_r = z | Z_s = z_s,\ \ s \in \mathcal{N}_r, s \neq r\big).$$

In other words, the conditional distribution at a site given the rest depends only on the values at the neighboring sites.

We would like to identify a Markov chain $\{X_n\}$ with state space Λ^S that has a given target stationary distribution π. The so-called *Gibb's sampler* is perhaps the most well-known algorithm for performing this task. The Gibb's sampler makes use of a strictly positive probability distribution q on S. This distribution chooses site s with probability q_s. The transition from $X_n = x$ to $X_{n+1} = y$ is as follows.

We only allow transitions from x to states y that differ from x by the value at one site. When we are at state $X_n = x$, we choose a site from the q distribution; call it s. Let us denote the configuration of x as $x(S)$ and the configuration of x except at site s as $x(S\backslash s)$. The value of the configuration at site s is denoted as $x(s)$. Thus in our notation, $x = x(S) = \big(x(s), x(S\backslash s)\big)$. We then change from the configuration x to y as follows. $y(S\backslash s) = x(S\backslash s)$, and the new value at s, $y(s)$ is selected with probability $\pi\big(y(s)|x(S\backslash s)\big)$. Thus, configuration x is changed into $\big(y(s), x(S\backslash s)\big)$ with probability $\pi\big(y(s)|x(S\backslash s)\big)$. Hence we have for all nonzero entries of the transition matrix for $\{X_n\}$,

$$P\big(X_{n+1} = y | X_n = x\big) = q_s \pi\big(y(s)|x(S\backslash s)\big) 1_{\{y(S\backslash s) = x(S\backslash s)\}}.$$

To prove that π is the stationary distribution we show that the detailed balance equation is satisfied. Consider any two states x, y that differ from each other by only one site s. Then

$$\begin{aligned}
\pi(x)q_s\frac{\pi\big(y(s),x(S\backslash s)\big)}{P\big(X(S\backslash s)=x(S\backslash s)\big)} &= \pi\big(y(s),x(S\backslash s)\big)q_s\frac{\pi(x)}{P\big(X(S\backslash s)=x(S\backslash s)\big)}\\
\pi(x)q_s\pi\big(y(s)|x(S\backslash s)\big) &= \pi(y)q_s\pi\big(x(s)|x(S\backslash s)\big)\\
\pi(x)P\big(X_{n+1}=y|X_n=x\big) &= \pi(y)P\big(X_{n+1}=x|X_n=y\big).
\end{aligned}$$

Thus $\{X_n\}$ is a reversible chain with stationary distribution π.

Example 2.3.2. Suppose $\Lambda = \{black, white\}$. At the chosen site, $x_s =$ *black* or *white*, so

$$\begin{aligned}
\pi(x)+\pi(y) &= P\big(X(s)=white, X(S\backslash s)=x(S\backslash s)\big)\\
&\quad + P\big(X(s)=black, X(S\backslash s)=x(S\backslash s)\big)\\
&= P\big(X(S\backslash s)=x(S\backslash s)\big)\\
\frac{\pi(y)}{\pi(x)+\pi(y)} &= \frac{P\big(X(s)=newcolor, X(S\backslash s)=x(S\backslash s)\big)}{P\big(X(S\backslash s)=x(S\backslash s)\big)}\\
&= P\big(X(s)=newcolor|X(S\backslash s)=x(S\backslash s)\big).
\end{aligned}$$

We note that this methodology is merely Barker's algorithm. The process of selecting a candidate configuration y from the current configuration x is clearly symmetric. In this case the acceptance probability for a Barker algorithm is exactly of the form given above. This example can be extended to fields other than binary ones. Geman and Geman [32] call this variant the "Gibb's sampler."

3. Large Deviation Theory

How dare we speak of the laws of chance? Is not chance the antithesis of all law?

Bertrand Russell

The probability of anything happening is in inverse ratio to its desirability.

One of the many variants of Murphy's Law

We will find that many of the rare events that we wish to simulate are the result of the occurrence of a *large deviation* event. In order to develop good simulation strategies for these events, we need to understand something of the probability theory associated with them. In this chapter, we provide an introduction to two of the basic concepts of this important area: Cramér's Theorem and the Gärtner–Ellis Theorem.

Before beginning the mathematical arguments, we should in an informal fashion try to understand what large deviation theory is about. In the simplest settings, the basic problem and results are quite easily presented. Suppose we have some sample average S_n of i.i.d. random variables $\{X_i\}$ with mean value m. By the law of large numbers, S_n is converging to m. Suppose we ask what is the probability that S_n is found outside some interval of width $2r$ about m? In other words, what is $P\big(|S_n - m| \geq r\big)$? This probability of course is going to zero as n gets large. Cramér's Theorem states that this probability behaves like

$$P\big(|S_n - m| \geq r\big) = f(n)\exp\big(-nI(r)\big),$$

where $I(r) > 0$ is the exponential rate constant and $f(n)$ is a sequence going to zero more slowly than the exponential. (Typically $f(n)$ is going to zero like some constant divided by n to some positive power.) This sequence $f(n)$ is difficult to compute and so large deviation results are typically given as limit theorems in the form

$$\lim_{n\to\infty} \frac{1}{n} \log P\big(|S_n - m| \geq r\big) = -I(r).$$

3.1 Cramér's Theorem

The simplest setting in which to obtain large deviation results is that of considering sums of i.i.d. $\mathcal{R}$-valued random variables. For example, we consider the large excursion probabilities of sums such as the sample average

$$S_n = \frac{X_1 + X_2 + \cdots + X_n}{n},$$

where the $\{X_i\}$, $i = 1, 2, \ldots, n$ are i.i.d. and n is approaching infinity. Suppose that $\mathbb{E}[X_1] = m$ exists and is finite. By the law of large numbers, we know that S_n is approaching m. Consider the function

$$I(x) = \sup_\theta[\theta x - \log M(\theta)],$$

where θ is a real variable and

$$M(\theta) = \mathbb{E}[\exp(\theta X_1)].$$

$M(\theta)$ is called the *moment generating function.* $I(x)$, the *large deviation rate function,* will play a crucial role in the development of the theory. Let us list some properties of the large deviation rate function and give a couple of examples.

Property 1. $I(x)$ is convex.

Proof. I is the pointwise supremum of a collection of affine (and hence convex) functions. Hence I is convex. More directly, let $\lambda \in [0, 1]$; then

$$\begin{aligned}
&I(\lambda x_1 + (1-\lambda)x_2) \\
&= \sup_\theta[\theta(\lambda x_1 + (1-\lambda)x_2) - \log M(\theta)] \\
&= \sup_\theta[\theta\lambda x_1 - \lambda \log M(\theta) + \theta(1-\lambda)x_2 - (1-\lambda)\log M(\theta)] \\
&\le \sup_\theta[\lambda(\theta x_1 - \log M(\theta))] + \sup_\theta[(1-\lambda)(\theta x_2 - \log M(\theta))] \\
&= \lambda I(x_1) + (1-\lambda)I(x_2).
\end{aligned}$$

□

We remark that a convex function $I(\cdot)$ on the real line is continuous everywhere on $D_I = \{x : I(x) < \infty\}$, the *effective domain* of $I(\cdot)$.

Property 2. $I(\cdot)$ has its minimum value at $x = \mathbb{E}[X_1] = m$. Furthermore $I(m) = 0$.

Proof. Since $M(0) = 1$, $I(x) \ge 0x - \log M(0) = 0$. By Jensen's inequality $M(\theta) \ge \exp(\theta m)$, hence $\theta m - \log M(\theta) \le 0$ for all θ. Thus $I(m) = 0$ and $I(x) \ge I(m)$ for all x. □

Property 3. For $x > m$,

$$I(x) = \sup_{\theta \ge 0}[\theta x - \log M(\theta)], \tag{3.1}$$

and is a nondecreasing function for $x > m$. Similarly, for $x < m$,

$$I(x) = \sup_{\theta \le 0}[\theta x - \log M(\theta)], \tag{3.2}$$

and is nonincreasing for $x < m$.

Proof. For all $\theta \in \mathcal{R}$, by Jensen's inequality, we have

$$\log M(\theta) = \log \mathbb{E}[\exp(\theta X_1)] \geq \mathbb{E}[\log \exp(\theta X)] = \theta m.$$

From Property 2, we know that $I(m) = 0$, and thus for $x \geq m$ and all $\theta < 0$,

$$\theta x - \log M(\theta) \leq \theta m - \log M(\theta) \leq I(m) = 0.$$

Hence the supremum in the definition for $I(x)$ for $x > m$, must occur over the nonnegative values of θ. Equation (3.1) implies the monotonicity of $I(x)$ on $[m, \infty)$, since the function $\theta x - \log M(\theta)$ is nondecreasing as a function of x. The statements for $x < m$ follow by considering the moment generating function of $-X$. □

Property 4. $M(\cdot)$ is differentiable on the interior of its effective domain (denoted as $\overset{\circ}{D}_M$),

$$M'(\theta) = \mathbb{E}[X_1 \exp(\theta X_1)] \tag{3.3}$$

and

$$\frac{M'(\theta_x)}{M(\theta_x)} = x \Rightarrow I(x) = \theta_x x - \log M(\theta_x). \tag{3.4}$$

Proof. Equation (3.3) is a standard result for moment generating functions and basically follows by interchanging the order of differentiation and integration. Now suppose $M'(\theta_x)/M(\theta) = x$, and consider the concave function of θ, $g(\theta) = \theta x - \log M(\theta)$. Since g is concave and $g'(\theta_x) = 0$, it follows that $g(\theta_x)$ is the maximum of $g(\theta)$ and (3.4) is established. □

Example 3.1.1. Suppose that X_1 is a Gaussian random variable with mean zero and variance 1. The moment generating function is well known to be $M(\theta) = \exp(\theta^2/2)$. We then have $I(x) = \sup_\theta[\theta x - \theta^2/2] = x^2/2$.

Example 3.1.2. X_1 is Bernoulli with parameter p; that is, $P\big(X_1 = 1\big) = p$ and $P\big(X_1 = 0\big) = 1 - p$. Then $M(\theta) = p \exp(\theta) + (1 - p)$. Calculation gives $I(x) = x \log(x/p) + (1 - x) \log[(1 - x)/(1 - p)]$ for $0 \leq x \leq 1$ (take $0 \log 0 = 0$ in this expression). $I(x) = \infty$ for x outside this range.

Example 3.1.3. Suppose X_1 is exponential (with density $p(x) = \lambda \exp(-\lambda x)$). Then $M(\theta) = \lambda/(\lambda - \theta)$ for $\theta < \lambda$. We then find straightforwardly that $I(x) = \lambda x - 1 - \log(\lambda x)$ for $x > 0$, and ∞ otherwise.

Example 3.1.4. Suppose X_1 is Cauchy (with density $p(x) = (\pi(1 + x^2))^{-1}$. $M(\theta) = \infty$ for $\theta \neq 0$. Hence $I(x) = 0$ for all x.

Theorem 3.1.1 (Cramér). *For every closed subset $F \subset \mathcal{R}$,*

$$\limsup_{n\to\infty} \frac{1}{n} \log P\big(S_n \in F\big) \le -\inf_{x\in F} I(x) = -I(F)$$

and for every open subset $G \subset \mathcal{R}$,

$$\liminf_{n\to\infty} \frac{1}{n} \log P\big(S_n \in G\big) \ge -\inf_{x\in G} I(x) = -I(G).$$

Proof. We prove only a special case of the theorem. We only consider interval subsets of $\mathcal{R}$. We also (for the lower bound) assume that for every $y \in (a,b)$, there exists a θ_y such that $I(y) = \theta_y y - \log M(\theta_y)$.

Upper bound. We suppose that our set $F = [a,b]$. If $a < m < b$, then by the law of large numbers and the fact that $I(m) = 0$ (which implies $\inf_{x\in F} I(x) = I(F) = 0$), we have that the upper bound theorem statement is true. Now, without loss of generality, suppose that $a \ge m$ (the case $a < b \le m$ is handled by replacing X_k with $-X_k$). Then for $\theta \ge 0$,

$$\begin{aligned} P\big(S_n \in [a,b]\big) &= \int_{[a,b]} dP_n(x) \\ &\le \exp[-\theta a] \int_{[a,b]} \exp(\theta x) dP_n(x) \\ &\le \exp[-\theta a] \int \exp(\theta x) dP_n(x) \\ &= \exp(-\theta a) M\left(\frac{\theta}{n}\right)^n, \end{aligned}$$

where $P_n(\cdot)$ is the probability distribution function of S_n. This implies that

$$\frac{1}{n} \log P\big(S_n \in [a,b]\big) \le -\frac{\theta a}{n} + \log M(\frac{\theta}{n}).$$

One may now replace θ by $n\theta$ to obtain

$$\frac{1}{n} \log P\big(S_n \in [a,b]\big) \le -[\theta a - \log M(\theta)].$$

Since this is true for all $\theta > 0$, we have

$$\begin{aligned} \frac{1}{n} \log P\big(S_n \in [a,b]\big) &\le \inf_{\theta>0} -[\theta a - \log M(\theta)] \\ &= -\sup_{\theta>0}[\theta a - \log M(\theta)] \\ &= -I(a) \\ &= -\inf_{x\in[a,b]} I(x). \end{aligned}$$

One now merely takes the limit superior over n to finish the upper bound part of the proof. □

Lower bound. We suppose that $G = (a, b)$. Let $y \in (a, b)$. Let $\delta > 0$ be small enough so that $(y - \delta, y + \delta) \subset (a, b)$. It is then sufficient to show that

$$\liminf_{n\to\infty} \frac{1}{n} \log P\big(S_n \in (y - \delta, y + \delta)\big) \geq -I(y).$$

Let θ_y be the point such that $I(y) = \theta_y y - \log M(\theta_y)$. Without loss of generality, we can suppose that y is a number greater than or equal to the mean value m. This then implies that θ_y is greater than or equal to zero by the arguments we made at the beginning of the proof. Define a new random variable $X^{(\theta_y)}$ with probability distribution

$$P_{\theta_y}(z) = P\big(X^{(\theta_y)} \leq z\big) = \frac{\int_{-\infty}^{z} \exp(\theta_y x) dP_1(x)}{M(\theta_y)}.$$

It is easy to see that the mean value of this new distribution is y. Therefore, by the law of large numbers, we must have for any $\delta_1 > 0$,

$$\lim_{n\to\infty} \int_{|(x_1+x_2+\cdots+x_n)/n-y|<\delta_1} dP_{\theta_y}(x_1) dP_{\theta_y}(x_2) \cdots dP_{\theta_y}(x_n) = 1.$$

Now note that for $\delta_1 < \delta$ we have,

$$\begin{aligned}
&P\big(S_n \in (y - \delta, y + \delta)\big) \\
&= \int_{|(x_1+x_2+\cdots+x_n)/n-y|<\delta} dP_1(x_1) dP_1(x_2) \cdots dP_1(x_n) \\
&\geq \int_{|(x_1+x_2+\cdots+x_n)/n-y|<\delta_1} dP_1(x_1) dP_1(x_2) \cdots dP_1(x_n) \\
&\geq \exp[(-ny - \delta_1 n)\theta_y] \int_{|(x_1+x_2+\cdots+x_n)/n-y|<\delta_1} \\
&\quad \exp[\theta_y(x_1 + x_2 + \cdots + x_n)] dP_1(x_1) dP_1(x_2) \cdots dP_1(x_n) \\
&\geq \exp[(-ny - \delta_1 n)\theta_y] M(\theta_y)^n \\
&\quad \int_{|(x_1+x_2+\cdots+x_n)/n-y|<\delta_1} dP_{\theta_y}(x_1) dP_{\theta_y}(x_2) \cdots dP_{\theta_y}(x_n).
\end{aligned}$$

Therefore,

$$\begin{aligned}
\liminf_{n\to\infty} \frac{1}{n} \log P\big(S_n \in (y - \delta, y + \delta)\big) &\geq (-y - \delta_1)\theta_y + \log M(\theta_y) \\
&= -I(y) - \delta_1 \theta_0.
\end{aligned}$$

Since δ_1 can be made arbitrarily small, we have finished the lower bound part of the proof. □

One of the more interesting aspects of this theorem is that the asymptotics of the probability of the sample average lying in some set away from the mean depend essentially on one point, that is, the logarithmic asymptotics of

$P(S_n \in (m+\epsilon, m+2\epsilon))$ are the same as for $P(S_n \in (m+\epsilon, \infty))$. A point of a set where $I(x)$ attains its minimum is called a *minimum rate point* of the set. The minimum rate points of the set essentially determine the asymptotic rate with which its probability goes to zero.

One should point out that the asymptotic upper bound of Cramér's Theorem actually holds for every n, not just for n large enough: $P(S_n \in [a,b]) \leq \exp(-nI(a))$ for all positive integers n (for the case $a > m$). This upper bound is usually called the *Chernoff bound* in communications theory.

In the lower bound part of the proof of the theorem we defined a new probability distribution P_{θ_y}. This distribution has a variety of names and uses. It is variously called the *twisted distribution*, the *tilted distribution*, the *exponential change of measure*, etc. The primary reason for this distribution is to shift the probability mass of the original random variables so that their mean value equals the minimum rate point of the set under consideration.

The theorem that we have given is by no means the most general form that is known even for the i.i.d. case. A complete version of the theorem is given in [21] with all the technicalities in place. Large deviation theorems are usually expressed as two separate limit theorems: an upper bound for closed sets and a lower bound for open sets. Only in the case of certain nice sets, such as intervals (or, more generally in higher-dimensional spaces, convex sets) can one be guaranteed that the upper bound equals the lower bound. For most of the applications that we have in mind, the interval case will be sufficient.

Recall the definitions $D_I = \{x : I(x) < \infty\}$, the effective domain of $I(\cdot)$, and similarly, $D_M = \{\theta : M(\theta) < \infty\}$. In our proof of the lower bound, we assumed that for every $y \in D_I$ there exists a θ_y such that $I(y) = \theta_y y - \log M(\theta_y)$. An alternate condition implies this property is to assume or require that $\log M(\theta)$ is steep. Steepness means simply that the magnitude of the derivative of $\log M(\theta)$ approaches infinity as it approaches the boundary of D_M, denoted ∂D_M. More precisely, a function $f : \mathcal{R}^d \to \mathcal{R}$ differentiable on its effective domain D_f is *steep* if $\{x_n\} \subset D_f$ and $x_n \to x_0 \in \partial D_f$ implies that $\|\nabla f(x_n)\| \to \infty$.

Example 3.1.5. We can derive the large deviation rate result for sums of i.i.d. Gaussian random variables directly. First note (for $w > 0$) that if we differentiate

$$\left(\frac{1}{w} - \frac{1}{w^3}\right) \exp(-\frac{1}{2}w^2) \qquad w > 0, \tag{3.5}$$

we get $-(1 - 3/w^4)\exp(-w^2/2)$ which is the derivative (by the Fundamental Theorem of the Calculus) of

$$\int_w^\infty (1 - \frac{3}{x^4}) \exp(-\frac{1}{2}x^2) dx \qquad w > 0. \tag{3.6}$$

Since both expressions tend to zero as $w \to \infty$, there are no constant terms and thus the expression in (3.5) equals that of (3.6). Now note that, for $w > 0$,

$$\begin{aligned}\int_w^\infty \exp(-\frac{1}{2}x^2)dx &\geq \int_w^\infty (1 - \frac{3}{x^4}) \exp(-\frac{1}{2}x^2)dx \\ &= (\frac{1}{w} - \frac{1}{w^3}) \exp(-\frac{1}{2}w^2)\end{aligned}$$

and

$$\begin{aligned}\int_w^\infty \exp(-\frac{1}{2}x^2)dx &\leq \int_w^\infty \frac{x}{w} \exp(-\frac{1}{2}x^2)dx \\ &= \frac{1}{w} \exp(-\frac{1}{2}w^2).\end{aligned}$$

Now let $S_n = (1/n)\sum_{i=1}^n X_i$, where X_1 is a standard Gaussian random variable. Then S_n is zero mean Gaussian with variance $1/n$. Then for $w > 0$,

$$\begin{aligned}P(S_n > w) &= P(\frac{X_1}{\sqrt{n}} > w) \\ &= P(X_1 > w\sqrt{n}) \\ &= \int_{w\sqrt{n}}^\infty \frac{\exp(-\frac{1}{2}x^2)}{\sqrt{2\pi}} dx.\end{aligned}$$

Using our upper and lower bounds then gives us

$$\lim_{n\to\infty} \frac{1}{n} \log P(S_n > w) = -\frac{w^2}{2},$$

the same result as Cramér's Theorem.

Example 3.1.6. In this example we compare rate results obtained by careless use of the Central Limit Theorem with those obtained via Cramér's Theorem. Let $\{X_j\}$ be an i.i.d. collection of Gaussian, mean zero, variance 1, random variables. Define $S_n = (1/n)\sum_1^n X_i^2$. S_n converges to 1 by the law of large numbers. We are interested in the asymptotics of $P(S_n > \gamma)$ for some γ greater than 1. Carelessly, one might say that S_n is approximately Gaussian with mean value 1 and variance $2/n$. Let $N(m, \sigma^2)$ denote a Gaussian random variable of mean m and variance σ^2. Hence

$$\begin{aligned}P(S_n > \gamma) &= P(N\left(1, \frac{2}{n}\right) > \gamma) \\ &= P(N\left(0, \frac{1}{n}\right) > \frac{\gamma - 1}{\sqrt{2}}) \\ &= P(N(0,1) > \sqrt{n}\frac{\gamma-1}{\sqrt{2}}) \\ &= k(n)\exp(-n(\gamma-1)^2/4),\end{aligned}$$

where, from the previous example we have $k(n)$ bounded as

$$\frac{1}{\sqrt{n}\frac{\gamma-1}{\sqrt{2}}} - \left(\frac{1}{\sqrt{n}\frac{\gamma-1}{\sqrt{2}}}\right)^3 \leq k(n) \leq \frac{1}{\sqrt{n}\frac{\gamma-1}{\sqrt{2}}}.$$

We thus conclude that $\lim_n (1/n) \log P(S_n > \gamma) = -(\gamma - 1)^2/4$. However, X_1^2 has a chi-squared distribution. $\log M(\theta)$ is calculated (by completing a square) to be $-(1/2)\log(1-2\theta)$ for θ in the range $(-\infty, (1/2)) = D_M$. $I(\gamma)$ is then easily found to be $(1/2)(\gamma - \log\gamma - 1)$. It is trivial to check that $\log M(\theta)$ is steep. Hence we have that $\lim_n (1/n)\log P(S_n > \gamma) = -I(\gamma)$. One makes more than a factor of 2 error in the rate (!) when $\gamma = 3$ using the Central Limit Theory approximations.

Note that in the previous example as γ approaches 1, the mean value of x_1^2, the two rate calculations are asymptotically equivalent. Of course, they both go to zero as γ goes to 1; but also, and more important, they approach zero at the same rate; that is, $\lim_{\gamma\to 1}[(\gamma - 1)^2/4]/(\gamma) = 1$. This phenomenon of the Central Limit Theorem approximations and the large deviation calculations coinciding as the minimum rate points of the sets under consideration approach the mean value is the usual behavior.

3.2 Gärtner–Ellis Theorem

Cramér first presented his theorem at a probability symposium in 1937 [18]. Since then, the theory has been broadened substantially. One of the most useful and surprising generalizations is the one due to Gärtner [31] and, more recently, Ellis [24]. These authors dispense with assumptions about the dependency structure of the random variable sequence and work entirely with assumptions on a sequence of moment generating functions. Their methods thus allow large deviation results to be derived for dependent random processes such as Markov chains, functionals of Gaussian random processes, etc. The main idea of their proof is to look closely at the proof of Cramér's Theorem and see which aspects are really essential to the upper and lower bound derivations. To point out in an informal fashion what the differences are between Cramér's Theorem and the Gärtner–Ellis theorem, consider the following table where we try to show the relevant quantities and results for each theorem in the scalar random variable setting.

Cramér	**Gärtner–Ellis**
$S_n = \sum_{i=1}^n X_i \quad \{X_i\}$ i.i.d.	Y_n
$\log(M(\theta)) = \log(\mathbb{E}[\exp(\theta X_1)])$	$\phi(\theta) = \lim_{n\to\infty} \frac{1}{n}\log(\mathbb{E}[\exp(\theta Y_n)])$
$= \frac{1}{n}\log(\mathbb{E}[\exp(\theta S_n)])$	
$I(x) = \sup_\theta[\theta x - \log(M(\theta))]$	$I(x) = \sup_\theta[\theta x - \phi(\theta)]$
$P\left(\frac{S_n}{n} \in E\right) \approx f_c(n)\exp\left(-nI(E)\right)$	$P\left(\frac{S_n}{n} \in E\right) \approx f_{ge}(n)\exp\left(-nI(E)\right)$

We now give a more formal treatment of the Gärtner–Ellis theorem for $\mathcal{R}^d$-valued random variables. Suppose that we have some infinite sequence of $\mathcal{R}^d$-valued random variables $\{Y_n\}$. No assumptions are made about the dependency structure of this sequence (at least directly). Define for $\theta \in \mathcal{R}^d$,

$$\phi_n(\theta) = \frac{1}{n}\log\mathbb{E}[\exp(\langle\theta, Y_n\rangle)].$$

First of all note that ϕ_n is a convex function, since for $\lambda \in [0,1]$ we have via Hölder's inequality

$$\begin{aligned}
&\phi_n\left(\theta_1\lambda + \theta_2(1-\lambda)\right)\\
&= \frac{1}{n}\log\mathbb{E}[\left(\exp(\langle\theta_1, Y_n\rangle)\right)^\lambda \left(\exp(\langle\theta_2, Y_n\rangle)\right)^{(1-\lambda)}]\\
&\le \frac{1}{n}\log\left(\mathbb{E}[\exp(\langle\theta_1, Y_n\rangle)]\right)^\lambda + \frac{1}{n}\log\left(\mathbb{E}[\exp(\langle\theta_2, Y_n\rangle)]\right)^{(1-\lambda)}\\
&= \lambda\phi_n(\theta_1) + (1-\lambda)\phi_n(\theta_2).
\end{aligned}$$

We now list three assumptions that we call the **standard assumptions** for any sequence of convex functions ϕ_n.

Assumption A1. $\phi(\theta) = \lim_n \phi_n(\theta)$ exists for all $\theta \in \mathcal{R}^d$, where we allow ∞ both as a limit value and as an element of the sequence $\{\phi_n(\theta)\}$.

In other words, the limit exists and is defined to be $\phi(\theta)$ for all $\theta \in \mathcal{R}^d$. Note that the effective domain of ϕ, D_ϕ is convex and thus ϕ itself is convex since it is the limit of convex functions on a convex set.

Assumption A2. The origin belongs to the interior of the effective domain of ϕ, that is, $0 \in \overset{\circ}{D}_\phi$ and ϕ itself is lower semi-continuous.

We always have that $0 \in D_\phi$ since $\phi(0) = 0 < \infty$. (This, of course, wouldn't be true for an arbitrary sequence of convex functions.) The convexity of ϕ follows from the convexity of each ϕ_n (the limit of convex functions is convex). The lower semi-continuity of ϕ is equivalent to ϕ being a *closed* convex function which in turn is equivalent to the condition that if for each real α the set $\{\theta \in \mathcal{R}^d : \phi(\theta) \leq \alpha\}$ is closed in $\mathcal{R}^d$.

A convex function ϕ is *essentially smooth* if three conditions hold, the set $\overset{\circ}{D}_\phi$ is nonempty, ϕ is differentiable everywhere in $\overset{\circ}{D}_\phi$, and ϕ is steep.

Assumption A3. ϕ is essentially smooth.

Note that if ϕ is essentially smooth, it will be automatically lower semi-continuous.

Assumption A4. ϕ is strictly convex.

Now we may define the (Gärtner–Ellis) large deviation rate function

$$I(x) = \sup_\theta[\langle \theta, x\rangle - \phi(\theta)].$$

In convex analysis, I is denoted the *convex conjugate function* of ϕ. If ϕ satisfies A2, A3, A4, then by Theorem A.0.10 in the Appendix, we have that I must also be essentially smooth and strictly convex. Furthermore for every $x \in \overset{\circ}{D}_I$, we have $I(x) = \langle \theta_x, x\rangle - \phi(\theta_x)$, where θ_x is the unique solution to $\nabla\phi(\theta_x) = x$.

Furthermore, if ϕ satisfies A2, A3, A4, then ϕ is the convex conjugate function of I and thus

$$\phi(\theta) = \sup_x[\langle x, \theta\rangle - I(x)].$$

(Actually, all that is required is that ϕ be closed which for a proper convex function is equivalent to lower semi-continuous.) Furthermore, $\phi(\theta_x) = \langle x, \theta_x\rangle - I(x)$, where x is the unique solution to $\nabla I(x) = \theta_x$.

Before proceeding to the main theorem, we first prove a simple but important lemma.

Lemma 3.2.1 (Exponential Overbound). *Let Z be a reals-valued random variable. Then for $t \geq 0$, $z \in \mathcal{R}$, $P(Z > z) \leq \mathbb{E}[\exp(tZ)]\exp(-tz)$.*

Proof. The proof follows immediately from Markov's inequality. More directly, let F be the distribution function of Z. $P(Z > z) = \int_z^\infty 1dF(y)$. Now for $y \in [z, \infty)$, $1 \leq \exp[t(y - z)]$ for all $t \geq 0$. Hence

$$\begin{aligned} P(Z > z) &\leq \int_z^\infty \exp[t(y-z)]dF(y) \\ &\leq \int_{-\infty}^\infty \exp[t(y-z)]dF(y) \\ &= \mathbb{E}[\exp(tZ)]\exp(-tz). \end{aligned}$$

□

Theorem 3.2.1 (Gärtner–Ellis). *Assume A1. For every compact subset* $K \subset \mathcal{R}^d$,

$$\limsup_{n\to\infty} \frac{1}{n} \log P\big(\frac{Y_n}{n} \in K\big) \le -\inf_{x\in K} I(x) = -I(K).$$

Assume A1 and A2. For every closed subset $F \subset \mathcal{R}^d$,

$$\limsup_{n\to\infty} \frac{1}{n} \log P\big(\frac{Y_n}{n} \in F\big) \le -\inf_{x\in F} I(x) = -I(F).$$

Assume A1 and A3. For every open subset $G \subset \mathcal{R}$,

$$\liminf_{n\to\infty} \frac{1}{n} \log P\big(Y_n \in G\big) \ge -\inf_{x\in G} I(x) - I(G).$$

Lemma 3.2.2. *Assume A1. Fix* $a \in \mathcal{R}$ *and* $\theta_1, \theta_2, \ldots, \theta_k$ *each in* $\mathcal{R}^d$. *For each* $\theta \in \mathcal{R}^d$, *define the half-space* $H_\theta(a) = \{x \in \mathcal{R}^d : \langle x, \theta\rangle - \phi(\theta) \le a\}$. *Let* $C = \cap_{i=1}^k H_{\theta_i}(a)$. *Then*

$$\limsup_{n\to\infty} \frac{1}{n} \log P\big(\frac{Y_n}{n} \notin C\big) \le -a.$$

Proof (Lemma 3.2.2).

$$\begin{aligned}
&P\big(\frac{Y_n}{n} \notin C\big) \\
&= P\big(\frac{Y_n}{n} \not\in \cup_{i=1}^k H_{\theta_i}^c(a)\big) \\
&\le \sum_{i=1}^k P\big(\langle \frac{Y_n}{n}, \theta_i\rangle - \phi(\theta_i) > a\big) \\
&= \sum_{i=1}^k P\big(\langle Y_n, \theta_i\rangle > n\phi(\theta_i) + an\big) \\
&\le \sum_{i=1}^k \exp(-n[a + \phi(\theta_i)])\mathbb{E}[\exp(\langle Y_n, \theta_i\rangle)] \quad \text{by Lemma 3.2.1} \\
&= \sum_{i=1}^k \exp\big(-n[a + \phi(\theta_i)]\big) \exp\big(n\phi_n(\theta_i)\big).
\end{aligned}$$

We note that $\limsup \phi_n(\theta) = \phi(\theta)$. Now, taking logarithms, multiplying by $1/n$, and taking the limit superior, gives us the result. □

Proof (Theorem 3.2.1, upper bound for compact K). Assume A1. Recall $I(K) = \inf_{x \in K} I(x)$ and now define the level sets $L_a = \{x : I(x) \leq a\}$. Fix an $\epsilon > 0$. Now note that

$$\begin{aligned} K &\subset L^c_{I(K)-\epsilon} \\ &= \{x : I(x) > I(K) - \epsilon\} \\ &= \{x : \sup_\theta[\langle\theta, x\rangle - \phi(\theta)] > I(K) - \epsilon\} \\ &= \cup_\theta \{x : \langle\theta, x\rangle - \phi(\theta) > I(K) - \epsilon\}. \end{aligned}$$

But this is an open cover of a compact set. Hence by the Heine–Borel Theorem, there exists a finite subcover, or in other words, there exists $\theta_1, \theta_2, \ldots, \theta_k$ such that

$$\begin{aligned} K &\subset \cup_{i=1}^k \{x : \langle\theta_i, x\rangle - \phi(\theta_i) > I(K) - \epsilon\} \\ &= \cup_{i=1}^k H^c_{\theta_i}(I(K) - \epsilon) \\ &= (\cap_{i=1}^k H_{\theta_i}(I(K) - \epsilon))^c. \end{aligned}$$

We now invoke Lemma 3.2.2 to obtain,

$$\limsup_{n\to\infty} \frac{1}{n} \log P\big(\frac{Y_n}{n} \in K\big) \leq -[I(K) - \epsilon],$$

for every $\epsilon > 0$. We can take ϵ as small as we wish and obtain the theorem statement for compact K. □

Lemma 3.2.3. *Assume A1 and A2. Then, for every $a \in \mathcal{R}$, L_a is compact.*

Proof (Lemma 3.2.3). Take $v \in L_a$. Then $I(v) \leq a$. Since 0 is in the interior of the effective domain, we can find an $\epsilon > 0$ such that a ball $B_{2\epsilon}$ of radius 2ϵ and center 0 is contained in $\overset{\circ}{D}_\phi$. Hence

$$\begin{aligned} \infty > a &\geq \sup_\theta[\langle\theta, v\rangle - \phi(\theta)] \\ &\geq \sup_{\theta \in B_\epsilon}[\langle\theta, v\rangle - \phi(\theta)] \\ &\geq \sup_{\theta \in B_\epsilon}[\langle\theta, v\rangle] - \sup_{\theta \in B_\epsilon}[\phi(\theta)] \\ &= \epsilon\|v\|^2 - \sup_{\theta \in B_\epsilon}[\phi(\theta)]. \end{aligned}$$

Hence,

$$\|v\|^2 \leq \frac{a}{\epsilon} + \frac{1}{\epsilon} \sup_{\theta \in B_\epsilon}[\phi(\theta)] < \infty.$$

Therefore L_a is bounded. It is closed since by Assumption A2, ϕ is a lower semi-continuous (closed) convex function. Bounded and closed in $\mathcal{R}^d$ is equivalent to compact. The proof of the lemma is complete. □

Lemma 3.2.4. *Assume A1 and A2. Let L_a^δ be the δ neighborhood of L_a, that is, $\{x : \|x - y\| < \delta$ for some $y \in L_a\}$. Then*

$$\limsup_{n\to\infty} \frac{1}{n} \log P\Big(\frac{Y_n}{n} \notin L_a^\delta\Big) \le -a.$$

Proof (Lemma 3.2.4).

$$\begin{aligned} L_a &= \{x : \sup[\langle\theta, x\rangle - \phi(\theta)] \le a\} \\ &= \cap_\theta \{x : \langle\theta, x\rangle - \phi(\theta) \le a\} \\ L_a^c &= \cup_\theta \{x : \langle\theta, x\rangle - \phi(\theta) > a\}. \end{aligned}$$

Note that $S = (L_a^\delta)^c$ is closed. $S \subset L_a^c$. The boundary of S, ∂S is closed, bounded, and is a subset of S. It is compact; therefore there exist $\theta_1, \theta_2, \ldots, \theta_r$ such that

$$\partial S \subset \cup_{i=1}^r \{x : \langle\theta_i, x\rangle - \phi(\theta_i) > a\} = U. \tag{3.7}$$

Suppose there exists an $x \in S$ that is not in U. Then $x \in U^c = \cap_{i=1}^r \{x : \langle\theta_i, x\rangle - \phi(\theta_i) \le a\}$. Choose some $y \in L_a$. Note also that $y \in U^c$ (remember that y is in the infinite intersection of all such sets). Since the intersection of convex sets is convex, we know that U^c is convex. Hence a straight line drawn between any two points in the set must lie entirely in the set. But a straight line drawn between x and y would have to intersect the boundary ∂S at some point b. Hence b must lie in U^c. However, this is a contradiction of (3.7). Hence

$$S \subset \cup_{i=1}^r \{x : \langle\theta_i, x\rangle - \phi(\theta_i) > a\}.$$

The lemma statement then follows by Lemma 3.2.2. □

Proof (Theorem 3.2.1 upper bound for closed sets). We assume A1 and A2. Let $F \subset \mathcal{R}^d$ be closed. Choose some some positive number a. Note that L_a^δ is bounded and hence $L_a^\delta \cap F$ is compact. $P(Y_n/n \in F) = P(Y_n/n \in F \cap L_a^\delta) + P(Y_n/n \in F \cap (L_a^\delta)^c)$. Note that

$$\limsup_{n\to\infty} \frac{1}{n} \log P\Big(\frac{Y_n}{n} \in F \cap L_a^\delta\Big) \le -I(F)$$

and

$$\limsup_{n\to\infty} \frac{1}{n} \log P\Big(\frac{Y_n}{n} \in F \cap (L_a^\delta)^c\Big) \le -a.$$

Hence

$$\limsup_{n\to\infty} \frac{1}{n} \log P\Big(\frac{Y_n}{n} \in F\Big) \le -[\max(I(F), a)].$$

We can take a arbitrarily large and then get the theorem statement. The proof of the upper bound for closed sets is complete. □

We now embark upon the proof of the lower bound for open sets. We first state and prove some building block lemmas.

Lemma 3.2.5. *Let F_n denote the distribution function of Y_n and define the distribution function of the random variable $Y_n^{(\theta)}$ as*

$$dF_n^{(\theta)}(x) = \frac{dF_n(x)\exp(\langle\theta, x\rangle)}{\int \exp(\langle\theta, x\rangle)dF_n(x)} = \frac{dF_n(x)\exp(\langle\theta, x\rangle)}{\exp(n\phi_n(\theta))}.$$

Let $B_\delta(v) = \{x : \|x - v\| < \delta\}$. Suppose $v \in \nabla\phi(D_\phi)$, that is, $\nabla\phi(\theta) = v$ has a solution θ_v. Then

$$\lim_{n\to\infty} P\Big(\frac{Y_n^{(\theta_v)}}{n} \notin B_\delta(v)\Big) = 0.$$

Proof (Lemma 3.2.5). We actually show the stronger result that this convergence is exponentially fast. Define $\phi_n^{(\theta_v)}(\cdot)$ as

$$\begin{aligned}\phi_n^{(\theta_v)}(\beta) &= \frac{1}{n}\log\int \exp(\langle\beta, x\rangle)F_n^{(\theta_v)}(dx)\\ &= \frac{1}{n}\log\int \exp(\langle\theta_v + \beta, x\rangle)F_n(dx) - \phi_n(\theta_v)\\ &\to_{n\to\infty} \phi(\theta_v + \beta) - \phi(\theta).\end{aligned}$$

Now note that $\phi_n^{(\theta_v)}(\beta)$ is differentiable at the origin. Hence the $\phi_n^{(\theta_v)}$ for the random variables $Y_n^{(\theta_v)}$ satisfy conditions A1 and A2. Denote the rate function of these random variables as $I^{(\theta_v)}(\cdot)$. Let $L_{\theta_v,\epsilon}$ be the ϵ level set of $I^{(\theta_v)}(\cdot)$. Then (by the differentiability of $\phi_n^{(\theta_v)}(\cdot)$ at the origin, there exists an $\epsilon > 0$ so that $L_{\theta_v,\epsilon}^{\delta/2} \subset B_\delta(v)$. Hence

$$\begin{aligned}P\Big(\frac{Y_n}{n} \notin B_\delta(v)\Big) &\leq P\Big(\frac{Y_n}{n} \notin L_{\theta_v,\epsilon}^{\delta/2}\Big)\\ &\leq \exp(-\epsilon' n)\end{aligned}$$

by invoking Lemma 3.2.4 for n sufficiently large and for some $\epsilon' < \epsilon$. This finishes the proof of the lemma. □

Lemma 3.2.6. *Assume $v = \nabla\phi(\theta)$ for $\theta = \theta_v$. Then*

$$\liminf_{n\to\infty} \frac{1}{n} \log P\big(Y_n/n \in B_\delta(v)\big) \geq -I(v) - \delta\|\theta_v\|.$$

Proof (Lemma 3.2.6).

$$
\begin{aligned}
&P\big(\frac{Y_n}{n} \in B_\delta(v)\big) \\
&= P\big(Y_n \in nB_\delta(v)\big) \quad \big(\text{note that } nB_\delta(v) = B_{n\delta}(nv)\big) \\
&= \int_{nB_\delta(v)} \exp[-\langle\theta, x\rangle + n\phi_n(\theta)] dF_n^{(\theta)}(x) \\
&= \exp[-n\big(\langle\theta, v\rangle + \phi_n(\theta)\big)] \int_{nB_\delta(v)} \exp[-\langle\theta, x - nv\rangle] dF_n^{(\theta)}(x) \\
&= \exp[-n\big(\langle\theta_v, v\rangle + \phi_n(\theta_v)\big)] \exp[-n\|\theta_v\|\delta] F_n^{(\theta_v)}\big(nB_\delta(v)\big),
\end{aligned}
$$

where the last line follows by setting $\theta = \theta_v$ and where $F_n^{(\theta_v)}(A)$ denotes the probability of the set A under the probability distribution $F_n^{(\theta_v)}$. Note that $(1/n)\log(\cdot)$ of the first exponential product term converges to $-I(v)$ as n approaches infinity. $F_n^{(\theta_v)}\big(nB_\delta(v)\big)$ converges to 1 by Lemma 3.2.5, and hence $(1/n)\log(\cdot)$ of it converges to zero. We have thus finished the proof of the lemma. □

Now we consider open sets that are in the range of the gradient map of ϕ.

Proof (Theorem 3.2.1 lower bound for open sets G, $\ni G \subset \nabla\phi(D_\phi)$). Take any $v \in G$. Then

$$
\begin{aligned}
\liminf_{n\to\infty} \frac{1}{n} \log P\big(\frac{Y_n}{n} \in (a, b)\big) &\geq \liminf_{n\to\infty} \frac{1}{n} \log P\big(\frac{Y_n}{n} \in B_\delta(v)\big) \\
&\geq -I(v) - \delta|\theta_v|,
\end{aligned}
$$

for all sufficiently small $\delta > 0$. Since δ can be made arbitrarily small and since $v \in G$ is arbitrary, we have shown the lower bound portion of the theorem statement. □

Proof (Theorem 3.2.1 lower bound). Again take G to be open in $\mathcal{R}^d$. Take some $v \in G$. We would like to show that

$$
\liminf_{n\to\infty} \frac{1}{n} \log P\big(\frac{Y_n}{n} \in G\big) \geq -I(v). \tag{3.8}
$$

It is known [68, Theorem 26.5] from convex function theory that $\overset{\circ}{D}_I \subset \nabla\phi(\overset{\circ}{D}_\phi)$. Hence for $v \in G$, there are three cases. If $v \notin D_I$, then (3.8) holds trivially since $-I(v) = -\infty$. Now suppose that $v \in \overset{\circ}{D}_I$. Now we are in the same situation as in our proof of the lower bound for the case of $G \subset \nabla\phi(D_\phi)$. Hence for this case (3.8) holds. The third and last case is if $v \in D_I$ but not in the interior. In this setting, one can show that any tiny ball taken about v will contain another point $v' \in \overset{\circ}{D}_I$ such that $I(v') \leq I(v)$. Hence when we are minimizing the rate function over the set G, we need only worry about the points that are in the interior of the set D_I; the boundary points need not be considered. Our proof of the Gärtner–Ellis Theorem is now complete. □

Remark 3.2.1. Under the standard Assumptions A1, A2, and A3, ϕ is differentiable at zero, with value $\nabla\phi(0) = m$. By an argument similar to that for Property 2 of the rate function in Cramér's Theorem, we have that m is the unique minimum of I ($I(x) > I(m) = 0$ for $x \neq m$). Consider some ball of radius $\delta > 0$ centered at m, $B_\delta(m)$. Then $P\big(Y_n/n \notin B_\delta(m)\big)$ goes to zero exponentially fast for all $\delta > 0$. This is enough to deduce that Y_n/n converges almost surely to m. (The convergence in probability is immediate. The fact that it converges exponentially fast allows us to deduce via the use of the Borel–Cantelli lemma that the convergence is almost sure.)

Let us now consider an example.

Example 3.2.1. Consider the autoregressive process $x_n = \alpha x_{n-1} + w_n$, $n = 1, 2, \dots$; $|\alpha| < 1$ and $x_0 = 0$. $\{w_n\}$ $n = 1, 2, \dots$ is an i.i.d. sequence such that $P\big(w_1 \in [-1/2, 1/2]\big) = 1$, and $\mathbb{E}[w_1] = 0$. Define $Y_n = \sum_1^n x_i$. We are interested in the large deviation asymptotics of $P\big(Y_n/n > \beta\big)$ where $\beta > 0$. Now $x_1 = w_1, x_2 = \alpha w_1 + w_2, \dots, x_n = \sum_1^n \alpha^{n-i} w_i$. Therefore,

$$Y_n = \frac{\alpha^n - 1}{\alpha - 1} w_1 + \frac{\alpha^{n-1} - 1}{\alpha - 1} w_2 + \cdots + w_n.$$

Hence

$$\exp[n\phi_n(\theta)] = \prod_{i=1}^{n} M_w\Big(\frac{1-\alpha^i}{1-\alpha}\theta\Big),$$

where $M_w(\cdot)$ is the moment generating function of w_1.

This implies that

$$\phi_n(\theta) = \frac{1}{n}\sum_{i=1}^{n} \log M_w\Big(\frac{1-\alpha^i}{1-\alpha}\theta\Big) \to_{n\to\infty} \log M_w\Big(\frac{\theta}{1-\alpha}\Big).$$

Hence the asymptotics of the sum Y_n are the same as the asymptotics of a sum of i.i.d. random variables with the same distribution as $w_1/(1-\alpha)$. To be more specific, consider $P\big(w_1 = 1/2\big) = 1/2 = P\big(w_1 = -1/2\big)$. Then $M_w(\theta) = [\exp(\theta/2) + \exp(-\theta/2)]/2 = \cosh(\theta/2)$. Thus

$$\begin{aligned} I(x) &= \sup_\theta \Big[\theta x - \log\big(\cosh\big(\frac{\theta}{2(1-\alpha)}\big)\big)\Big] \\ &= x(1-\alpha)\log\frac{m^+(x)}{m^-(x)} - \log\left[\frac{1}{2}\Big(\sqrt{\frac{m^+(x)}{m^-(x)}} + \sqrt{\frac{m^-(x)}{m^+(x)}}\Big)\right], \end{aligned}$$

where the functions are defined as $m^+(x) = 1 + 2(1-\alpha)x$ and $m^-(x) = 1 - 2(1-\alpha)x$, for $0 < |x| < 1/[2(1-\alpha)]$ and infinity for x outside this range. Note that the assumptions that we have placed on the distribution of w_1 guarantee that $\mathcal{R} \subset \phi'(D_\phi)$ and $\limsup |Y_n/n| \leq 1/[2(1-\alpha)] < \infty$.

The differentiability of ϕ on $\overset{\circ}{D}_\phi$ is necessary for the lower bound, as we see in the following example.

Example 3.2.2. Suppose that Y_n is an independent sequence of random variables with $P(Y_n = n) = 1/2 = P(Y_n = -n)$. $\exp[n\phi_n(\delta)] = \cosh(n\theta)$. Therefore $\lim_n \phi_n(\theta) = |\theta|$ and Assumption A2 does not hold. Note that $I(x) = 0$ for $-1 \leq x \leq 1$ and is equal to ∞ otherwise. However $P\big(Y_n/n \in (-1,1)\big) = 0$, which implies that

$$\liminf \frac{1}{n} \log P\big(\frac{Y_n}{n} \in (-1,1)\big) = -\infty \ngeq - \inf_{x\in(-1,1)} I(x) = 0.$$

In the previous example the theorem failed because Assumption A2 didn't hold at the origin. Other examples can be constructed in which the theorem fails because the assumption fails at other points.

Example 3.2.3. Let $\{x_n\}$ be a d-dimensional i.i.d. sequence of Gaussian random vectors with mean value m and covariance matrix K. Assume that K is full rank (i.e., K is invertible). The moment generating function for a Gaussian random vector is well known to be $\mathbb{E}[\exp(\langle\theta, x_1\rangle)] = \exp(\langle\theta, m\rangle + (1/2)\theta^T K\theta)$ where the superscript T indicates transpose (for vectors and matrices). $I(x) = \sup_\theta[\langle\theta, x\rangle - \langle\theta, m\rangle - (1/2)\theta^T K\theta]$. Setting the gradient with respect to θ in the term in the braces to zero results in $x - m - K\theta = 0$. This implies that $\theta_x = K^{-1}(x - m)$. Substituting this value of θ back into the supremum expression yields $I(x) = (1/2)(x-m)^T K^{-1}(x-m)$.

Example 3.2.4. Let E be an exponential random variable with mean value of one. Suppose X_i are i.i.d. normal random variables of mean one and variance one and also jointly independent of E. We are interested in the quantity

$$P\big(E + \sum_{i=1}^{n} X_i > nA\big),$$

for some real-valued set A. We define $Y_n = E + \sum_{i=1}^n X_i$ and thus since E and X_i are independent, we can write

$$\begin{aligned}
\mathbb{E}[\exp(\theta Y_n)] &= \mathbb{E}[\exp\big(\theta(E + \sum_{i=1}^{n} X_i)\big)] \\
&= \mathbb{E}[\exp(\theta E)]\mathbb{E}[\exp(\theta X_1)]^n \\
\frac{1}{n}\log(\mathbb{E}[\exp(\theta Y_n)]) &= \frac{1}{n}\log(\mathbb{E}[\exp(\theta E)]) + \log(\mathbb{E}[\exp(\theta X_1)]).
\end{aligned}$$

We know

$$\mathbb{E}[\exp(\theta E)] = \begin{cases} \frac{1}{1-\theta} & \text{if } \theta < 1 \\ \infty & \text{otherwise} \end{cases}$$

and also

$$\mathbb{E}[\exp(\theta X_1)] = \exp(\theta + \frac{\theta^2}{2}).$$

Therefore

$$\phi(\theta) = \log[\mathbb{E}[\exp(\theta X_1)]] \qquad \text{for} \quad \theta < 1,$$

and takes on the value of ∞ for $\theta \geq 1$ since for such values of θ, we have $\phi_n(\theta) = \infty$ for all n. Therefore

$$\phi(\theta) = \begin{cases} \theta + \frac{\theta^2}{2} & \text{if } \theta < 1 \\ \infty & \text{otherwise} \end{cases}$$

Hence Assumption A1 is satisfied. Note that Assumption A2 fails. 0 *is* in the interior of the effective domain but ϕ is not closed since for $t > 3/2$, the set $\{\theta \in \mathcal{R} : \phi(\theta) \leq t\} = \{\theta \in [-1 - \sqrt{2t+1}, 1)$ is not closed. *A*3 fails, since ϕ is not steep. *A*4 does hold.

Now we compute the rate function as follows,

$$\begin{aligned} I(x) &= \sup_{\theta}[\theta x - \phi(\theta)] \\ &= \sup_{\theta<1}[\theta x - \theta - \frac{\theta^2}{2}] \\ &= \begin{cases} \frac{(x-1)^2}{2} & x < 2 \\ x - \frac{3}{2} & x \geq 2 \end{cases} \end{aligned}$$

We remark that since A1 holds, we can use this rate function as an *upper* bound to the true rate for compact sets A. For general closed sets, to make use of Theorem 3.2.1, we need A2 to hold also. To use Theorem 3.2.1 to lower bound the rate for open sets, we would need A3 to hold.

This example shows that adding even one heavy-tailed random variable to a sample average can significantly distort the large deviation asymptotics from what one might expect. In this example, the usual reaction is to think that the addition of just one exponential random variable to a very large sum of i.i.d. normal random variables will not make much difference in the overall scope of things. Counterintuitively this turns out to be false.

Example 3.2.5. Suppose $\{X_n^{(j)}\}$ is a collection of independent random variables. $X_n^{(j)}$ is uniform on the j points $\{0, 1/j, 2/j, \dots, (j-1)/j\}$. Define

$$Y_n = \sum_{i=1}^{n} X_i^{(n)};$$

we are interested in $P(Y_n/n > 0.7)$. Note that Cramér's Theorem can't be used in this setting since the distribution of the random variables in the sum is changing with n. Hence we must try to employ the Gärtner–Ellis framework.

Note that

$$\exp[n\phi_n(\theta)] = \prod_{j=1}^{n} M_{X_j}(\theta) = \left[\frac{1}{n}\frac{\exp(\theta)-1}{\exp(\frac{\theta}{n})-1}\right]^n$$

$$\phi_n(\theta) = \frac{1}{n}\log\left[\frac{1}{n}\frac{\exp(\theta)-1}{\exp(\frac{\theta}{n})-1}\right]^n$$

$$= \log\left[\frac{1}{n}\frac{\exp(\theta)-1}{\exp(\frac{\theta}{n})-1}\right].$$

The limit of the above, as $n \to \infty$, can be determined by L'Hospital's rule, giving

$$\phi(\theta) = \lim_{n\to\infty}\phi_n(\theta) = \log\frac{\exp(\theta)-1}{\theta}.$$

Interestingly we have found that $\phi(\theta) = \log(M(\theta))$ where $M(\theta)$ is the moment generating function of a uniform random variable on the interval $[0,1]$. Note also that this $\phi(\theta)$ is not steep.

The rate function is thus

$$I(x) = \sup_\theta\left[\theta x - \log(\frac{\exp(\theta)-1}{\theta})\right]$$

which unfortunately will not have a closed form solution. Thus, we numerically compute $I(.7)$ and find $I(0.7) \approx 0.2528$. The maximizing θ turns out to be ≈ 2.67.

Example 3.2.6 (Square Law Detector). Suppose that we have a signal detection (hypothesis testing) device that performs the following operation on samples $\{X_i\}$ coming into a communications receiver

$$\frac{1}{n}\sum_{i=1}^{n} X_i^2 \underset{H_0}{\overset{H_1}{\gtrless}} \gamma,$$

where the symbol $\underset{H_0}{\overset{H_1}{\gtrless}}$ signifies that if the left-hand side of the equation is greater than γ; we decide that hypothesis H_1 is true and if it is less than γ, we decide that hypothesis H_0 is true. We take H_1 to be the hypothesis that the samples are composed of a known deterministic signal sequence $\{s_i\}$ with deterministic power spectrum $P(\omega)$ plus a noise sequence $\{N_i\}$ which is assumed zero mean Gaussian with power spectrum $f(\omega)$. Define the vectors $N = (N_1 N_2 \cdots N_n)^T$ and $s = (s_1 s_2 \cdots s_n)^T$. N is Gaussian with covariance matrix V. We can diagonalize this matrix via $V = U^T \Lambda U$ where Λ is a diagonal matrix consisting of the eigenvalues of V and U is a unitary matrix

$(U^TU = I)$ whose columns are the eigenvectors of V. Define the transformations $N^* = \Lambda^{-1/2}U^TN$ and $s^* = \Lambda^{-1/2}U^Ts$. N^ast has covariance matrix I and hence the transformation we use is called a "whitening transformation."

Suppose we are interested in the probability of an error when hypothesis H_1 is true. Hence we need to compute the probability

$$\begin{aligned}
&P(\sum_{i=1}^{n}(s_i + N_i)^2 \le n\gamma) \\
&= P\big((s+N)^T(s+N) \le n\gamma\big) \\
&= P\big((s^*+N^*)^T\Lambda^{1/2}U^TU\Lambda^{1/2}(s^*+N^*) \le n\gamma\big) \\
&= P\big((s^*+N^*)^T\Lambda(s^*+N^*) \le n\gamma\big) \\
&= P\big(\sum_{i=1^n}\lambda_i(s_i^*+N_i^*)^2 \le n\gamma\big).
\end{aligned}$$

We now try to use the machinery of Theorem 3.2.1.

$$\begin{aligned}
\phi_n(\theta) &= \frac{1}{n}\log\mathbb{E}[\exp\big(\theta\sum_{i=1}^{n}\lambda_i(s_i^*+N_i^*)^2\big)] \\
&= \frac{1}{n}\sum_{i=1}^{n}\log\mathbb{E}[\exp\big(\theta\lambda_i(N_i^*-s_i^*)^2\big)],
\end{aligned}$$

where we can write the final minus sign since the $\{N_i\}$ are symmetric independent random variables. Continuing, we find

$$\begin{aligned}
\phi_n(\theta) &= \frac{1}{n}\log\left\{\frac{1}{\sqrt{1-2\theta\lambda_i}}\exp\Big(\frac{\theta\lambda_i(s_i^*)^2}{1-2\theta\lambda_i}\Big)\right\} \\
&= \frac{1}{n}\sum_{i=1}^{n}-\frac{1}{2}\log(1-2\theta\lambda_i^{(n)}) + \frac{1}{n}\sum_{i=1}^{n}\frac{\theta(s_i^*)^2\lambda_i^{(n)}}{1-2\theta\lambda_i^{(n)}},
\end{aligned}$$

where in the final expression the dependence of the eigenvalues on n is made explicit. Now as n approaches infinity (by the Toeplitz Distribution Theorem) we obtain

$$\lim_{n\to\infty}\frac{1}{n}\sum_{i=1}^{n}-\frac{1}{2}\log(1-2\theta\lambda_i^{(n)}) = -\frac{1}{4\pi}\int_0^{2\pi}\log(1-2\theta)f(\omega)d\omega.$$

Recall that the s^* vector was obtained by applying the whitening transformation $\Lambda^{-1/2}U^T$ to s. According to Toeplitz distribution theory (with the right technical conditions), the U matrix acts more and more like the discrete Fourier Transform (DFT) matrix; that is, the matrix acts as if its columns were samples from a complex sinusoid. Hence, we expect $(s_i^*)^2$ to be asymptotically like $P(\omega_i)/f(\omega_i)$, where ω_i is the frequency corresponding to the ith column of the DFT matrix ($\omega_i = 2\pi i/n$). By this argument and the Toeplitz Distribution Theorem, we have that

$$\lim_{n\to\infty}\frac{1}{n}\sum_{i=1}^{n}\frac{\theta(s_i^*)^2\lambda_i^{(n)}}{1-2\theta\lambda_i^{(n)}} = \frac{\theta}{2\pi}\int_0^{2\pi}\frac{P(\omega)}{1-2\theta f(\omega)}d\omega.$$

Hence

$$\phi(\theta) = -\frac{1}{4\pi}\int_0^{2\pi}\log(1-2\theta)f(\omega)d\omega + \frac{\theta}{2\pi}\int_0^{2\pi}\frac{P(\omega)}{1-2\theta f(\omega)}d\omega.$$

By Theorem 3.2.1,

$$\lim_{n\to\infty}\frac{1}{n}\log P\big(\frac{1}{n}\sum_{i=1}^{n}(s_i+N_i)^2 \le \gamma\big) = -\sup_\theta[\theta\gamma - \phi(\theta)].$$

To find the optimum θ, θ_γ, we differentiate with respect to θ and set the result to zero to find that θ_γ solves

$$\gamma = \frac{1}{2\pi}\int_0^{2\pi}\frac{P(\omega)+f(\omega)(1-2\theta_\gamma f(\omega))}{(1-2\theta_\gamma f(\omega))^2}d\omega.$$

In general this equation can't be solved in closed form. However, if the background noise is white (i.e., $f(\omega=1)$ we can indeed solve the system in closed form. In this case (letting $P=\int_0^{2\pi}P(\omega)d\omega$),

$$\begin{aligned}
\gamma &= \frac{1}{2\pi}\int_0^{2\pi}\frac{P(\omega)+(1-2\theta_\gamma)}{(1-2\theta_\gamma)^2}d\omega \\
&= \frac{P}{(1-2\theta_\gamma)^2}+\frac{1}{1-2\theta_\gamma} \\
0 &= P+(1-2\theta_\gamma)-\gamma(1-2\theta_\gamma)^2 \\
\theta_\gamma &= \frac{1}{2}\left(1-\frac{1+\sqrt{1+4\gamma P}}{2\gamma}\right).
\end{aligned}$$

Now substituting this back into the expression for the rate function obtains

$$I(\gamma) = \frac{P+\gamma}{2} - \frac{1+\sqrt{1+4\gamma P}}{4} - \frac{P\gamma}{1+\sqrt{1+4\gamma P}} + \frac{1}{2}\log\frac{1+\sqrt{1+4\gamma P}}{2\gamma}.$$

3.3 Level Crossing Times

Our motivation in this section is to derive some large deviation type results for the times to a level crossing for certain sequences of random variables. We use this material when we discuss the simulation of waiting times in certain types of queueing systems.

We consider the usual Gärtner–Ellis Theorem setup. Let $\{Y_n\}$ be a sequence of $\mathcal{R}$-valued random variables. Define

$$\phi_n(\theta) = \frac{1}{n} \log \mathbb{E}[\exp(\theta Y_n)],$$

where $\theta \in \mathcal{R}$. We assume A1, A2, and A3 of the previous section. A further new assumption is that we suppose that there exists a $\theta^* \in \overset{\circ}{D}_\phi$ such that $\theta^* > 0$, $\phi(\theta^*) = 0$, and $\phi'(\theta^*) > 0$. In the setting where $Y_n = \sum_{i=1}^n X_i$ where $\{X_i\}$ are i.i.d. random variables (in this section we refer to this as the i.i.d. case), this new assumption would imply that $\mathbb{E}[X_1] < 0$ and $P(X_1 > 0) > 0$.

As usual we define

$$I(x) = \sup_\theta [\theta x - \phi(\theta)].$$

Now let $T_M = \inf\{n : Y_n > M\}$; that is, T_M is the time of the first level crossing of size M. We are interested in the probabilities

$$\rho_M = P(T_M < \infty).$$

If one considers the i.i.d. case, the sum Y_n has a negative drift. In general there is positive probability that $T_M = \infty$.

Theorem 3.3.1.

$$\lim_{M \to \infty} \frac{1}{M} \log P(T_M < \infty) = -\theta^*.$$

Proof. Upper bound proof. Let F_n denote the probability distribution function of Y_n. As usual, we denote its exponential shifted version as $Y_n^{(\theta)}$ with probability distribution function

$$dF_n^{(\theta)}(x) = \frac{\exp(\theta x)}{\exp(n\phi_n(\theta))} dF_n(x).$$

Now for $j \geq 1$,

$$\begin{aligned}
P(T_M = j) &\leq P(Y_j > M) \\
&= \int 1_{\{x > M\}} dF_j(x) \\
&= \int 1_{\{x > M\}} \exp(-\theta x) \exp(j\phi_j(\theta)) dF_j^{(\theta)}(x) \\
&\leq \exp(-\theta M) \exp(j\phi_j(\theta)) \int 1_{\{x > M\}} dF_j^{(\theta)}(x) \quad \text{for } \theta > 0 \\
&= \exp(-\theta M) \exp(j\phi_n(\theta)) P(Y_j^{(\theta)} > M) \quad \text{for } \theta > 0.
\end{aligned}$$

Thus,

$$\begin{aligned}
P(T_M < \infty) &= \sum_{j=1}^\infty P(T_M = j) \\
&\leq \exp(-\theta M) \sum_{j=1}^\infty \exp(j\phi_j(\theta)) P(Y_j^{(\theta)} > M) \quad \text{for } \theta > 0.
\end{aligned}$$

Now since the functions $\{\phi_n(\theta)\}$ are convex and converge to $\phi(\theta)$, that convergence is *uniform* on any compact set contained in $\overset{\circ}{D}_\phi$. Furthermore these functions are uniformly bounded on any such compact set. (See the appendix for a proof of these properties.) Now $\phi(0) = 0 = \phi(\theta^*)$. Hence, by the uniform convergence, for any $\theta^* > \epsilon > 0$, there exists $\delta_\epsilon, j_\epsilon$ such that

$$\phi_j(\theta^* - \epsilon) < \delta_\epsilon < 0 \quad \forall\ j > j_\epsilon.$$

Also, by the uniform bound, we have $|\exp\big(j\phi_j(\theta^* - \epsilon)\big)| < B_\epsilon$ for all $j < j_\epsilon$. Thus

$$\begin{aligned} P\big(T_M < \infty\big) &\leq \exp\big(-(\theta^* - \epsilon)M\big)\sum_{j=1}^{\infty}\exp\big(j\phi_j(\theta^* - \epsilon)\big)P\big(Y_j^{\theta^*-\epsilon} > M\big) \\ &\leq \exp\big(-(\theta^* - \epsilon)M\big)[j_\epsilon B_\epsilon + \sum_{j=0}^{\infty}\exp(j\delta_\epsilon)] \\ &\leq \exp\big(-(\theta^* - \epsilon)M\big)[j_\epsilon B_\epsilon + \frac{1}{1-\exp(\delta_\epsilon)}]. \end{aligned}$$

Hence

$$\limsup_{M\to\infty}\frac{1}{M}\log P\big(T_M < \infty\big) \leq -\theta^* + \epsilon,$$

and since $\epsilon > 0$ can be taken to be arbitrarily small, we have completed the upper bound part of the theorem. □

Surprisingly enough the lower bound part of the theorem is easier (it's usually the other way around with large deviation theorems). Let $y \in (0, \infty)$ be fixed and denote $N = \lceil My \rceil$ ($\lceil a \rceil$ is the smallest integer greater than or equal to a). Then

$$\begin{aligned} P\big(T_M < \infty\big) &\geq P\big(Y_N > M\big) \\ &\geq P\big(\frac{Y_N}{N} \geq \frac{1}{y}\big) \end{aligned}$$

and so

$$\begin{aligned} \frac{1}{M}\log P\big(T_M < \infty\big) &\geq \frac{1}{M}\log P\big(\frac{Y_N}{N} \geq \frac{1}{y}\big) \\ &= \frac{N}{M} \cdot \frac{1}{N}P\big(Y_N/N \geq \frac{1}{y}\big) \\ &\to_{M\to\infty} -yI\big(\frac{1}{y}\big). \end{aligned}$$

We now wish to choose the best value of y for the lower bound. To do this we take the infimum over all y of $yI(1/y)$. Now note that $I(1/y) = \sup_\theta[\theta/y -$

$\phi(\theta)]$. So by choosing $\theta = \theta^*$ we have an easy lower bound of $I(1/y) \geq \theta^*/y$ (with equality if $1/y = \phi'(\theta^*)$). Thus

$$\inf_{y \in (0,\infty)} yI(\frac{1}{y}) \geq \inf_{y \in (0,\infty)} y\theta^* \frac{1}{y} = \theta^*,$$

which completes the proof of the lower bound. □

3.4 Functionals of Finite State Space Markov Processes

Markov models are probably the simplest systems with memory that confront engineers and statisticians. We utilize the general theory of the previous sections to investigate in detail what is entailed in computing the large deviation rate functions for systems with memory. In this section we consider only finite state space ($k < \infty$) Markov chains. Functions on this state space may be thought of in two ways. They may be thought of as functions mapping from the finite state space into $\mathcal{R}$. Alternatively, the function values may be arranged into a vector with dimension k, the cardinality of the state space. Similarly, linear operators may be thought of as mapping functions of the state space into other functions on the state space or, alternatively, as $k \times k$ matrices operating on vectors.

Let

$$p_{ij} = P\big(X_{n+1} = j | X_n = i\big)$$

be the transition probabilities of a Markov chain with state space $S = \{0, 1, \dots, k-1\}$. Let f be a fixed function that maps from S into $\mathcal{R}$. Let us suppose that the chain is irreducible (and hence ergodic). Then

$$S_n = \frac{1}{n} \sum_{i=1}^{n} f(X_i)$$

converges to

$$\mathbb{E}_\mu[f] = \sum_{j=0}^{k} f(j)\mu(j),$$

where μ is the stationary probability distribution of the chain. To calculate the large deviation behavior of S_n, we use the Gärtner–Ellis Theorem directly.

Define

$$\phi_n(\theta) = \frac{1}{n} \log \mathbb{E}[\exp(\theta n S_n)].$$

Now,

$$\begin{aligned}
&\mathbb{E}[\exp(\theta S_n)] \qquad (3.9)\\
&= \sum_{x_1=0}^{k-1} \sum_{x_2=0}^{k-1} \cdots \sum_{x_n=0}^{k-1} \exp(\theta \sum_{i=1}^{n} f(x_i)) p(x_1, x_2, \ldots, x_n)\\
&= \sum_{x_1=0}^{k} \sum_{x_2=0}^{k} \cdots \sum_{x_n=0}^{k} \exp(\theta \sum_{i=1}^{n} f(x_i)) (\prod_{i=1}^{n-1} p(x_{i+1}|x_i)) p(x_1), \qquad (3.10)
\end{aligned}$$

where $p(x_1, x_2, \ldots, x_n)$ is the multidimensional probability mass function of the first n steps of the chain. Let $\mathcal{G}$ be the collection of all functions mapping from S into $\mathcal{R}$. Define the linear operator T_θ, $\{T_\theta : \mathcal{G} \to \mathcal{G}\}$ as

$$T_\theta(g)(x) = \sum_{y=0}^{k-1} \exp(\theta f(y)) g(y) p_{xy}. \qquad (3.11)$$

Thus we may rewrite (3.10) as

$$\mathbb{E}[\exp(\theta S_n)] = \sum_{x_1=0}^{k-1} T_\theta(1)^{(n-1)}(x_1) \exp\big(\theta f(x_1)\big) p(x_1),$$

where the notation $T_\theta^{(n-1)}(1)$ means that the operator is iterated n times starting from the constant function 1. In this finite-dimensional setting, this corresponds exactly to multiplying the matrix representation of T_θ by itself $n-1$ times and then operating that result on the all ones vector. Since this matrix is strictly positive, the Perron–Frobenius Theorem states that T_θ has a largest eigenvalue $\lambda(\theta)$ such that

$$\lambda(\theta) = \|T_\theta\|_s = \sup_{g:\|g\| \le 1} [\|T_\theta(g)\|],$$

where $\|\cdot\|_s$ is called the strong matrix norm and $\|\cdot\|$ indicates the usual Euclidean norm for $\mathcal{R}^k$. Furthermore, this eigenvalue has an associated unique (up to scaling) eigenvector ψ. This implies then that a first-order expansion exists for the iterates of T_θ as follows,

$$T_\theta^{n-1}(g)(x) = c\lambda(\theta)^{n-1}\psi(x) + o(n)$$

with the $o(n)$ terms being exponentially insignificant as n approaches infinity, and $c > 0$ is a constant that depends on g. Therefore

$$\begin{aligned}\lim_{n\to\infty}\phi_n(\theta) &= \lim_{n\to\infty}\frac{1}{n}\log\sum_{x_1=0}^{k-1}T_\theta(1)^{(n-1)}(x_1)\exp\big(\theta f(x_1)\big)p(x_1)\\ &= \lim_{n\to\infty}\frac{1}{n}\log\sum_{x_1=0}^{k-1}c\lambda(\theta)^{n-1}\psi(x_1)\exp\big(\theta f(x_1)\big)p(x_1)\\ &= \log\lambda(\theta)+\lim_{n\to\infty}\frac{1}{n}\log\sum_{x_1=0}^{k-1}c\psi(x_1)\exp\big(\theta f(x_1)\big)p(x_1)\\ &= \log\lambda(\theta).\end{aligned}$$

We may then define

$$I(x)=\sup_\theta[\theta x-\log\lambda(\theta)].$$

In this finite state space setting, it is easy to verify that $\log\lambda(\theta)$ is a closed, convex, and steep function with effective domain all of $\mathcal{R}$, and hence from the Gärtner–Ellis Theorem we know that I is the rate function of the sum.

Example 3.4.1. Suppose we have a two-state $\{0,1\}$ Markov chain that stays where it is with probability p and flips state with probability $1-p=q$. Thus it has transition matrix

$$P=\begin{pmatrix}p\,q\\ q\,p\end{pmatrix}.$$

Let our functional f be $f(x)=1-x$, where x is the state of the chain. Then, in order to find $\lambda(\theta)$, we must find the largest eigenvalue of the matrix T_θ:

$$T_\theta=\begin{pmatrix}p\exp(\theta) & q\\ q\exp(\theta) & p\end{pmatrix}.$$

To do this we must set the determinant of λI_2-T_θ equal to zero, where I_2 is the two-dimensional identity matrix. This gives us

$$\lambda^2+\lambda(-p\exp(\theta)-p)+\exp(\theta)(p^2-q^2)=0$$

or via the quadratic formula and finding the largest root

$$\lambda(\theta)=\frac{p(\exp(\theta)+1)+\sqrt{p^2(1+\exp(\theta))^2+4\exp(\theta)(1-2p)}}{2}.$$

One can then plug this result into the equation $I(x)=\sup_\theta[\theta x-\log\lambda(\theta)]$ to obtain the final form of the rate function.

By this simple example and the resulting algebraic mess, one can appreciate that usually it is desirable to compute the eigenvalues via standard numerical techniques. We note that much more general Markov models other than finite-state space chains can be handled by virtually the same techniques [13].

3.5 Contraction Principle

Let E be a complete separable metric space. A *good rate function* is a lower semi-continuous mapping $I : E \to [0, \infty]$ such that for all $\alpha \in [0, \infty)$, the level set $\{x \in E : I(x) \leq \alpha\}$ is compact. We note that for a sequence of $\mathcal{R}^d$-valued random variables $\{f_n(Y_n)\}$ that satisfy the standard assumptions A1 to A4, the rate function given by Theorem 3.2.1 is a good rate function.

Now suppose we have a sequence of E-valued random variables $\{Z_n\}$ (on some probability space) that satisfy the large deviation principle:

$$\limsup_{n\to\infty} \frac{1}{n} \log P\big(Z_n \in nA\big) \leq -\inf_{x\in A} I(x) \quad \text{for all closed sets } A,$$

and

$$\liminf_{n\to\infty} \frac{1}{n} \log P\big(Z_n \in nG\big) \geq -\inf_{x\in G} I(x) \quad \text{for all open sets } G,$$

where $I(\cdot)$ is a good rate function. Let $F(\cdot)$ be a continuous function from E into another complete separable metric space E'. Then

Theorem 3.5.1. *Under the above assumptions, the sequence of E'-valued random variables $\{F(Z_n/n)\}$ satisfies a large deviation principle with good rate function $J(\cdot)$, where*

$$J(y) = \begin{cases} \inf_{x:F(x)=y} I(x) & \text{if } \{x : F(x) = y\} \neq \emptyset \\ \infty & \text{if } \{x : F(x) = y\} = \emptyset \end{cases}.$$

Proof. First we show that J is good. The lower semi-continuity of J follows from the continuity of F. Thus we need only worry about the compactness of the level sets of J. Since I is good, for all $y \in F(E)$ the infimum in the definition of J is obtained at some point of E. Thus for the level sets of J we have

$$\begin{aligned} \{y \in E'\} &= \{F(x) : I(x) \leq \alpha\} \\ &= F(\{x : I(x) \leq \alpha\}). \end{aligned}$$

Thus $\{y \in E'\}$ is the continuous image of a compact set and thus is compact. Upper Bound. Let $A \subset E'$ be a closed set. Let $C = \{x : F(x) \in A\}$. Then $P\big(F(Z_n/n) \in A\big) = P\big(Z_n/n \in C\big)$. Note that C is a closed set since $F(\cdot)$ is continuous. Then

$$\begin{aligned} \limsup_{n\to\infty} \frac{1}{n} \log P\big(F(Z_n/n) \in A\big) &= \limsup_{n\to\infty} \frac{1}{n} \log P\big(Z_n/n \in C\big) \\ &\leq -\inf_{x\in C} I(x) \\ &= -\inf_{x:F(x)=y \text{ and } y\in A} I(x) \\ &= -\inf_{y\in A} \inf_{x:F(x)=y} I(x) \\ &= -\inf_{y\in A} J(y). \end{aligned}$$

Lower Bound. Let $G \subset E'$ be open and let $y \in G$ and $x \in E$ such that $F(x) = y$. Since $F(\cdot)$ is continuous, we can find a neighborhood V of x such that $F(V) \subset G$. Therefore,

$$\begin{aligned}\liminf_{n\to\infty} \frac{1}{n} \log P\big(F(Z_n/n) \in G\big) &= \liminf_{n\to\infty} \frac{1}{n} \log P\big(Z_n/n \in V\big) \\ &\geq -I(x).\end{aligned}$$

This is true for every x such that $F(x) \in G$, and hence the proof is complete. □

Example 3.5.1. Suppose $S_n = \sum_{i=1}^n X_i/n$ where the $\{X_i\}$ are i.i.d. real valued standard Gaussian random variables. $\{S_n\}$ obeys a large deviation principle with rate function $I(x) = x^2/2$. Suppose we have $F(x) = x^2$. Then

$$J(y) = \inf_{x:x^2=y} \frac{x^2}{2} = \frac{y}{2} \quad y \geq 0,$$

and equals ∞ for $y < 0$. Therefore we would have (for an interval (a, b) on the positive half line)

$$\lim_{n\to\infty} \frac{1}{n} \log P\big(S_n^2 \in (a,b)\big) = -\inf_{y\in(a,b)} \frac{y}{2} = -\frac{a}{2}.$$

3.6 Notes and Comments

Cramér [18] is usually credited with presenting the first large deviation asymptotic expansion. He gave not only the exponential rate, but an expansion for the slowly varying constants in front of the exponential for the case of i.i.d. random variables with a continuous component and for the case of infinite intervals (such as $[a, \infty)$). Chernoff [19] showed that whatever the type of random variable (discrete, continuous, or mixed) the large deviation rate function gives the correct exponential rate in the i.i.d. case. This original work was extended in various directions.

J. Gärtner [31] is credited with being the first to consider looking only at the convergence properties of the moment generating function of a sequence of random variables. He assumed throughout that $M(\theta) < \infty$ for all θ. The form of the theorem given in the text should be credited almost entirely to Ellis [24], who was one of the pioneers in the use of convex analysis in the theory of large deviations.

The material in the section on level crossing times is a reworking of the classical results of Siegmund [74] for i.i.d. random variables. Our presentation follows the line given by Lehtonen and Nyrhinen [52] for the i.i.d. setting.

The first place that the operator theory for the large deviations of functionals of a Markov chain appears is in a paper by H. Miller [61]. The presentation given in this section is of course very different from that historical reference. The contraction principle is actually a collection of theorems dealing with the mappings of large deviation principles from one domain to another through continuous (or nearly continuous) maps. These theorems spring from an original result due to M. Donsker and S.R.S. Varadhan [23].

4. Importance Sampling

Applying simulation methodology is simply finding the right wrench to pound in the correct screw. *Anon.*

4.1 The Basic Problem of Rare Event Simulation

Large and/or nonlinear stochastic systems, due to analytic intractability, must often be simulated in order to obtain estimates of the key performance parameters. Typical situations of interest could be a buffer overload in a queueing network or an error event in a digital communication system. In many system designs or analyses a low probability event is a key parameter of the system's efficacy.

Since the test statistic's probability distribution is usually very difficult or impossible to compute in closed form, one is faced with the problem of computer simulation in order to find the probability of interest.

How does one go about simulating a rare event in order to find its probability? This is a more difficult problem than one may expect. Consider a sequence $\{X_j\}$ of i.i.d. Bernoulli random variables with

$$P\big(X_1 = 1\big) = \rho = 1 - P\big(X_1 = 0\big).$$

Suppose that we wish to estimate from the observed sequence the parameter ρ. We wish to have at most a 5% error on ρ with 95% confidence. This means that we must have

$$P\big(|\rho - \hat{\rho}| \leq .05\rho\big) = .95,$$

where $\hat{\rho}$ is the estimate of ρ. For example, the maximum likelihood estimate would be

$$\hat{\rho} = \frac{1}{k}\sum_{i=1}^{k} X_i.$$

The variance of the Bernoulli random variable X_1 is $\rho(1-\rho)$. If ρ is very small (i.e., $\{X_1 = 1\}$ is a rare event), the variance of X_1 is approximately ρ. Hence the variance of $\hat{\rho}$ is

$$\frac{\rho(1-\rho)}{k} \approx \frac{\rho}{k}.$$

The mean value of $\hat{\rho}$ is ρ which (by definition) means that the estimate is unbiased. Hence using a Central Limit Theorem approximation, we have that

$$\begin{aligned} P\big(|\rho - \hat{\rho}| \leq .05\rho\big) &= P\big(|\frac{1}{\sqrt{k}} \sum_{i=1}^{k} \frac{X_i - \rho}{\sqrt{\rho}}| \leq .05\sqrt{\rho k}\big) \\ &\approx P\big(|Z| \leq .05\sqrt{\rho k}\big), \end{aligned}$$

where Z is a Gaussian, mean zero, variance 1 (i.e., standard Gaussian) random variable. Now for a standard Gaussian we have that $P\big(|Z| \leq z\big) = .95$ implies (from tables) that $z \approx 2$; that is, two standard deviations about the mean captures 95% of the probability of a Gaussian distribution. Thus we must have that $.05\sqrt{\rho k} = 2$ which in turn implies that $k = 1600/\rho$. Therefore, if ρ is somewhere on the order of 10^{-6}, we would need something like 1.6×10^9 number of samples to estimate it to the desired level of precision. This is a very large number of simulation runs and will impose severe demands on our random number generator. Unfortunately (for the simulation designer) error probabilities of this order are typical in digital communication systems and many other systems of engineering and scientific interest.

4.2 Importance Sampling

The principal method that we use to attack the rare event simulation problem is to utilize a variance reduction technique from simulation theory known as *importance sampling*. Suppose that we wish to estimate

$$\rho = \mathbb{E}[\eta(X)],$$

where X is a random variable (or vector) describing some observation on a random system. If η is the indicator function of some set (a typical case), then ρ would be the probability of that set. Suppose that the observation random variable X is controlled by a probability density function $p(\cdot)$. The direct simulation method would be to generate a sequence of i.i.d. random numbers $X^{(1)}, X^{(2)}, \ldots, X^{(k)}$ from the density $p(\cdot)$ and form the estimate

$$\hat{\rho}_p = \frac{1}{k} \sum_{i=1}^{k} \eta(X^{(i)}).$$

Alternatively, we could generate a sequence of i.i.d. random numbers $Y^{(1)}, Y^{(2)}, \ldots, Y^{(k)}$ distributed with density $q(\cdot)$. The density $q(\cdot)$ is called the *importance sampling biasing distribution*. We then form the *importance sampling estimator* or *estimate* as

$$\hat{\rho}_q = \frac{1}{k} \sum_{i=1}^{k} \eta(Y^{(i)}) \frac{p(Y^{(i)})}{q(Y^{(i)})}.$$

Immediately we see that we could have problems unless $q(x)$ is never zero for any value of x where $p(x)$ is positive. Mathematically, this means that the support of $q(\cdot)$ must include the support of $p(\cdot)$. However, we see that there can only be a problem if $\eta(x)$ is nonzero at x and the ratio of $p(\cdot)$ and $q(\cdot)$ blows up. This is equivalent to saying that support of $\eta(x)p(x)$ is included in the support of $\eta(x)q(x)$. Thus our requirement may be stated as

$$support(p \cdot \eta) \subset support(q \cdot \eta).$$

The expected value of $\hat{\rho}_q$ under the density $q(\cdot)$ is just

$$\begin{aligned} \mathbb{E}_q[\hat{\rho}_q] &= \frac{1}{k} \sum_{i=1}^{k} \int \eta(y^{(i)}) \frac{p(y^{(i)})}{q(y^{(i)})} q(y^{(i)}) dy^{(i)} \\ &= \int \eta(y) p(y) dy \\ &= \mathbb{E}_p[\eta(X)] \\ &= \rho. \end{aligned}$$

Therefore, the estimate $\hat{\rho}_q$ is unbiased and, as $k \to \infty$, we expect it to be converging by the law of large numbers to its mean value ρ.

Example 4.2.1. Suppose we are interested in

$$\rho = P\big(\frac{1}{n} \sum_{j=1}^{n} Z_j > T\big),$$

where $\{Z_j\}$ are i.i.d. $\mathcal{R}$-valued random variables with mean value zero, density function $p^*(\cdot)$, and T is a positive constant. We simulate with some other random variables $\{R_j\}$ with density function $q^*(\cdot)$ and form the importance sampling estimator,

$$\hat{\rho}_q = \frac{1}{k} \sum_{i=1}^{k} 1_{\{\frac{1}{n}\sum_{j=1}^{n} R_j^{(i)} > T\}} \frac{\prod_{j=1}^{n} p^*(R_j^{(i)})}{\prod_{j=1}^{n} q^*(R_j^{(i)})}.$$

To fit in with the theoretical framework given in the introduction, note that we have the correspondences

$$
\begin{aligned}
(Z_1, Z_2, \ldots, Z_n) &\Rightarrow X \\
1_{\{\frac{1}{n}\sum_{i=1}^{n} Z_i > T\}} &\Rightarrow \eta(X) \\
(R_1, R_2, \ldots, R_n) &\Rightarrow Y \\
\prod_{j=1}^{n} p^*(Z_j^{(i)}) &\Rightarrow p(X^{(i)}) \\
\prod_{j=1}^{n} q^*(R_j^{(i)}) &\Rightarrow q(Y^{(i)}).
\end{aligned}
$$

In this example we can see some of the possible utility in using an importance sampling estimator. Consider again the quantity ρ. It involves the sum of n random variables. Even if we know $p(\cdot)$, the distribution of the sum would involve the n-fold convolution of $p(\cdot)$. For large n, this could be a very difficult task even numerically. After that, then one would be forced to try to integrate over the tail of the resulting distribution, another task that could be very difficult analytically or numerically. With importance sampling, we see that knowledge of the one-dimensional densities is sufficient to come up with an unbiased estimator of ρ.

One key question of importance sampling is: Are there better choices for $q(\cdot)$ than just $p(\cdot)$ (the direct Monte Carlo choice)? Let us consider the variance of $\hat{\rho}_q$. Since this estimator is the average of k i.i.d. terms, the variance will be $1/k$ times the variance of one of the terms. Thus

$$
\begin{aligned}
k\,Var(\hat{\rho}_q) &= \int [\eta(x)\frac{p(x)}{q(x)} - \rho]^2 q(x)dx \\
&= \int [\frac{\eta(x)^2 p(x)^2}{q(x)} - 2\rho p(x)\eta(x) + \rho^2 q(x)]dx \\
&= \int \frac{\eta(x)^2 p(x)^2}{q(x)} dx - \rho^2 \\
&= F_q - \rho^2. \qquad (4.1)
\end{aligned}
$$

In the above expression, we have emphasized that $k\,Var(\hat{\rho}_q)$ may be written as a difference of a first term, F_q, and a second term ρ^2.

We now try to choose $q(\cdot)$ to minimize this expression,

$$
\begin{aligned}
F_q &= \int \frac{\eta(x)^2 p(x)^2}{q(x)} dx \\
&= \int \frac{\eta(y)^2 p(y)^2}{q(y)^2} q(y) dy \\
&= \mathbb{E}[\frac{\eta(Y)^2 p(Y)^2}{q(Y)^2}] \\
&= \geq \left(\mathbb{E}[\frac{|\eta(Y)|p(Y)}{q(Y)}]\right)^2 \\
&= \int \frac{|\eta(y)|p(y)}{q(y)} q(y) dy \\
&= \int |\eta(y)|p(y) dy,
\end{aligned}
$$

where we have used Jensen's inequality or the fact that for any random variable Z, $\mathbb{E}[Z^2] \geq (\mathbb{E}[Z])^2$. Furthermore we have strict equality if and only if Z is almost surely a constant. Thus F_q is minimized when $|\eta(Y)|p(Y)/q(Y)$ is almost surely a constant. However, this can hold only if $|\eta(x)|p(x)/q(x)|$ is a constant. Thus the optimal choice for q is

$$
q_{opt}(x) = \frac{p(x)|\eta(x)|}{\int |\eta(y)p(y)dy}. \tag{4.2}
$$

Let us investigate $q_{opt}(\cdot)$. Suppose, for simplicity that $\eta(\cdot)$ is a non-negative function. Then we note that $F_{q_{opt}} = \rho \int \eta(x)p(x)dx = \rho^2$. Thus $k\,Var(\hat{\rho}_{q_{opt}}) = 0$ for all k. Unfortunately, this is not as wonderful as it might seem at first glance. In the first place, $p(\cdot)$ in many cases is not specified in closed form. It could be (for example) the distribution of a large sum of i.i.d. (as in the Example 4.2.1) or Markov distributed random variables. We may generate samples from it easily enough but explicit expressions for it are generally not available. Secondly, even if $p(\cdot)$ were known, the constant of proportionality is exactly ρ^{-1}, precisely the parameter that we are trying to estimate! Hence in computing the weighting factor for the importance sampling estimator of ρ, we first must know what ρ is. Clearly, we must search for other methods or criteria by which to choose a good simulation distribution $q(\cdot)$. This need to find good criteria for choosing the simulation distribution has sparked the vast amount of research in this area for the past two decades.

Let us see if we can elucidate some guidelines for choosing good practical simulation distributions. Consider the "optimal" choice in (4.2), for the case that $\eta(\cdot) = 1_{\{E\}}(\cdot)$. We think of the set E as being some "rare event." We can gain some insight from the optimal choice on what properties a good practical simulation distribution should have. First note that $q_{opt}(\cdot)$ puts all of its probability mass on the set or event E (i.e., its support is contained in or equal to E). Thus, intuitively, we want to choose the simulation distribution so that more events of interest occur. The second observation is that $q_{opt}(\cdot)$

has the same shape over the set E as the original distribution (in fact it *is* the same distribution except just scaled by ρ^{-1}). Thus if a region of E has more probability mass than another region of E under the original distribution, then the optimal choice will also have this property. We can summarize these two principles as

P1) Choose the simulation distribution so that we "hit" the rare event E of interest more often.

P2) Choose the simulation distribution so that the more likely or higher probability regions of E are hit more often during the simulation than the lower probability or less likely regions of E.

These properties have spawned a variety of ad hoc techniques for choosing the simulation distribution. By far the two most popular methods are variance scaling and mean translation.

The variance scaling method increases the "hit" probability by choosing as the simulation random variables, the original random variables multiplied by a constant. Typically the constant is greater than one and thus we are merely increasing the variance of the original distribution. Thus for some rare event E, typically this would put more probability mass on it, causing us to "hit" it more often during the simulation, which would satisfy our first property quite nicely. Whether the second property is satisfied depends on the problem. Typically, we would try to choose the variance scaling parameter to satisfy as much as possible, the second property.

Example 4.2.2. In Example 4.2.1, the variance scaling method would correspond to

$$q^*(x) = \frac{1}{a}p^*(\frac{x}{a}),$$

where a is the variance scaling parameter (if the original density $p^*(\cdot)$ has variance σ^2, then $q^*(\cdot)$ has variance $a^2\sigma^2$).

The variance scaling method has been largely superseded by the mean translation method. This method seeks to increase the "hit" probability by adding a mean value to the input random variables.

Example 4.2.3. In Example 4.2.1, the mean translation method would correspond to

$$q^*(x) = p^*(x - m)$$

where m is the mean shift parameter (if the original density $p^*(\cdot)$ has mean m_o, then $q^*(\cdot)$ has mean $m_o + m$). In this method, almost always m is just chosen to be T, which partially explains its popularity. The scaling parameter in the variance scaling method has no such "default" choice available.

Recall that for the importance sampling estimator, we require that the support of $\eta(\cdot)q(\cdot)$ include the support of $\eta(\cdot)p(\cdot)$. We must always take this into account when choosing simulation distributions.

Example 4.2.4. Suppose for example that $\eta(x) = 1_{\{E\}}(x)$; that is, $\eta(\cdot)$ has support E. If the original density has support $[0, s]$ $(s > 0)$, and $E = [s/2, s]$, then the variance scaling simulation distribution has support $[0, sa]$ (assuming $a > 0$). Thus

$$\begin{aligned} support(p \cdot \eta) &= [0, s] \cap E = E \\ &\subset support(q \cdot \eta) \\ &= [0, sa] \cap E = E, \end{aligned}$$

and thus satisfies our support requirement.

The mean shift method on the other hand has

$$\begin{aligned} support(q \cdot \eta) &= [m, s + m] \cap E \\ &= [m, s + m] \cap [s/2, s] \end{aligned}$$

which will violate the support requirement for any choice of $m > s/2$.

4.3 The Fundamental Theorem of System Simulation

Before we begin our study of how to choose good importance sampling biasing distributions, we need to consider some fundamental properties of importance sampling estimators. In this section, we consider a very general question in the field of system simulation: that is, "Should we bias the random variables at the input, at the output, or at some intermediate point of a system?." To be a bit more specific, consider the following example.

Example 4.3.1. Suppose we are interested in estimating

$$\rho = P\big(S + N > a\big),$$

where S and N are two independent random variables with densities $p_s(\cdot)$ and $p_n(\cdot)$, respectively. Denote the sum of these two random variables as R with density denoted as $p_r(\cdot)$. We use importance sampling to estimate the value of ρ. We generate simulation random variables $S^{(1)\prime}, S^{(2)\prime}, \ldots, S^{(k)\prime}$ i.i.d. with marginal density $p_{s'}$, i.i.d $N^{(1)\prime}, N^{(2)\prime}, \ldots, N^{(k)\prime}$ i.i.d. with marginal density $p_{n'}$. The sequences are also independent of each other. We also consider $R^{(1)\prime}, R^{(2)\prime}, \ldots, R^{(k)\prime}$ i.i.d. with marginal density $p_{r'}$ which are generated by the relation $R^{(j)\prime} = S^{(j)\prime} + N^{(j)\prime}$.

We consider two types of estimators, an input estimator and an output estimator. The "input" estimator is explicitly given as

$$\hat{p}_i = \frac{1}{k}\sum_{j=1}^{k} 1_{\{S^{(j)\prime}+N^{(j)\prime}>a\}} \frac{p_s(S^{(j)\prime})p_n(N^{(j)\prime})}{p_{s'}(S^{(j)\prime})p_{n'}(N^{(j)\prime})}$$

and the "output" estimator as

$$\hat{p}_o = \frac{1}{k}\sum_{j=1}^{k} 1_{\{R^{(j)\prime}>a\}} \frac{p_r(R^{(j)\prime})}{p_{r'}(R^{(j)\prime})}.$$

these estimators are unbiased. Which has lower variance?

We should note that in many situations, an output formulation of the bias distribution is impossible. If the system is very complicated, it may very well be virtually impossible to calculate the biasing distributions at the output of the system. However, it may very well be possible to calculate the distributions at some intermediate point of the system. In this chapter we take the first steps toward attempting to understand how much there is to gain or lose by using an input over an output formulation. It is essential to the theory of importance sampling in system simulation that we try to gain some understanding of the role of the bias point in Monte Carlo simulation.

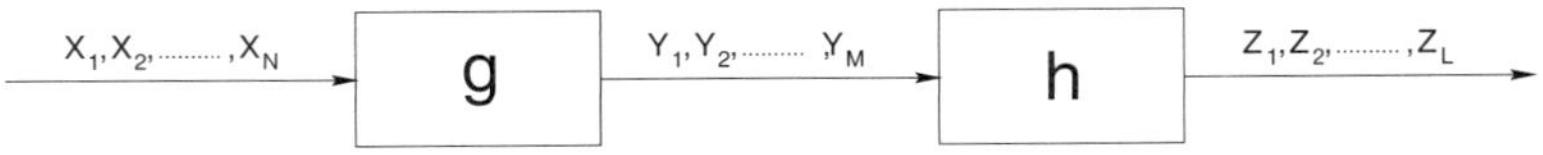

Fig. 4.1. A multi-input, multi-output system.

We are given two (Borel measurable) functions $g : \mathcal{R}^N \to \mathcal{R}^M$ and $h : \mathcal{R}^M \to \mathcal{R}^L$, which define our system as shown in Fig. 4.1. Let $(X_1, X_2, \ldots, X_N)$ be an arbitrary random vector. We consider these to be our "input random variables." We denote their joint probability measure as P_x. Define $(Y_1, Y_2, \ldots, Y_M) = g(X_1, X_2, \ldots, X_N)$ which we consider to be our "intermediate random variables" with joint probability measure P_y and lastly $(Z_1, Z_2, \ldots, Z_L) = h(Y_1, Y_2, \ldots, Y_M)$ our "output random variables" with joint measure P_z.

Let f be a (Borel measurable) function mapping $\mathcal{R}^L$ to $\mathcal{R}^d$. Suppose we are interested in the quantity

$$\begin{aligned}\rho &= \mathbb{E}[f(Z_1, \ldots, Z_L)],\\ &= \mathbb{E}[f\big(h(Y_1, Y_2, \ldots, Y_M)\big)],\\ &= \mathbb{E}[f\ \big(h\big(g(X_1, X_2, \ldots, X_N)\big)\big)].\end{aligned}$$

$\rho = (\rho_1, \rho_2, \ldots, \rho_d)$ is of course a d-dimensional vector. The bias probability measures are always denoted with the symbol Q with a subscript to indicate which random variables are being biased, for example, Q_x, Q_y, Q_z. We assume that the original probability measures are absolutely continuous with respect to these measures. It is enough to assume that $P_x \ll Q_x$ since this automatically implies $P_y \ll Q_y$ and $P_z \ll Q_z$. This of course guarantees the existence of the Radon–Nikodym derivatives $dP_x/dQ_x, dP_y/dQ_y, dP_z/dQ_z$ needed for our importance sampling estimators. We assume that biasing measures have the same relationship between them as do the actual measures. (We denote the biased random variables as the original random variable written with a tilde over it.) Thus, if $\tilde{X}_1, \tilde{X}_2, \ldots, \tilde{X}_N$ are generated to have measure Q_x, then $g(\tilde{X}_1, \tilde{X}_2, \ldots, \tilde{X}_n)$ will have measure Q_y and $h\big(g(\tilde{X}_1, \tilde{X}_2, \ldots, \tilde{X}_n)\big)$ will have measure Q_z.

Depending on at which point of the system we wish to bias, we can define various estimators of ρ. The possibilities are

$$\hat{\rho}_i = \frac{1}{k}\sum_{j=1}^{k} f\Big(h\big(g(\tilde{X}_1^{(j)}, \tilde{X}_2^{(j)}, \ldots, \tilde{X}_N^{(j)})\big)\Big)\frac{dP_x}{dQ_x}(\tilde{X}_1^{(j)}, \ldots, \tilde{X}_N^{(j)})$$

$$\hat{\rho}_m = \frac{1}{k}\sum_{j=1}^{k} f\big(h(\tilde{Y}_1^{(j)}, \tilde{Y}_2^{(j)}, \ldots, \tilde{Y}_M^{(j)})\big)\frac{dP_y}{dQ_y}(\tilde{Y}_1^{(j)}, \ldots, \tilde{Y}_M^{(j)})$$

and

$$\hat{\rho}_o = \frac{1}{k}\sum_{j=1}^{k} f(\tilde{Z}_1^{(j)}, \ldots, \tilde{Z}_L^{(j)})\frac{dP_z}{dQ_z}(\tilde{Z}_1^{(j)}, \ldots, \tilde{Z}_L^{(j)})$$

as the input, intermediate, and output estimators, respectively, and where the superscript on a random variable indicates which one of k independent simulation runs is under consideration. Each of these estimates are d-dimensional vectors; $\hat{\rho}_i = (\hat{\rho}_{i,1}, \hat{\rho}_{i,2}, \ldots, \hat{\rho}_{i,d})$, $\hat{\rho}_m = (\hat{\rho}_{m,1}, \hat{\rho}_{m,2}, \ldots, \hat{\rho}_{m,d})$, and $\hat{\rho}_o = (\hat{\rho}_{o,1}, \hat{\rho}_{o,2}, \ldots, \hat{\rho}_{o,d})$.

We now state the following fundamental theorem of importance sampling Monte Carlo system simulation:

Theorem 4.3.1.

$$Var(\hat{\rho}_{i,r}) \geq Var(\hat{\rho}_{m,r}) \geq Var(\hat{\rho}_{o,r}) \quad r = 1, 2, \ldots, d$$

with equality for the first inequality if and only if

$$\frac{dP_x}{dQ_x}(\tilde{X}_1^{(j)}, \ldots, \tilde{X}_N^{(j)}) = s_i(\tilde{Y}_1^{(j)}, \ldots, \tilde{Y}_M^{(j)})$$

for some function s_i, and with equality for the second inequality if and only if

$$\frac{dP_y}{dQ_y}(\tilde{Y}_1^{(j)}, \ldots, \tilde{Y}_M^{(j)}) = s_o(\tilde{Z}_1^{(j)}, \ldots, \tilde{Z}_L^{(j)})$$

for some function s_o.

We first give a simple lemma for the importance sampling weight functions.

Lemma 4.3.1.

$$\frac{dP_y}{dQ_y}(\tilde{Y}) = \mathbb{E}[\frac{dP_x}{dQ_x}(\tilde{X})|\tilde{Y}].$$

Proof (Lemma 4.3.1). To characterize

$$\mathbb{E}[\frac{dP_x}{dQ_x}(\tilde{X})|\tilde{Y}],$$

or more precisely,

$$\mathbb{E}_{Q_y}[\frac{dP_x}{dQ_x}(\tilde{X})|\tilde{Y}],$$

first of all note that, for every bounded (measurable) function $h(y)$, we have

$$\mathbb{E}_{Q_x}[\frac{dP_x}{dQ_x}(\tilde{X})h(\tilde{Y})] = \mathbb{E}_{Q_y}[\mathbb{E}_{Q_x}[\frac{dP_x}{dQ_x}(\tilde{X})|\tilde{Y}]h(\tilde{Y})].$$

On the other hand, we also have

$$\begin{aligned}\mathbb{E}_{Q_x}[\frac{dP_x}{dQ_x}(\tilde{X})h(\tilde{Y})] &= \mathbb{E}_{Q_x}[\frac{dP_x}{dQ_x}(\tilde{X})h(g(\tilde{X}))] \\ &= \mathbb{E}_{P_x}[h(g(\tilde{X}))] \\ &= \mathbb{E}_{P_y}[h(\tilde{Y})] \\ &= \mathbb{E}_{Q_y}[\frac{dP_y}{dQ_y}(\tilde{Y})h(\tilde{Y})].\end{aligned}$$

Thus,

$$\mathbb{E}_{Q_y}[\mathbb{E}_{Q_x}[\frac{dP_x}{dQ_x}(\tilde{X})|\tilde{Y}]h(\tilde{Y})] = \mathbb{E}_{Q_y}[\frac{dP_y}{dQ_y}(\tilde{Y})h(\tilde{Y})]$$

for all functions h. The only way that this can occur is

$$\mathbb{E}_{Q_x}[\frac{dP_x}{dQ_x}(\tilde{X})|\tilde{Y}] = \frac{dP_y}{dQ_y}(\tilde{Y}) \quad Q_y - \text{a.s.}$$

□

Proof (Proof of Theorem 4.3.1). Without loss of generality, we can just consider the relationship between the input and intermediate estimators. Also without loss of generality, we just suppose that $d = 1$, otherwise we could just work with the rth component of the estimators and have the same supposition.

Since the two estimators have the same mean, it suffices to compare the second moments of typical terms. For simplicity we write

$$\tilde{X} = (\tilde{X}_1^{(j)}, \tilde{X}_2^{(j)}, \ldots, \tilde{X}_N^{(j)})$$

and

$$\tilde{Y} = (\tilde{Y}_1^{(j)}, \tilde{Y}_2^{(j)}, \ldots, \tilde{Y}_M^{(j)}).$$

Thus, the typical term of the input estimator has second moment

$$\mathbb{E}[f\Big(h\big(g(\tilde{X})\big)\Big)^2 \frac{dP_x}{dQ_x}(\tilde{X})^2] = \mathbb{E}[f\big(h(\tilde{Y})\big)^2 \frac{dP_x}{dQ_x}(\tilde{X})^2] \tag{4.3}$$

$$= \mathbb{E}[f(h(\tilde{Y}))^2 \mathbb{E}[\frac{dP_x}{dQ_x}(\tilde{X})^2|\tilde{Y}]] \tag{4.4}$$

while the typical term for the intermediate estimator has second moment

$$\mathbb{E}[f(h(\tilde{Y}))^2 \frac{dP_y}{dQ_y}(\tilde{Y})^2] = \mathbb{E}[f(h(\tilde{Y}))^2 \mathbb{E}[\frac{dP_x}{dQ_x}(\tilde{X})|\tilde{Y}]^2],$$

where we have used Lemma 4.3.1.

Now observe that by Jensen's inequality,

$$\mathbb{E}[\frac{dP_x}{dQ_x}(\tilde{X})|\tilde{Y}]^2 \le \mathbb{E}[\frac{dP_x}{dQ_x}(\tilde{X})^2|\tilde{Y}].$$

Hence the general term for the input estimator has greater than equal second moment (and hence greater than or equal variance) than that of the intermediate estimator. We note also that we have equality in the Jensen's inequality if and only if $(dP_x/dQ_x)(\tilde{X})$ conditioned on $\tilde{Y}$ is almost surely a constant (dependent possibly on $\tilde{Y}$). This is equivalent to $(dP_x/dQ_x)(\tilde{X}) = s(\tilde{Y})$ for some deterministic function s. This completes the proof of the theorem. □

Remark 4.3.1. In writing (4.4), we appealed to two elementary properties of conditional expectation: Equations (34.6) and (34.4) of [8]. Equation (34.6) requires that the right-hand side of (4.3) be finite. However, if (4.3) is infinite, the theorem is trivial. To use (34.4) further requires that $\mathbb{E}[(dP_x/dQ_x(\tilde{X}))^2] < \infty$. If this is not the case, put $L_n(\cdot) = \min(dP_x/dQ_x(\cdot), n)$, and write

$$
\begin{aligned}
&\mathbb{E}[f\big(h(\tilde{Y})\big)^2 \frac{dP_x}{dQ_x}(\tilde{X})^2] \\
&= \lim_{n\to\infty} \mathbb{E}[f\big(h(\tilde{Y})\big)^2 L_n(\tilde{X})^2] \\
&= \lim_{n\to\infty} \mathbb{E}[f\big(h(\tilde{Y})\big)^2 \mathbb{E}[L_n(\tilde{X})^2|\tilde{Y}]] \\
&\geq \lim_{n\to\infty} \mathbb{E}[f\big(h(\tilde{Y})\big)^2 \mathbb{E}[L_n(\tilde{X})|\tilde{Y}]^2] \\
&= \mathbb{E}[f\big(h(\tilde{Y})\big)^2 \lim_{n\to\infty} \mathbb{E}[L_n(\tilde{X})|\tilde{Y}]^2] \\
&= \mathbb{E}[f\big(h(\tilde{Y})\big)^2 \mathbb{E}[\frac{dP_x}{dQ_x}(\tilde{X})|\tilde{Y}]^2]
\end{aligned}
$$

where we use the monotone convergence theorem along with a conditional dominated convergence theorem [8, Theorem 34.2(v)]. Note also that the quantity

$$
\frac{dP_x}{dQ_x}(\tilde{X})
$$

is integrable since its expectation is one.

Remark 4.3.2. It is possible that the inequalities be met with equality. For example, consider the case of $h\big(g(x_1, \ldots, x_N)\big) = \sum_{i=1}^N r(x_i)$, for some arbitrary function $r : \mathcal{R} \to \mathcal{R}$. We can suppose that the output estimator and the intermediate estimator are the same. Suppose $dP_x(x_1, x_2, \ldots, x_N) = \prod_{i=1}^N p(x_i)$, where $p(\cdot)$ is the input probability density (or mass function if we are dealing with discrete random variables). Suppose we choose the biasing distributions to be *exponential shifts*:

$$
dQ_{x,\theta}(x_1, \ldots, x_N) = \prod_{i=1}^N q_\theta(x_i) = \frac{\prod_{i=1}^N p(x_i)\exp\big(\theta r(x_i)\big)}{M(\theta)^N},
$$

where $M(\theta) = \int p(x)\exp\big(\theta r(x)\big)dx$ is the moment generating function of the scalar random variable $r(X)$. Now note that

$$
\begin{aligned}
\frac{dP_x}{dQ_x}(\tilde{X}_1, \ldots, \tilde{X}_N) &= \frac{\prod_{i=1}^N p(\tilde{X}_i)}{\Big(\prod_{i=1}^N p(\tilde{X}_i)\exp\big(\theta r(\tilde{X}_i)\big)\Big) M(\theta)^{-N}} \\
&= \exp\big(-\theta \sum_{i=1}^N r(\tilde{X}_i)\big) M(\theta)^N \\
&= \exp\big(-\theta \tilde{Y}\big) M(\theta)^N \\
&= s_i(\tilde{Y}).
\end{aligned}
$$

Thus, in this sum of i.i.d. random variables setting with exponential shift bias distributions, no performance loss is incurred by using the simpler input formulation.

4.4 Conditional Importance Sampling

Suppose we are interested in

$$\rho = \mathbb{E}[f(Z_1, Z_2)],$$

where Z_i is an $\mathcal{R}^{n_i}$-valued random variable for $i = 1, 2$ and $f : \mathcal{R}^{n_1} \times \mathcal{R}^{n_2} \rightarrow \mathcal{R}^d$. Denote the probability measure on $\mathcal{R}^{n_1} \times \mathcal{R}^{n_2}$ associated with the random variables (Z_1, Z_2) as P. We suppose that we wish to use importance sampling to estimate ρ and thus we have a biasing probability measure on $\mathcal{R}^{n_1} \times \mathcal{R}^{n_2}$ which we denote as Q. The usual importance sampling estimator is given by

$$\hat{\rho}_{IS} = \frac{1}{k} \sum_{j=1}^{k} f(\tilde{Z}_1^{(j)}, \tilde{Z}_2^{(j)}) \frac{dP}{dQ}(\tilde{Z}_1^{(j)}, \tilde{Z}_2^{(j)}).$$

Now note that by the smoothing property of conditional expectation,

$$\rho = \mathbb{E}_P[f(Z_1, Z_2)] = \mathbb{E}[\mathbb{E}_P[f(Z_1, Z_2)|Z_2]] = \mathbb{E}[g(Z_2)],$$

where we denote $\mathbb{E}_P[f(Z_1, Z_2)|Z_2]$ which is only a function of Z_2, as $g(Z_2)$. It could very well be in certain situations that this conditional expectation g is known or is easily computable. Of course ρ is the just the expectation of the g function. Thus we can use importance sampling to estimate the expectation of the g function. This leads us to the so-called *conditional importance sampling estimate* (also known in certain situations as the g-method),

$$\begin{aligned} \hat{\rho}_g &= \frac{1}{k} \sum_{j=1}^{k} \mathbb{E}_P[f(Z_1^{(j)}, Z_2^{(j)})|Z_2^{(j)}] \frac{dP}{dQ}(\tilde{Z}_2^{(j)}) \\ &= \frac{1}{k} \sum_{j=1}^{k} g(Z_2^{(j)}) \frac{dP}{dQ}(\tilde{Z}_2^{(j)}), \end{aligned}$$

where $(dP/dQ)(z_2)$ is just the Radon–Nikodym derivative of the marginal distribution of Z_2 under P with respect to the marginal distribution of Z_2 under Q. For example, it is easy to verify that

$$\mathbb{E}_Q[\frac{dP}{dQ}(\tilde{Z}_1, \tilde{Z}_2)|\tilde{Z}_2] = \frac{dP}{dQ}(\tilde{Z}_2).$$

The main result here is

Theorem 4.4.1.

$$Var(\hat{\rho}_{g,i}) \leq Var(\hat{\rho}_{IS,i}) \qquad i = 1, 2, \ldots, d.$$

Proof. For simplicity, we just take $d = 1$, otherwise without loss of generality, we can just consider the ith component of the estimator in isolation. As always $k\,Var(\hat{\rho}_g) = F_g - \rho^2$ and of course $k\,Var(\hat{\rho}_{IS}) = F_{IS} - \rho^2$. We now have

$$\begin{aligned} F_g &= \int g(z_2)^2 \big(\frac{dP}{dQ}(z_2)\big)^2 dQ(z_2) \\ &= \int g(z_2)^2 \frac{dP}{dQ}(z_2) dP(z_2). \end{aligned}$$

Consider the first term in the integrand above,

$$\begin{aligned} g(z_2)^2 &= \Big(\int f(z_1, z_2) dP(z_1|z_2)\Big)^2 \\ &= \Big(\int f(z_1, z_2) \frac{dP(z_1|z_2)}{dQ(z_1|z_2)} dQ(z_1|z_2)\Big)^2 \end{aligned}$$

now applying Schwarz' inequality

$$\begin{aligned} &\le \Big(\int dQ(z_1|z_2)\Big) \int f^2(z_1, z_2) \big(\frac{dP(z_1|z_2)}{dQ(z_1|z_2)}\big)^2 dQ(z_1|z_2) \\ &= \int f^2(z_1, z_2) \frac{dP(z_1|z_2)}{dQ(z_1|z_2)} dP(z_1|z_2). \end{aligned}$$

Hence

$$\begin{aligned} F_g &= \int g(z_2)^2 \frac{dP}{dQ}(z_2) dP(z_2) \\ &\le \int\int f^2(z_1, z_2) \frac{dP(z_1|z_2)}{dQ(z_1|z_2)} dP(z_1|z_2) \frac{dP}{dQ}(z_2) dP(z_2) \\ &= \int\int f^2(z_1, z_2) \frac{dP}{dQ}(z_1, z_2) dP(z_1, z_2) \\ &= \int\int f^2(z_1, z_2) \big(\frac{dP}{dQ}(z_1, z_2)\big)^2 dQ(z_1, z_2) \\ &= F_{IS}. \end{aligned}$$

This completes the proof of the theorem. □

4.5 Simulation Diagnostics

We typically want our importance sampling estimate $\hat{\rho}$ of a probability ρ to be within x percent accuracy with probability y. Usually this leads us to consider some measure of the *relative precision* of the estimate. Let Z be a standard Gaussian random variable. Denote the two-sided quantile of Z by $P\big(|Z| \le t_y\big) = y$. We control the relative accuracy of our importance sampling estimates by controlling the number of simulation runs $\tilde{k}$. Recall that since the importance sampling estimate $\hat{\rho}$ is unbiased, we can write $\tilde{k}\, Var(\hat{\rho}) = F - \rho^2$. Thus,

$$
\begin{aligned}
y &= P\big(|\hat{\rho} - \rho| \le \frac{x}{100}\rho\big) \\
&\approx P\big(|Z\sqrt{Var(\hat{\rho})}| \le \frac{x}{100}\rho\big) \\
&= P\big(|Z\sqrt{\frac{F-\rho^2}{\tilde{k}}}| \le \frac{x}{100}\rho\big) \\
&= P\big(|Z| \le \frac{x\rho\sqrt{\tilde{k}}}{100\sqrt{F-\rho^2}}\big) \\
t_y &= \frac{x\rho\sqrt{\tilde{k}}}{100\sqrt{F-\rho^2}} \\
\tilde{k} &= \left(\frac{t_y 100}{x}\right)^2 \left(\frac{F}{\rho^2} - 1\right). \qquad (4.5)
\end{aligned}
$$

We should always set a desired level of precision and confidence before we begin a simulation. Equation (4.5) requires that we know F and ρ in order to set the number of simulation runs $\tilde{k}$ beforehand. Obviously, we don't have this information and so we must do something else. In fact no procedure in which the run length is fixed before the simulation begins can be relied upon to produce a confidence interval that covers the true value with the desired probability level. In the author's opinion, the only practical solution to this problem is to develop some sort of sequential procedure. In other words we will, as the simulation progresses, use the simulation outputs themselves to decide when we have collected enough data and can stop the simulation.

Suppose that $\rho = P\big(f(Z_p) \in E\big)$ where Z_p is an $\mathcal{S}$-(a complete separable metric space[1])valued random variable (with associated probability measure P) and f is a (measurable) function mapping from S into $\mathcal{R}^d$.

To implement an importance sampling estimator, we generate an i.i.d. sequence of $\mathcal{S}$-valued random variables $\{Z_q^{(1)}, Z_q^{(2)}, \ldots\}$, with associate probability measure Q. As the simulation progresses, we compute an estimate of F (in addition to the importance sampling estimate of ρ). Thus

$$
\begin{aligned}
\hat{\rho}(k) &= \frac{1}{k}\sum_{j=1}^{k} \frac{dP}{dQ}(Z_q^{(j)}) \;\; 1_{\{f(Z_q^{(j)}) \in E\}} \\
\hat{F}(k) &= \frac{1}{k}\sum_{j=1}^{k} \left[\frac{dP}{dQ}(Z_q^{(j)})\right]^2 \;\; 1_{\{f(Z_{q,n}^{(j)}) \in E\}}.
\end{aligned}
$$

We can then compute

[1] We have chosen to make the domain of f a complete separable metric space (also called a Polish space). We could allow a more general topological space; all that we really require is that f be a measurable mapping.

$$k^*(k) = \left(\frac{t_y 100}{x}\right)^2 \left(\frac{\hat{F}(k)}{\hat{\rho}(k)^2} - 1\right). \tag{4.6}$$

Some common useful values for t_y are: $t_{.80} = 1.2816$, $t_{.90} = 1.6440$, $t_{.95} = 1.96$, $t_{.99} = 2.5758$. Common values for x would be $1, 5, 10$, or 20. Our criterion for stopping the simulation is as follows.

Sequential Stopping Criterion: Stop after k simulation runs if

$$k \geq k^*(k).$$

Sometimes (usually out of laziness), we wish just to run a simulation for a certain number of times k and look at the output. We can use (4.5) in another way. We might want to know what level of precision we have attained. We can then estimate an x percent level of precision with y percent confidence, by computing

$$x = 100 t_y \sqrt{\frac{1}{k}\left(\frac{\hat{F}(k)}{\hat{\rho}(k)^2} - 1\right)}.$$

Example 4.5.1. Suppose we have the following observation model for an observed sequence of random variables in a signal detection problem

$$r_i = s + N_i \qquad i = 1, \ldots, n,$$

where $s = -1$ under hypothesis zero (H_0), $s = +1$ under hypothesis one (H_1), and $\{N_i\}$ is an i.i.d. sequence of standard normals under either hypothesis. We process these data by computing

$$R = \frac{1}{n}\sum_{i=1}^{n} r_i = \frac{1}{n}\sum_{i=1}^{n} s + N_i = s + \bar{N}.$$

If $R > 0$, we announce H_1 is true; otherwise we announce H_0 is true. The probability of error for this receiver is

$$P(error) = P(\frac{1}{n}\sum_{i=1}^{n} N_i > 1) = P(\bar{N} > 1).$$

Since the N_i are i.i.d. standard normals, $\bar{N}$ is normal mean zero, variance $1/n$. Using tables of the error function, we can easily evaluate this probability for a given n. For example, $n = 24$ gives $P(error) \approx 4.8 \times 10^{-7}$.

The importance sampling method we choose to simulate this system is mean shifting. Instead of directly simulating the standard normal noise samples $\{N_i\}$, we use the mean shifted $\{\tilde{N}_i\}$ (taken to be mean one, variance one) random variables in an input formulation. For an output formulation we have $p_{\bar{N}}$ is normal mean zero, variance $1/n$ and $q_{\bar{N}}$ is normal mean one, variance $1/n$. Both formulations will give the same variance (the mean shift is also an exponential shift in the Gaussian setting); we use the simpler (to simulate) output formulation.

We want to investigate a bit our results for k^* in the setting of this very simple system (where we can calculate everything in closed form). We compute $\hat{\rho}_n$ and $\hat{F}_n$ as the simulation progresses to determine the number of simulation runs we need to achieve x percent accuracy with probability y. For this example, we use $y = .9$ and x taking on the range of values $\{2.5, 5, 10, 20, 40\}$. We find that as our desired accuracy x decreases, our empirical probability $\hat{y}$ of achieving accuracy x increases to its expected value $y = .9$ using the k^* stopping criterion. The following table shows the results of this experiment for 1000 trials.

Accuracy (x)	Achieved Probability ($\hat{y}$)
40	.840
20	.873
10	.887
5	.893
2.5	.909

Table 4.1. Simulation results

4.6 Notes and Comments

The importance sampling idea, that of focusing on the region(s) of importance so as to save computational resources, evidently springs from a 1956 paper due to A. Marshall [59]. Importance sampling is but one of a variety of variance reduction techniques known to simulation practitioners. A very readable introduction to the subject of variance reduction in simulation is found in [67].

The notion of input versus output estimators and bias point selection was first posed (at least in the engineering literature) by P. Hahn and M. Jeruchim [37]. The basic theorem is from [14].

Conditional importance sampling estimators first appear (in the engineering literature) under the name of the g-method of R. Srinavasan [78]. The basic theorem given in the text is a generalization of his result for i.i.d. sums.

A rigorous (asymptotic) analysis of the sequential stopping rule given in the section on simulation diagnostics can be found in [62].

5. The Large Deviation Theory of Importance Sampling Estimators

Probability theory is nothing but common sense reduced to calculation.
Pierre Simon Laplace
If we complicate things, they get less simple. *Anon.*

5.1 The Variance Rate of Importance Sampling Estimators

In this section we first prove a very general theorem regarding the variance rate of importance sampling estimators. At first sight the setting may appear to be fairly abstract but we argue that the level of generality presented here will pay dividends later on. We encourage the reader to follow the problem setup given here with some care and then compare it with the first example given below to see how this framework is actually used.

For every integer n, let $Z_{p,n}$ be a random variable taking values in some complete separable metric space $\mathcal{S}_n$. Let P_n be the probability measure induced by $Z_{p,n}$ on $\mathcal{S}_n$. Instead of directly simulating $Z_{p,n}$, we choose to simulate with another $\mathcal{S}_n$-valued random variable $Z_{q,n}$ which in turn induces a probability measure on $\mathcal{S}_n$, Q_n. Let f_n be an $\mathcal{R}^d$-valued measurable function on the space $\mathcal{S}_n$ (i.e. $f_n : \mathcal{S}_n \to \mathcal{R}^d$).

To create the importance sampling estimators, we must assume that P_n is absolutely continuous with respect to Q_n for all n (and hence the Radon–Nikodym derivative dP_n/dQ_n exists). We suppose we are interested in $\rho_n = P\big(f_n(Z_{p,n})/n \in E\big)$, for some Borel set $E \subset \mathcal{R}^d$. We assume that a large deviation principle is controlling this probability as n gets large. The importance sampling estimator of ρ is given as

$$\hat{\rho}_n = \frac{1}{k} \sum_{j=1}^{k} \frac{dP_n}{dQ_n}(Z_{q,n}^{(j)}) \; 1_{\left\{\frac{f_n(Z_{q,n}^{(j)})}{n} \in E\right\}}.$$

The bottom line in most simulation parameter estimation is to evaluate the variance of the estimator. Thus, we would like to investigate the variance performance of the above importance sampling estimator as we vary k and n. Since $\hat{\rho}$ is an average of k i.i.d. terms, $k\,Var(\hat{\rho})$ is a constant. Thus we know exactly how the variance changes as k, the number of simulation runs, is varied. A much more interesting question is how does the variance change as we vary the large deviation parameter n? This section is devoted to this question.

Example 5.1.1. Let $\{X_i\}$ be a sequence of $\mathcal{R}$-valued random variables. The density of $(X_1, X_2, \ldots, X_n)$ is denoted as $p_n(x_1, x_2, \ldots, x_n)$. We are interested in estimating $\rho_n = P(\frac{1}{n}\sum_{i=1}^n X_i \in E)$, for some set $E \subset \mathcal{R}$. We simulate using another sequence of $\mathcal{R}$-valued random variables $\{Y_i\}$. We denote the probability density of $(Y_1, Y_2, \ldots, Y_n)$ as $q(y_1, y_2, \ldots, y_n)$. Our estimator is thus

$$\hat{\rho}_n = \frac{1}{k}\sum_{j=1}^{k} \frac{p(Y_1^{(j)}, Y_2^{(j)}, \ldots, Y_n^{(j)})}{q(Y_1^{(j)}, Y_2^{(j)}, \ldots, Y_n^{(j)})} 1_{\{\frac{1}{n}\sum_{i=1}^n Y_i^{(j)} \in E\}}.$$

To match up with the notation given above,

$$\begin{aligned}
d &= 1 \\
\mathcal{S}_n &= \mathcal{R}^n \\
Z_{p,n} &= (X_1, X_2, \ldots, X_n) \\
\text{for } B \subset \mathcal{R}^n, \quad P_n(B) &= \int_B p(z_1, z_2, \ldots, z_n) dz_1 dz_2 \cdots dz_n \\
Z_{q,n} &= (Y_1, Y_2, \ldots, Y_n) \\
\text{for } B \subset \mathcal{R}^n, \quad Q_n(B) &= \int_B q(z_1, z_2, \ldots, z_n) dz_1 dz_2 \cdots dz_n \\
\text{for } z = (z_1, z_2, \ldots, z_n) \in \mathcal{R}^n, \; f_n(z) &= \sum_{i=1}^n z_i \\
\frac{dP_n}{dQ_n}(z) &= \frac{p(z_1, z_2, \ldots, z_n)}{q(z_1, z_2, \ldots, z_n)}.
\end{aligned}$$

We first need to make a few definitions. For $\theta \in \mathcal{R}^d$, define (if the integral exists)

$$c_n(\theta) = \frac{1}{n}\log\Big(\int \frac{dP_n}{dQ_n}(z)\exp\big(\langle\theta, f_n(z)\rangle\big)dP_n(z)\Big).$$

As we showed in the last chapter for ϕ_n, we have that c_n is a convex function for each n.

For each n, define a probability measure μ_n (if $c_n(0) < \infty$) by defining for every Borel set $B \subset \mathcal{S}_n$,

$$\mu_n(B) = \exp\big(-nc_n(0)\big)\int_B \frac{dP_n(z)}{dQ_n(z)} dP_n(z).$$

Let Y_n be an $\mathcal{S}_n$-valued random variable with associated probability measure μ_n. Then $\{f_n(Y_n)\}$ is a sequence of $\mathcal{R}^d$-valued random variables. For $\theta \in \mathcal{R}^d$, we have the usual log moment generating function sequence for these random variables as

$$\begin{aligned}\phi_n(\theta) &= \frac{1}{n}\log \mathbb{E}[\exp(\langle\theta, f_n(Y_n)\rangle)] \\ &= \frac{1}{n}\log(\int \frac{dP_n}{dQ_n}(z)\exp(\langle\theta, f_n(z)\rangle)\exp(-nc_n(0))dP_n(z)) \\ &= c_n(\theta) - c_n(0).\end{aligned}$$

If the standard assumptions A1, A2, and A3 hold for the c_n function sequence, then they will also hold for the ϕ_n sequence. Hence Theorem 3.2.1 holds for the random variables $\{f_n(Y_n)\}$.

We now define the *variance rate function* $R(\cdot)$ as

$$R(x) = \sup_{\theta\in\mathcal{R}^d} [\langle\theta, x\rangle - c(\theta)] \quad (\text{for } x \in \mathcal{R}^d),$$

where c of course is the limit of the $\{c_n\}$ sequence. For every Borel set E, we define

$$R(E) = \inf_{x\in E} R(x),$$

and denote $\overset{\circ}{E}$ = the interior of E, $\bar{E}$ = the closure of E, and ∂E = the boundary of $E = \bar{E}/\overset{\circ}{E}$.

As stated above, we are interested in determining the rate of the variance expression as a function of n. The first term (which usually determines the rate of the variance) of that expression may be written as

$$F_n = \int (\frac{dP_n}{dQ_n}(z))^2 1_{\{\frac{f_n(z)}{n}\in E\}} dQ_n(z).$$

We now prove the following theorem.

Theorem 5.1.1. *Let E be any Borel set such that $\overset{\circ}{E} \neq \emptyset$, $\bar{E} = \bar{\overset{\circ}{E}}$, and $0 < R(E) < \infty$. Then*

$$\lim_{n\to\infty}\frac{1}{n}\log(F_n) = -R(E)$$

Proof.

$$F_n = \int (\frac{dP_n}{dQ_n}(z))^2 1_{\{\frac{f_n(z)}{n}\in E\}} dQ_n(z)$$

$$
\begin{aligned}
&= \exp(nc_n(0)) \int_{\frac{f_n(z)}{n}\in E} d\mu_n(z) \\
\frac{1}{n}\log F_n &= c_n(0) + \frac{1}{n}\log \int_{\frac{f_n(z)}{n}\in E} d\mu_n(z) \\
&\to_{n\to\infty} c(0) - \inf_{x\in E}\sup_{\theta\in\mathcal{R}^d} [\langle\theta, x\rangle - c(\theta) - c(0)] \\
&= -\inf_{x\in E}\sup_{\theta\in\mathcal{R}^d} [\langle\theta, x\rangle - c(\theta)] \\
&= -R(E).
\end{aligned}
$$

□

Remark 5.1.1. We stated the theorem for sets E with nonempty interior such that $\bar{E} = \bar{\mathring{E}}$. The principal reason for doing this was so that we could state a limit theorem for the variance asymptotics cleanly without recourse to an upper bound for closed sets and a lower bound for open sets.

Remark 5.1.2. This theorem is a slight generalization of Theorem 3.2.1 of the previous chapter. We see that if we take $Q_n = P_n$ (which is the direct Monte Carlo choice), we have

$$
\begin{aligned}
F_n &= \int 1_{\{\frac{f_n(z)}{n}\in E\}} dQ_n(z) \\
&= \int 1_{\{\frac{f_n(z)}{n}\in E\}} dP_n(z) \\
&= P\left(\frac{f_n(z)}{n} \in E\right) \\
&= \rho_n
\end{aligned}
$$

and our theorem reduces directly to Theorem 3.2.1 which deals with the asymptotic rate of probabilities to zero. This is not surprising since k times the variance of the Monte Carlo estimator is just $\rho_n(1-\rho_n) = \rho_n - \rho_n^2 \approx \rho_n$ for very small ρ_n. The asymptotic rate of the variance for the Monte Carlo estimator should be (and is) the same as the asymptotic rate of the probabilities ρ_n.

Example 5.1.2. Suppose we are interested in estimating

$$
\rho = P\left(\left|\frac{1}{10}\sum_{i=1}^{10} X_i\right| > 3\right),
$$

where $\{X_i\}$ are i.i.d. Laplacian random variables with density

$$
p(x) = \frac{1}{2}\exp(-|x|).
$$

We use the variance scaling method to choose an i.i.d. simulation distribution $q(x)$; that is, $q(x)$ will have the density of aX_1, where $a > 0$ is the variance scale parameter. In order to find a good value of a, we compute the variance rate function $R_a(3)$ for a range of values for a. The value of a that maximizes $R_a(3)$ is the best choice at least in the sense of maximizing the rate with which the asymptotic variance approaches zero.

Thus for each value of a, the corresponding simulation distribution is given by

$$q_a(x) = \frac{1}{2a} \exp(\frac{-|x|}{a}).$$

Therefore

$$\begin{aligned} c_a(\theta) &= \log\left(\int \frac{p(x)^2}{q(x)} \exp(\theta x)\, dx\right) \\ &= \log(2a-1) - \log((\frac{1}{a}-2)^2 - \theta^2). \end{aligned}$$

Now, we must compute

$$R_a(3) = \sup_\theta [\theta 3 - c_a(\theta)].$$

Taking the derivative to find the supremum in the above expression leads us to the optimizing value for θ as

$$\theta_{\{3,a\}} = \frac{-1 + \sqrt{1 + 9(\frac{1}{a} - 2)^2}}{3},$$

and

$$R_a(3) = 3\theta_{\{3,a\}} - \log(1-2a) + \log(\theta^2_{\{3,a\}} - (\frac{1}{a} - 2)^2).$$

In Fig. 5.1, we have a plot of $R_a(3)$ versus a. The optimal value is empirically found to be $a_o = 3.05897$. The estimator form is thus

$$\begin{aligned} \hat{\rho} &= \frac{1}{k} \sum_{j=1}^{k} 1_{\{|\sum_{i=1}^{10} Y_i^{(j)}|>30\}} \prod_{i=1}^{10} \frac{p(Y_i^{(j)})}{q_{a_o}(Y_i^{(j)})} \\ &= \frac{1}{k} \sum_{j=1}^{k} 1_{\{|\sum_{i=1}^{10} Y_i^{(j)}|>30\}} a_o^{10} \exp((\frac{1}{a_o} - 1) \sum_{i=1}^{10} |Y_i^{(j)}|). \end{aligned}$$

In Fig. 5.2, we show a typical simulation that stopped after $k^* = 172503$ simulation runs giving us an estimate of 10% error with 90% confidence.

Example 5.1.3. Suppose we are interested in estimating ρ given by

$$\rho = P(\frac{1}{10} \sum_{i=1}^{10} (\cos(\theta_i)) > 0.95),$$

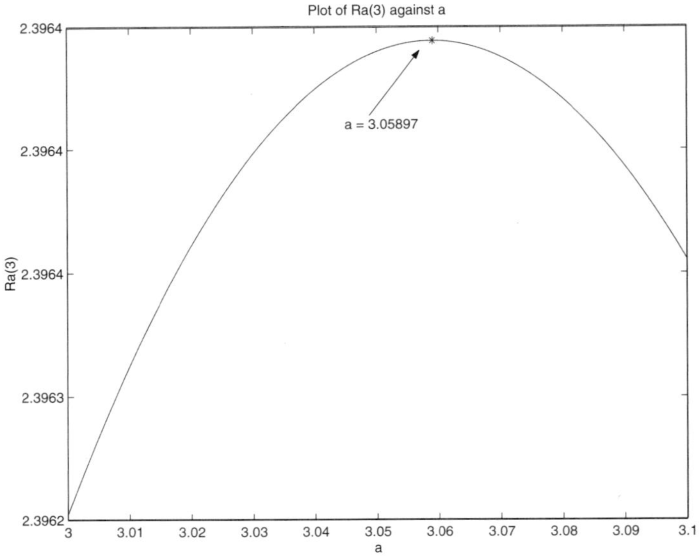

Fig. 5.1. Plot of $R_a(3)$ as a function of a in the neighborhood of the optimal a value, $a_o = 3.05897$.

where the $\{\theta_i\}$ are i.i.d. uniform $[0, 2\pi]$ random variables. The probability density function of $Y = \cos(\theta_1)$ is given by

$$p(y) = \frac{1}{2\pi\sqrt{1-y^2}}.$$

Our problem is to find a good biasing distribution in this setting. We decide to use an adhoc piecewise uniform strategy as follows,

$$q(y) = (1-a)q_1(y) + aq_2(y),$$

where q_1 is a uniform density on $[-1, a]$ and q_2 is uniform on $[a, 1]$. We still must make a good choice for the parameter a.

Our estimate for a given choice of a is thus

$$\hat{\rho} = \frac{1}{k}\sum_{j=1}^{k} 1_{\{\frac{1}{10}\sum_{i=1}^{10} y_i > 0.95\}} \frac{\prod_{i=1}^{10} p(y_i)}{\prod_{i=1}^{10} q(y_i)}.$$

We choose the a parameter by maximizing the variance rate function value at $y = .95$. The variance rate function of interest is given by

$$R_a(y) = \sup_\theta[\theta y - c_a(\theta)],$$

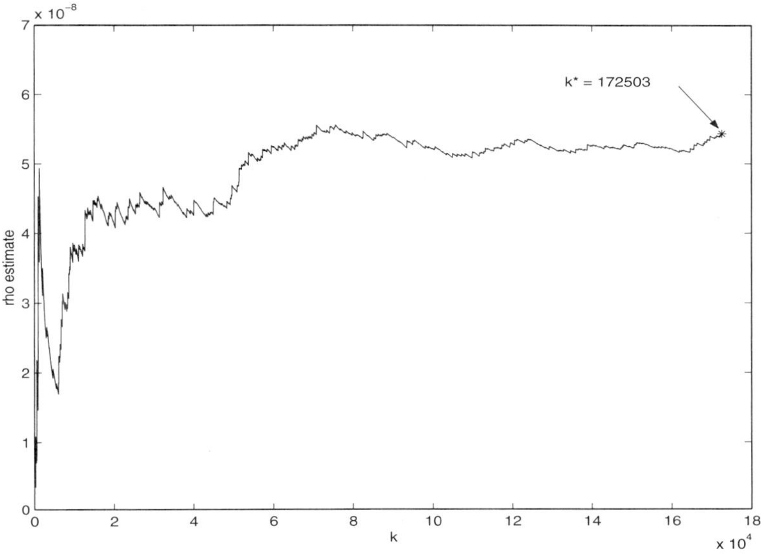

Fig. 5.2. Typical simulation run showing estimate as a function of the simulation runs. k^* was chosen to give 10% error with 90% confidence.

where

$$\begin{aligned} c_a(\theta) &= \log\left(\int_{-1}^{1} \frac{1/2\pi\sqrt{1-y^2}}{q(y)} \exp(\theta y) \frac{1}{2\pi\sqrt{1-y^2}} dy\right) \\ &= \log\left(\int_{-1}^{a} \frac{\exp(\theta y)(1+a)}{(2\pi)^2(1-a)(1-y^2)} dy + \int_{a}^{1} \frac{\exp(\theta y)(1-a)}{(2\pi)^2 a(1-y^2)} dy\right). \end{aligned}$$

Of course we are only interested in the values $R_a(.95)$. To perform the supremum over θ, we need to solve the equation (for each value of a)

$$0.95 = c'(\theta_{.95}(a)).$$

One can analytically take the derivative of c_a and numerically compute a solution to this equation. After doing so, we then compute $R_a(.95) = \theta_{.95}(a).95 - c_a(\theta_{.95}(a))$. In Fig. 5.3 we plot $R_a(.95)$ as a function of a. The maximum value occurs at $a = .845$, with variance rate of $R_{.845}(.95) = 2.8294$. In Example 5.2.7, we compute the large deviation rate of ρ_n and find that $I(.95) = 1.5572$. Note that $R_{.845}(.95) < 2I$ and so this estimator, even with an optimal choice of a, is not efficient.

5.2 Efficient Importance Sampling Estimators

Examination of (4.1) shows that $k\,Var(\hat{\rho}_n) = F_n - \rho_n^2$ which of course is greater than or equal to zero for all n. This implies in particular that the

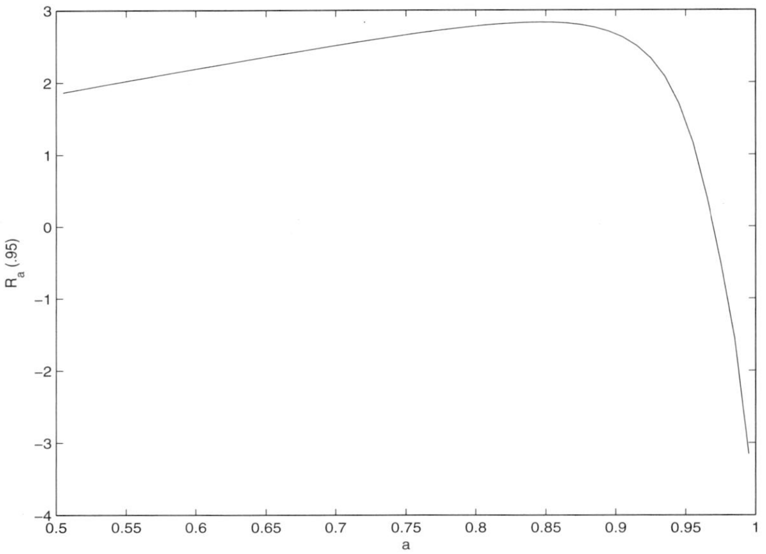

Fig. 5.3. Plot of $R_a(.95)$ as a function of a.

asymptotic rate (in n) with which F_n approaches zero can be no greater than the rate with which ρ_n^2 approaches zero.

Suppose we have the following limits existing

$$\lim_{n\to\infty} \frac{1}{n} \log \rho_n = -I$$
$$\lim_{n\to\infty} \frac{1}{n} \log F_n = -R.$$

Then we must have that $R \leq 2I$. If we have equality here (i.e. $R = 2I$), then we say that our simulation procedure is *efficient*. This notion of efficiency is crucial to the understanding of how to choose importance sampling simulation distributions. The philosophy that we adhere to in this book is that if a family of simulation distributions is efficient, then it is a good choice.

The notion of efficiency is fundamental to the understanding of rare event simulation. Recall the expression we derived for the number of simulation runs $\tilde{k}$ needed to achieve $x\%$ accuracy with $100y\%$ confidence,

$$\tilde{k} = \left(\frac{t_y 100}{x}\right)^2 \left(\frac{F_n}{\rho_n^2} - 1\right).$$

As n grows large, ρ_n is going to zero exponentially fast. Hence, $\tilde{k}$, the number of simulation runs needed to achieve a given precision, must then *grow* exponentially fast (unless and only unless) F_n approaches zero at exactly the same rate as ρ_n^2. In other words, unless the importance sampling simulation distributions are efficient, , the number of simulation runs needed must grow exponentially in n.

The philosophy of efficient simulation is based upon minimizing the asymptotic rate to zero of the estimator variance. We show that this philosophy will allow us to choose biasing strategies based upon very simple geometric interpretations of the error sets of the system under consideration. This chapter is more about presenting this simulation philosophy than it is about simulating any one type of rare event probability. We argue that in many rare event simulation problems, we should first consider the underlying large deviation theory of the problem. The reason for following this methodology is that by first embedding our problem as but one of a parametric sequence of problems (the parameter is n, the large deviation index), we can concern ourselves with maximizing the estimator variance *rate* to zero instead of minimizing the actual estimator variance itself. The mathematics of maximizing the variance rate is almost always far simpler than trying to minimize the actual variance over some class of simulation distributions. This is intuitive since our large deviation framework is only trying to maximize a rate parameter instead of the actual variance. Trying to minimize analytically the estimator variance directly almost always leads to a very complicated functional minimization problem. Of course, when this minimization can be carried out, it is very desirable to do so. In the vast majority of practical situations, it really can't be done.

We now present some models for which we can explicitly describe families of efficient importance sampling distributions. Suppose (as in the previous section) we are given a sequence of $\mathcal{S}_n$ valued random variables $\{Z_{p,n}\}$. $\{f_n(Z_{p,n})\}$ is a sequence of $\mathcal{R}^d$-valued random variables whose associated ϕ_n sequence satisfies the standard assumptions A1, A2, and A3.

We are interested in simulating to obtain estimates of the probability

$$\rho_n = P\left(\frac{f_n(Z_{p,n})}{n} \in E\right).$$

where E is a Borel set such that $\overset{\circ}{E} \neq \emptyset$, $\bar{E} = \bar{\overset{\circ}{E}}$, and $0 < I(E) < \infty$.

A point ν is a *minimum rate point* of the set E if $I(\nu) = I(E)$. A point ν is a *dominating point* of the set E if ν is a unique point such that

a) $\nu \in \partial E$,

b) there exists a unique $\theta_\nu \in \mathcal{R}^d$ such that $\nabla\phi(\theta_\nu) = \nu$, and

c) $E \subset \mathcal{H}(\nu) = \{x : \langle \theta_\nu, x - \nu \rangle \geq 0\}$.

The set $\mathcal{H}(\nu)$ is a *half space*. From convex function theory, it turns out that the hyperplane $\partial\mathcal{H}(\nu) = \{x : \langle \theta_\nu, x - \nu \rangle = 0\}$ is tangent to the rate function level set $\{x : I(x) = I(\nu)\}$ at the point ν. (This follows from the fact that $\nabla I(\nu) = \theta_\nu$.) As a result, a dominating point, if it exists, is always a unique minimum rate point. In Fig. 5.4, we have the happy situation of a dominating point existing. In the next Fig. 5.5, we have a situation where a unique minimum rate point exists but is *not* a dominating point.

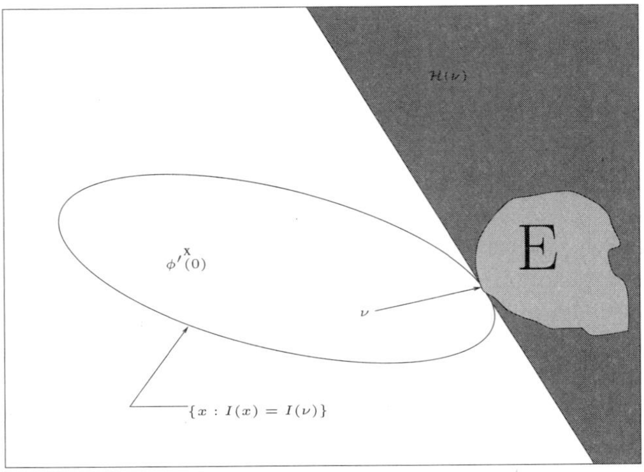

Fig. 5.4. A set E with dominating point ν.

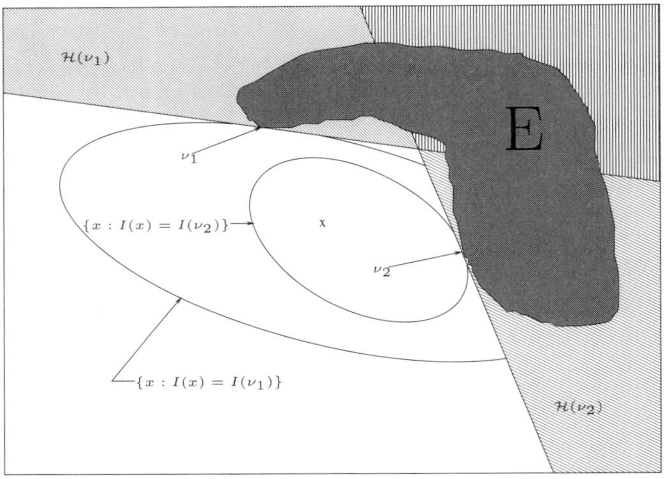

Fig. 5.5. A set E with minimum rate point ν_2 that is not a dominating point.

5.2.1 The Dominating Point Case

We assume in this subsection that E has a dominating point ν. Then

$$\lim_{n\to\infty} \frac{1}{n} \log \rho_n = -I(\nu) = \sup_\theta [\langle \theta, \nu \rangle - \phi(\theta)] = -[\langle \theta_\nu, \nu \rangle - \phi(\theta_\nu)].$$

Hence ρ_n^2 has rate $-2I(\nu)$. For *any* sequence of simulation distributions $\{Q_n\}$, we must have

$$\liminf_{n\to\infty} \frac{1}{n} \log F_n \geq -2I(\nu).$$

Remember, by definition, any sequence of simulation distributions that meets the above inequality with equality is efficient.

There are in general many efficient sequences. One universal (but impractical) choice is to merely choose for each n,

$$dQ_n = \frac{1_{\{\frac{f_n(Z_{p,n})}{n} \in E\}}}{\rho_n} dP_n,$$

which gives $F_n = \rho_n^2$ (and hence $k\,Var(\hat{\rho}_n) = 0$). Of course this choice of Q_n as a simulation distribution is of little interest since it requires exact knowledge of ρ_n, which is what we are trying to estimate in the first place. Are there other efficient sequences?

Consider the family of exponential shifts of P_n; that is,

$$dQ_n = \frac{\exp(\langle \psi, f_n(z)\rangle) dP_n(z)}{\int \exp(\langle \psi, f_n(z')\rangle) dP_n(z')} = \frac{\exp(\langle \psi, f_n(z)\rangle) dP_n(z)}{\exp(n\phi_n(\psi))}.$$

To compute the variance rate function associated with this choice, we must compute

$$\begin{aligned} c_n(\theta) &= \frac{1}{n}\log\Big(\int \frac{dP_n}{dQ_n}(z)\exp(\langle \theta, f_n(z)\rangle) dP_n(z)\Big) \\ &= \frac{1}{n}\log\Big(\int \exp(-\langle \psi, f_n(z)\rangle)\exp(\langle \theta, f_n(z)\rangle) dP_n(z)\Big) \\ &\quad + \frac{1}{n}\log\Big(\int \exp(\langle \psi, f_n(z')\rangle) dP_n(z')\Big) \\ &= \phi_n(\theta - \psi) + \phi_n(\psi) \\ &\to_{n\to\infty} \phi(\theta-\psi) + \phi(\psi) \\ &= c(\theta). \end{aligned}$$

Hence as long as $[-\psi, \psi]$ is in the interior of the domain of ϕ, we have hypotheses A1, A2, and A3 holding for c.

Hence, the variance rate function is

$$\begin{aligned} R(x) &= \sup_\theta[\langle \theta, x\rangle - c(\theta)] \\ &= \sup_\theta[\langle \theta, x\rangle - \phi(\theta-\psi) - \phi(\psi)] \\ &= \sup_\theta[\langle \theta - \psi, x\rangle - \phi(\theta-\psi)] + [\langle \psi, x\rangle - \phi(\psi)] \\ &= I(x) + [\langle \psi, x\rangle - \phi(\psi)]. \end{aligned}$$

Now, let's make the choice that $\psi = \theta_\nu$. Then

$$R(x) = I(x) + [\langle \theta_\nu, x\rangle - \phi(\theta_\nu)].$$

First, note that $R(\nu) = 2I(\nu)$. Since we assume that ν is the dominating point for E, we have for any $x \in E$, $\langle \theta_\nu, x - \nu\rangle \geq 0$. Also note that $I(x) \geq I(\nu)$ for all $x \in E$. Thus

$$R(x) - R(\nu) = I(x) - I(\nu) + \langle \theta_\nu, x - \nu \rangle \geq 0.$$

Thus

$$\inf_{x \in E} R(x) = R(E) = R(\nu) = 2I(\nu) = 2I(E), \tag{5.1}$$

and hence this sequence is efficient. We have proved the following theorem.

Theorem 5.2.1. *Suppose E has a dominating point ν. Then the sequence of simulation distributions given by*

$$dQ_n(z) = \exp\big(-n\phi_n(\theta_\nu)\big) \exp\big(\langle \theta_\nu, f_n(z) \rangle\big) dP_n(z),$$

is efficient.

A very important property of this efficient exponential shift family is its so-called *dominating point shift property.* Note that

$$\begin{aligned} &\phi_{Q,n}(\theta) \\ &= \frac{1}{n} \log\Big(\int \exp\big(\langle \theta, f_n(z) \rangle\big) dQ_n(z)\Big) \\ &= \frac{1}{n} \log\Big(\int \exp\big(\langle \theta, f_n(z) \rangle\big) \exp\big(-n\phi_n(\theta_\nu)\big) \exp\big(\langle \theta_\nu, f_n(z) \rangle\big) dP_n(z)\Big) \\ &= \phi_n(\theta + \theta_\nu) - \phi_n(\theta_\nu). \end{aligned}$$

This implies that $\phi_Q(\theta) = \phi(\theta + \theta_\nu) - \phi(\theta_\nu)$. Note that $\nabla \phi_Q(0) = \nabla \phi(\theta_\nu) = \nu$. Therefore by Remark 3.2.1, $f_n(Z_{q,n})/n$ converges almost surely to ν. Thus,

Dominating Point Property: The efficient shift parameter is characterized as that value which causes the sequence of simulation variables to converge almost surely to the dominating point.

In the next few subsections, we consider some specific probability models and some efficient simulation strategies for them.

Sum of i.i.d. $\mathcal{R}^d$ random variables. Let $\{X_i\}$ be a sequence of i.i.d. $\mathcal{R}^d$ random variables and probability density (or probability mass function if the random variables are discrete) $p(\cdot)$. For some (measurable) function $f : \mathcal{R}^d \to \mathcal{R}^{d'}$, suppose we are interested in the probability

$$\rho_n = P\Big(\frac{1}{n} \sum_{i=1}^{n} f(X_i) \in E\Big),$$

for some set $E \subset \mathcal{R}^{d'}$. Denote the moment generating function of $f(X_1)$ as $M_f(\theta) = \mathbb{E}[\exp(\langle \theta, f(X_1) \rangle)]$ for $\theta \in \mathcal{R}^{d'}$. We simulate with another sequence of i.i.d. $\mathcal{R}^d$-valued random variables $\{Y_n\}$ with density (or mass function) $q(\cdot)$. Thus the estimator is

$$\hat{\rho}_n = \frac{1}{k}\sum_{j=1}^{k} 1_{\{\sum_{i=1}^n f(Y_i^{(j)})\in nE\}} \prod_{i=1}^{n} \frac{p(Y_i^{(j)})}{q(Y_i^{(j)})}.$$

According to the setup for Theorem 5.1.1, we have $\mathcal{S}_n = \mathcal{R}^{nd}$, $Z_{p,n} = (X_1, X_2, \dots, X_n)$, $Z_{q,n} = (Y_1, Y_2, \dots, Y_n)$ and for $y_i \in \mathcal{R}^d$, $f_n(y_1, y_2, \dots, y_n) = \sum_{i=1}^n f(y_i)$. Thus for $\theta \in \mathcal{R}^{d'}$,

$$\begin{aligned} c_n(\theta) &= \frac{1}{n}\log\Big(\int \prod_{i=1}^{n} \frac{p(y_i)}{q(y_i)} \exp\big(\langle\theta, \sum_{i=1}^{n} f(y_i)\rangle\big) \prod_{i=1}^{n} p(y_i) dy_1 dy_2 \cdots dy_n\Big) \\ &= \log\Big(\int \frac{p(y)^2}{q(y)} \exp\big(\langle\theta, f(y)\rangle\big) dy\Big) \\ &= c(\theta). \end{aligned}$$

As always $k\,Var(\hat{\rho}_n) = F_n - \rho^2$. Then (under the usual assumptions)

$$\lim_{n\to\infty} \frac{1}{n} \log F_n = -R(E).$$

Also from Theorem 3.2.1, we have

$$\lim_{n\to\infty} \frac{1}{n} \log \rho_n = -I(E).$$

Suppose E has dominating point ν. In particular this means (among other things) that $I(E) = I(\nu) = \sup_\theta[\langle\theta,\nu\rangle - \log M_f(\theta)] = \langle\theta_\nu,\nu\rangle - \log M_f(\theta_\nu)$, where θ_ν satisfies $\nu = \nabla \log M_f(\theta_\nu)$.

Now let us choose q to be a certain exponential shift:

$$q_o(x) = \frac{p(x)\exp\big(\langle\theta_\nu, f(x)\rangle\big)}{M_f(\theta_\nu)}. \tag{5.2}$$

This choice corresponds exactly to choosing dQ_n as in (5.2.1) with the efficient parameter $\psi = \theta_\nu$. Hence, we know that it is efficient. We can also continue and compute the variance rate function directly as

$$\begin{aligned} R_{q_o}(\nu) &= \sup_\theta[\langle\theta,\nu\rangle - \log \int p(x)\exp\big(\langle\theta - \theta_\nu, f(x)\rangle\big)dx - \log M_f(\theta_\nu)] \\ &= \sup_\theta[\langle\theta,\nu\rangle - \log M_f(\theta - \theta_\nu) - \log M_f(\theta_\nu)] \\ &= 2I(E), \end{aligned}$$

where the maximization occurs for the value of $\theta = 2\theta_\nu$, and again we see that this choice of $q(\cdot)$ is efficient.

Interestingly enough, if we restrict our simulation random variables Y_n to be independent, then this particular exponential shift q_o is the *uniquely* efficient simulation distribution choice.

Theorem 5.2.2. *In the setting of this subsection, the simulation distribution given in (5.2) is the uniquely efficient i.i.d. simulation distribution.*

Proof. Suppose that we have another density choice, call it s, which is also efficient. Then $R_s(E) = R_{q_o}(E) = 2I(E)$.

Consider any point $x \in \mathcal{R}^d$. For any set B, $R_s(B) \le 2I(B)$. If we consider the ball of radius ϵ and center x, $B_x(\epsilon)$, we then have that $R_s(B_x(\epsilon)) \le 2I(B_x(\epsilon))$. Letting ϵ go to zero implies that $R_s(x) \le 2I(x)$. Also,

$$\inf_{x \in E} R_s(x) = 2I(\nu)$$

and

$$\inf_{x \in E} R_s(x) \le R_s(\nu) \le 2I(\nu)$$

implies

$$R_s(\nu) = 2I(\nu);$$

in other words, a minimum variance rate point for *any* efficient simulation strategy must be the dominating point of the set. Remember, the dominating point of the set is defined only through the large deviation asymptotics of the probabilities of the set. It has nothing to do with the particular choice of simulation distributions that we might make.

Therefore, since we are assuming the existence of a dominating point ν, and efficient simulation distributions, $R_s(\nu) = R_{q_o}(\nu) = 2I(\nu)$. Hence, we have two convex functions taking on the same value at a point ν and one lying entirely over the other one. This implies that the gradients (if they exist) at the point ν must also be equal, or $\nabla R_s(\nu) = 2\nabla I(\nu)$. The gradient of I exists due to Assumptions A1 to A4 holding. If the gradient of R_s at ν doesn't exist, we can choose another point in a small neighborhood of ν where the gradient does exist and will be arbitrarily close to $2I(\nu)$ in value. (The set where the gradient doesn't exist is at most countable.) We can then carry through the following argument with the appropriate analysis.

Assuming that we have $\nabla R_s(\nu) = 2\nabla I(\nu)$, we invoke a duality property for convex functions that says that $\nabla I(\nu) = \theta_\nu$. Therefore $\nabla R_s(\nu) = 2\theta_\nu$. Now another duality property says $c_s(\theta_x) = \langle \theta_x, x\rangle - R_s(x)$ if and only if $\nabla R_s(x) = \theta_x$. Hence $c_s(2\theta_\nu) = \langle 2\theta_\nu, \nu\rangle - R_s(\nu) = 2(\langle \theta_\nu, \nu\rangle - I(\nu)) = 2\log M_f(\theta_\nu)$. Thus $R_s(\nu) = 2I(\nu)$ implies that $c_s(2\theta_\nu) = 2\log M_f(\theta_\nu)$. Now by Jensen's inequality

$$\begin{aligned}
\exp\big(c(2\theta_\nu)\big) &= \int \frac{p(x)^2}{s(x)} \exp\big(2\langle \theta_\nu, f(x)\rangle\big)dx \\
&= \int \left(\frac{p(x)}{s(x)} \exp\big(\langle \theta_\nu, f(x)\rangle\big)\right)^2 s(x)dx \\
&\ge \left(\int \frac{p(x)}{s(x)} \exp\big(\langle \theta_\nu, x\rangle\big) s(x)dx\right)^2 \\
&= M_f(\theta_\nu)^2,
\end{aligned}$$

with equality if and only if $p(x)\exp\big(\langle\theta_\nu, f(x)\rangle\big)/s(x)$ is almost surely a constant. Hence if $s(x)$ is efficient, it must be the efficient exponential shift. □

The dominating point shift property tells us that the sequence of simulation random variables must converge to the dominating point of the set under consideration. In this i.i.d. setting, this is equivalent to the i.i.d. simulation distributions being chosen to have mean value equal to the dominating point. In this i.i.d. setting, we call this result the *mean value property*. To see how this works directly, consider the mean value of the density q_o,

$$\begin{aligned}
\mathbb{E}_{q_o}[f(X)] &= \int x q_o(x)dx \\
&= \int f(x)\frac{p(x)\exp\big(\langle\theta_\nu, f(x)\rangle\big)}{M_f(\theta_\nu)}dx \\
&= \frac{1}{M_f(\theta_\nu)}\nabla_\theta\Big(\int p(x)\exp\big(\langle\theta, f(x)\rangle\big)dx\Big)\Big|_{\theta=\theta_\nu}dx \\
&= \frac{1}{M_f(\theta_\nu)}\nabla_\theta M_f(\theta)|_{\theta=\theta_\nu}dx \\
&= \frac{\nabla M_f(\theta_\nu)}{M_f(\theta_\nu)} \\
&= \nabla\log M_f(\theta_\nu) \\
&= \nu.
\end{aligned}$$

Thus, in the i.i.d. sum setting, the unique efficient i.i.d. simulation distribution is the exponential shift whose shift parameter is chosen so that its mean value is at the dominating point of the set under consideration.

Example 5.2.1. Suppose X_1 is a one-dimensional Gaussian random variable with mean zero, variance σ^2. The moment generating function $M(\theta) = \exp(\theta^2\sigma^2/2)$. Suppose we are interested in $P\big(\sum_{i=1}^n X_i > nT\big)$. θ_T is the solution to

$$T = \frac{\theta_T\sigma^2\exp(\theta_T^2\sigma^2/2)}{\exp(\theta_T^2\sigma^2/2)}$$

or $\theta_T = T/\sigma^2$. Then the efficient exponential shift density is given by

$$\begin{aligned}
q(x) &= \frac{p(x)\exp(\theta_T x)}{M(\theta_T)} \\
&= \frac{\exp\big(-\frac{x^2}{2\sigma^2}\big)\exp\big(\frac{Tx}{\sigma^2}\big)\exp\big(-\frac{\theta_T^2\sigma^2}{2}\big)}{\sqrt{2\pi\sigma^2}} \\
&= \frac{\exp\big(-\frac{(x-T)^2}{2\sigma^2}\big)\exp\big(\frac{T^2}{2\sigma^2}\big)\exp\big(-\frac{\theta_T^2\sigma^2}{2}\big)}{\sqrt{2\pi\sigma^2}} \\
&= \frac{\exp\big(-\frac{(x-T)^2}{2\sigma^2}\big)}{\sqrt{2\pi\sigma^2}},
\end{aligned}$$

which means that the exponential shift efficient distribution corresponds to a mean shift of the original distribution.

Example 5.2.2. Suppose we are interested in $P\left(\sum_{i=1}^n X_i^2 > nT\right)$, where $\{X_i\}$ are i.i.d. standard Gaussian random variables. In this setting we have $f(x) = x^2$ and

$$\begin{aligned} M_f(\theta) &= \int_{-\infty}^{\infty} \exp(\theta x^2) \frac{\exp(-\frac{x^2}{2})}{\sqrt{2\pi}} dx \\ &= \frac{1}{\sqrt{1-2\theta}}. \end{aligned}$$

Therefore, the exponential shift (with $f(x) = x^2$) is

$$q(x) = \exp(\theta x^2) \exp(-\frac{x^2}{2}) \frac{\sqrt{1-2\theta}}{\sqrt{2\pi}},$$

which is a Gaussian density of mean zero and variance $1/(1-2\theta)$. To choose the optimal value of θ, we need that $\mathbb{E}_q[X^2] = T$. But this means that we want the variance of the simulation distribution to be T (or $\theta_T = (1 - 1/T)/2$). Hence a *variance shift* of the original Gaussian distribution is an efficient simulation strategy in this setting, where the efficient distribution is given by

$$q(x) = \frac{\exp(-\frac{x^2}{2T})}{\sqrt{2\pi T}}.$$

We can also compute the variance rate for this simulation choice to find that $R(T) = T - 1 - \log(T) = 2I(T)$.

These two examples above show that in some situations either mean shifting or variance scaling is indeed the intelligent choice to make for possible candidate simulation distributions.

Example 5.2.3. Suppose X_1 is Laplacian, mean zero, variance $2a^2$ with moment generating function $M(\theta) = 1/(1 - a^2\theta^2)$ for $|\theta| < 1/a$. (Note here that we need to verify that this generating function is steep (which it is) since it is not finite for all $\theta \in \mathcal{R}$.) Again we are interested in $P\left(\sum_{i=1}^n X_i > nT\right)$. Then θ_T is the solution to

$$T = \frac{M'(\theta_T)}{M(\theta_T)}$$

which gives after a little algebra (for $T > 0$),

$$\theta_T = \frac{\sqrt{a^2 + T^2} - a}{aT}.$$

Then the efficient exponential shift density is given by

$$
\begin{aligned}
q(x) &= \frac{p(x)\exp(\theta_T x)}{M(\theta_T)} \\
&= \frac{\exp(-\frac{|x|}{a})}{2a}\exp\Big(\frac{\sqrt{a^2+T^2}-a}{aT}x\Big)\frac{T^2-(\sqrt{a^2+T^2}-a)^2}{T^2}.
\end{aligned}
$$

Example 5.2.4. Suppose X_1 is Bernoulli, with $P\big(X_1 = 1\big) = p = 1 - P\big(X_1 = 0\big)$. The probability mass function is defined by $p(1) = p$, $p(0) = 1 - p$ and moment generating function $M(\theta) = 1 - p + p\exp(\theta)$. Suppose that $0 < p < T < 1$. We are interested in $P\big(\sum_{i=1}^n X_i > nT\big)$. Then

$$
\begin{aligned}
T &= \frac{M'(\theta)}{M(\theta)} \\
\theta_T &= \log\Big(\frac{(1-p)T}{p(1-T)}\Big).
\end{aligned}
$$

Hence, the efficient exponential shift density is given by

$$
\begin{aligned}
q(x) &= \frac{p(x)\exp(\theta_T x)}{M(\theta_T)} \\
&= \frac{p(x)\frac{(1-p)^x T^x}{p^x(1-T)^x}}{1-p+p\frac{(1-p)T}{p(1-T)}} \\
&= T^x(1-T)^{1-x},
\end{aligned}
$$

which is merely another Bernoulli distribution with success probability given by T.

We could have saved ourselves a bit of computation if we had noted that any other simulation distribution taking values only on $\{0, 1\}$ must also be a Bernoulli distribution. Then by the mean shift property, the mean value must be equal to the dominating point of the set (T in this case.) Thus the efficient i.i.d. simulation distribution must be Bernoulli with mean value T (or equivalently success probability T).

Example 5.2.5. Suppose X_1 is uniform $[0, 1]$ and we wish to estimate $\rho = P\big(\sum_{i=1}^n X_i > nT\big)$. The moment generating function for the uniform is computed as

$$
\begin{aligned}
M(\theta) &= \int_0^1 \exp(\theta x)dx \\
&= \frac{\exp(\theta)-1}{\theta},
\end{aligned}
$$

which implies

$$
M'(\theta) = \frac{\theta\exp(\theta) - (\exp(\theta)-1)}{\theta^2}.
$$

To be specific, we choose $T = .99$. Numerically we find that if $T = .99$, $\theta_T \approx 100$. The efficient simulation distribution choice is thus

$$q(x) = \frac{p(x)\exp(\theta_T x)}{M(\theta)} = \begin{cases} \frac{100\ \exp(100x)}{\exp(100)-1} & 0 \le x \le 1 \\ 0 & \text{otherwise.} \end{cases}$$

To generate random variables from this distribution, we employ the inverse method; we generate $Q^{-1}(U)$ where U is uniform $[0,1]$ and $Q^{-1}(u) = \frac{1}{100}\log\Big(u\big(\exp(100)-1\big)+1\Big)$.

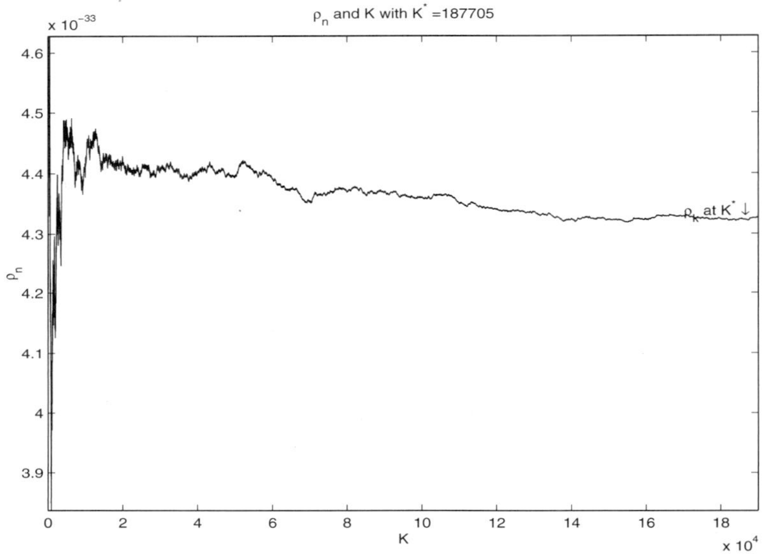

Fig. 5.6. Plot of an efficient estimate as a function of k, the number of simulation runs. $T = .99$, $n = 20$, and k^* is our estimate of the number of simulation runs that gives us 1% error with 95% confidence.

From the simulation (see Figs. 5.6 and 5.7), we obtain $\hat{\rho}_{20} = 4.324 \times 10^{-33}$ with 1% error and 95% confidence.

Example 5.2.6. Suppose X_1 is exponential, mean one, variance one, and with moment generating function $M(\theta) = 1/(1-\theta)$ for $\theta < 1$. (Note here that this generating function is steep.) We are interested in $P\big(\sum_{i=1}^n X_i > nT\big)$. We can attack this problem in the same fashion, but a simpler way is just to note that an exponential shift of an exponential is an exponential with a different parameter. The mean value property tells us that this exponential should have mean T. Thus the efficient simulation distribution would be an exponential with density

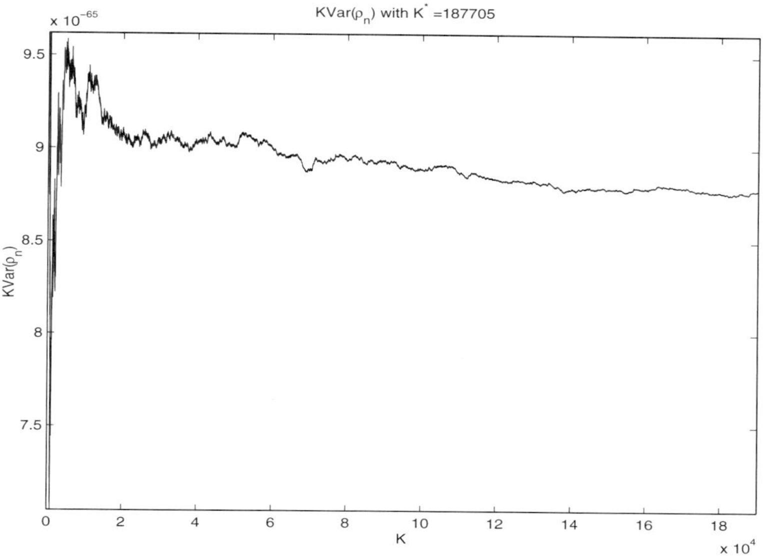

Fig. 5.7. Plot of k times sample variance as a function of k, the number of simulation runs. $T = .99$, $n = 20$, and k^* is our estimate of the number of simulation runs that gives us 1% error with 95% confidence.

$$q(x) = \frac{1}{T} \exp\left(-\frac{1}{T}x\right).$$

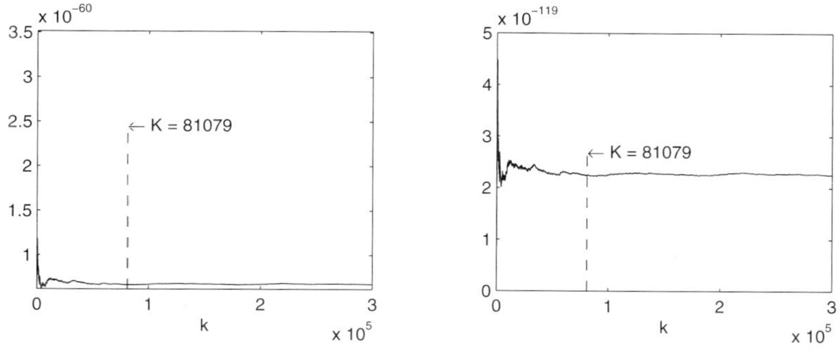

Fig. 5.8. Plot of an efficient estimate and k times sample variance as a function of k, the number of simulation runs. $T = 10$, $n = 20$, and K is our estimate of the number of simulation runs that gives us 5% error with 95% confidence.

In Fig. 5.8, we plot the estimate and k times the sample variance as a function of k for $T = 10$. The number of simulation runs needed for an

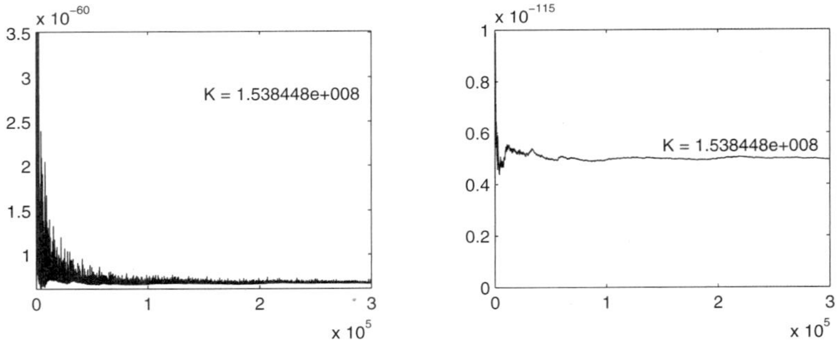

Fig. 5.9. Plot of an inefficient estimate (q is chosen to be exponential with parameter $1/(2T)$) and k times sample variance as a function of k, the number of simulation runs. $T = 10$, $n = 20$, and K is our estimate of the number of simulation runs that gives us 5% error with 95% confidence.

estimated 1% error with 95% confidence is derived from the simulation itself via the stopping criterion (4.5). To see the effects of choosing the biasing parameter incorrectly, we show in Fig. 5.9 the results of a simulation run where the mean value of the simulation distribution (again chosen to be exponential) is $2T$ instead of T. Of course this allows many more "hits" during the simulation of the set we are interested in, but one can see that the performance degrades substantially from that of the efficient choice. This is a very important point. Many adhoc choices for good simulation distributions are based upon the notion of causing more of the event of interest to happen. However, as this example shows, overbiasing can be just as significant a source of error as underbiasing (or even doing nothing at all).

Example 5.2.7. Suppose we are interested in

$$\rho = P\Big(\frac{1}{10}\sum_{i=1}^{10}\cos(\alpha_i) > 0.95\Big),$$

where α_i are i.i.d. uniform $[0, 2\pi]$ random variables. In this setting, an exponential shifted important sampling distribution of the following form will be efficient,

$$q(\beta) = \frac{\exp\big(\theta_T \cos(\beta)\big)}{2\pi M_f(\theta_T)},$$

where $M_f(\cdot)$ is the moment generating function corresponding to the random variable $\cos(\beta)$; that is,

$$M(\theta) = \frac{1}{2\pi}\int_0^{2\pi} \exp\big(\theta\cos(\beta)\big)\,d\beta.$$

$\theta_{.95}$ is given by the solution to the following equation,

$$\frac{M'(\theta)}{M(\theta)} = 0.95, \tag{5.3}$$

which we solve numerically to find,

$$\theta_{.95} \approx 10.2518, \qquad M(\theta_{.95}) \approx 3576.$$

We can then compute the large deviation rate function at the point of interest to be $I(.95) = \theta_{.95}.95 - \log(3576) = 1.5572$. Finally, the efficient important sampling distribution is given by:

$$q(\beta) = \frac{\exp\bigl(10.2518\,\cos(\beta)\bigr)}{2\pi \cdot 3576}. \tag{5.4}$$

We use an acceptance-rejection method for generating samples from $q(\beta)$. We can choose g to be uniform $[0, 2\pi]$, giving us $q(\beta) \le cg(\beta)$ with $c = [\exp(10.2518)]/3576 \approx 7.9232$. Our estimator is thus

$$\hat{\rho} = \frac{3576^{10}}{k} \sum_{j=1}^{k} 1_{\{\frac{1}{10}\sum_{i=1}^{10} \cos\beta_i^{(j)} > 0.95\}} \exp\bigl(-10.2518 \sum_{i=1}^{10} \cos(\beta_i^{(j)})\bigr).$$

In Fig. 5.10, we see a typical simulation run.

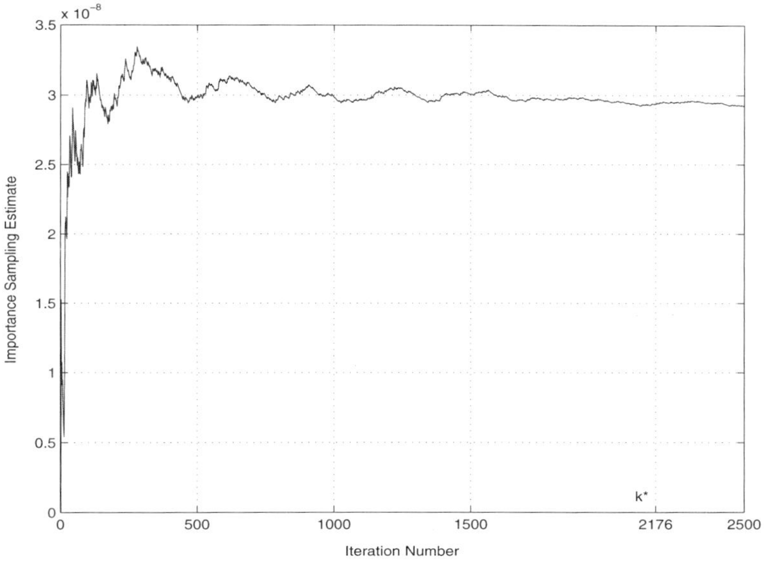

Fig. 5.10. Plot of an efficient estimate as a function of k, the number of simulation runs. k^* is our estimate of the number of simulation runs that gives us 5% error with 95% confidence.

Example 5.2.8. Suppose we are interested in the following probability,

$$\rho = P\left(\frac{1}{10}\sum_{i=1}^{10} X_i \in E\right),$$

where $\{X_i\}$ are i.i.d. random vectors taking values in $\mathcal{R}^5$. We denote $X_i = (X_{i1}X_{i2}X_{i3}X_{i4}X_{i5})$, where we assume that $\{X_{1i}\}$ are $\mathcal{R}$-valued i.i.d. exponential random variables with mean one. E, the set of interest is given by

$$E = \{x \in \mathcal{R}^5 : (x-\mu)^t K(x-\mu) \le 3\},$$

where $\mu = (7\ 5\ 8\ 4\ 8)^t$ and K is a symmetric, positive definite matrix given by

$$K = \begin{pmatrix} 5.116 & 4.736 & 4.058 & 1.821 & 1.109 \\ & 5.684 & 4.523 & 2.311 & 1.273 \\ & & 6.117 & 2.525 & 1.321 \\ & & & 4.432 & 2.481 \\ & & & & 2.134 \end{pmatrix}.$$

(We give only the top triangular part since it is symmetric.) We wish to come up with an efficient simulation scheme for this problem.

The large deviation rate function is given by

$$I(x) = \sum_{i=1}^{5} x_i - 1 - \log(x_i).$$

The set of interest $E = \{x \in \mathcal{R}^5 : (x-\mu)^t K(x-\mu) \le 3\}$ is a five-dimensional convex hyperellipsoid. Therefore there will exist a dominating point. The problem of finding the dominating point reduces to a convex optimization problem wherein we seek to minimize $I(x)$ over the set E.

At this point, one could use a standard nonlinear optimization routine from MatLab: fmincon('function',x0,[],[],[],[],[],[],'constraint'). Here 'function' was just $I(x)$ and 'constraint' was that x had to lie in the set E.

Another possibility is to consider the geometry of this problem and construct our own algorithm. The surface of E is a differentiable manifold given by $\partial E = \{x : (x-\mu)^t K(x-\mu) - 3 = 0\}$. From calculus we know that for any $x \in \partial E$, the gradient of $(x-\mu)^t K(x-\mu) - 3$ at that point is normal to the surface. Thus $\nabla(x-\mu)^t K(x-\mu) - 3 = 2K(x-\mu)$ is normal to the surface at x. Also at the dominating point ν_E, we know that the normal to the surface of E must point in the *opposite* direction from that of the gradient of I at that point; thus

$$\frac{\nabla I(\nu_E)}{\|\nabla I(\nu_E)\|} = -\frac{K(\nu_E - \mu)}{\|K(\nu_E - \mu)\|}.$$

We can generate an algorithm from this idea of the gradient of I being in the opposite direction of the surface normal to the set of interest as follows.

Gradient Search with Constraint:

Set ε and δ to be small numbers
Set $x_0 = \mu$ (or any point in the interior of E
REPEAT
REPEAT
$x_n \to x_{n-1} - \varepsilon \nabla I(x_{n-1})$
UNTIL $x_n \notin E$ i.e., $(x_n - \mu)^t K(x_n - \mu) > 3$
REPEAT $x_n \to x_{n-1} - \varepsilon 2K(x_{n-1} - \mu)$
UNTIL $x_n \in E$ i.e., $(x_n - \mu)^t K(x_n - \mu) \leq 3$
UNTIL $|(I(x_n)/\|I(x_n)\|) - (K(x_n - \mu)/\|K(x_n - \mu)\|)| < \delta$
RETURN x_n

The third REPEAT and the subsequent UNTIL is an attempt to "project" the sequence values back onto the surface of the set E whenever the gradient descent on I takes us out of the set E of interest. The algorithm works very well and is generalizable to many other problems of this same sort.

Using the above algorithm we obtain that the dominating point is approximately $\nu_E = (6.7362\ 5.1044\ 7.8271\ 4.5000\ 6.5487)^t$ with $I(\nu_E) = 16.7378$.

Therefore the biasing density is given by

$$q(y) = \frac{\exp(\langle \theta_E, y \rangle)p(y)}{M(\theta_E)} = \prod_{i=1}^{5} q(y_i),$$

where each $q(y_i)$ is an i.i.d. exponential random variable with parameter $\lambda = 1/(1 - \theta_E$. The estimator $\hat{\rho}$ was used where Y_i are random vectors generated from $q(y)$:

$$\hat{\rho} = \frac{1}{k} \sum_{j=1}^{k} 1_{\{(\frac{1}{10} \sum_{i=1}^{10} Y_i^{(j)} - \mu)^t K (\frac{1}{10} \sum_{i=1}^{10} Y_i^{(j)} - \mu) \leq 3\}} \prod_{i=1}^{10} \frac{p(Y_i^{(j)})}{q(Y_i^{(j)})}.$$

Fig. 5.11 shows the plot of the efficient estimate $\hat{\rho}$ as a function of k, the number of simulation runs. During the simulation run depicted, $k^* = 646{,}572$ which gave the estimate

$$\hat{\rho}(k^*) = 5.94 \times 10^{-11}.$$

Sum of a Functional of a Markov Chain. Let $\{X_i\}$ be samples from a finite state space irreducible Markov chain. Denote

$$p_{ij} = P(X_{n+1} = j | X_n = i)$$

as the transition probabilities of the chain where we take the state space (without loss of generality) to be $S = \{0, 1, \ldots, k-1\}$. Let f be a fixed function that maps from S into $\mathcal{R}$. Let us suppose that the chain is irreducible and aperiodic (and hence ergodic).

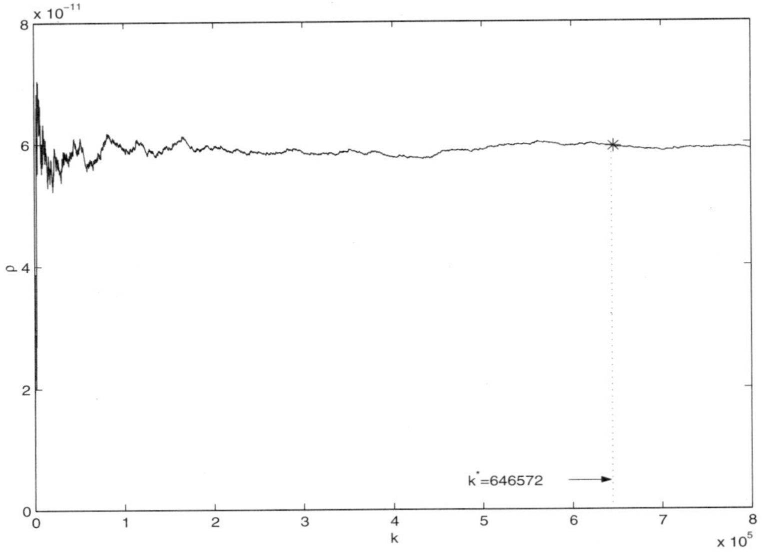

Fig. 5.11. Plot of an efficient estimate $\hat{\rho}$ as a function of k, the number of simulation runs. k^* is the number of simulation runs that gives us 5% error with 95% confidence. Note that $\hat{\rho}(k^*) = 6.66 \times 10^{-11}$ and $k^* = 646{,}572$.

Suppose we are interested in the probability

$$\rho_n = P\Big(\sum_{i=1}^{n} f(X_i) > nT\Big),$$

where $f(\cdot)$ is a given function $f : S \to \mathcal{R}$. We simulate from another Markov chain $\{Y_n\}$ on S with transition probabilities $\{q_{ij}\}$. Thus the estimator is

$$\begin{aligned}\hat{\rho}_n &= \frac{1}{k}\sum_{j=1}^{k} 1_{\{\sum_{i=1}^{n} f(Y_i^{(j)})>nT\}} \frac{p(Y_1^{(j)}, Y_2^{(j)}, \ldots, Y_n^{(j)})}{q(Y_1^{(j)}, Y_2^{(j)}, \ldots, Y_n^{(j)})} \\ &= \frac{1}{k}\sum_{j=1}^{k} 1_{\{\sum_{i=1}^{n} f(Y_i^{(j)})>nT\}} \frac{p(Y_1^{(j)}) \prod_{i=1}^{n-1} p_{Y_i^{(j)}, Y_{i+1}^{(j)}}}{q(Y_1^{(j)}) \prod_{i=1}^{n-1} q_{Y_i^{(j)}, Y_{i+1}^{(j)}}}.\end{aligned}$$

According to the setup of Section 5.1, we have $\mathcal{S}_n = \mathcal{R}^n$, $d = 1$, $Z_{p,n} = (X_1, X_2, \ldots, X_n)$, $Z_{q,n} = (Y_1, Y_2, \ldots, Y_n)$, and $f_n(x_1, x_2, \ldots, x_n) = \sum_{i=1}^{n} f(x_i)$.

Let $\mathcal{G}$ be the collection of all functions mapping from S into $\mathcal{R}$. Define the linear operator K_θ, $\{K_\theta : \mathcal{G} \to \mathcal{G}\}$ as

$$K_{q,\theta}(g)(x) = \sum_{y=0}^{k-1} \exp\big(\theta f(y)\big) g(y) \frac{p_{xy}^2}{q_{xy}}.$$

This is a positive operator and by the Perron–Frobenius Theorem will have a unique largest positive eigenvalue $\gamma_q(\theta)$. Thus

$$\begin{aligned}
c_n(\theta) &= \frac{1}{n}\log\Big(\sum_{y_1,y_2,\dots,y_n}\frac{p(y_1,y_2,\dots,y_n)^2}{q(y_1,y_2,\dots,y_n)}\exp\big(\theta\sum_{i=1}^{n} f(y_i)\big)\Big) \\
&= \frac{1}{n}\log\Big(\sum_{y_1,y_2,\dots,y_n}\frac{p(y_1)^2\prod_{i=1}^{n-1}p_{y_i,y_{i+1}}^2}{q(y_1)\prod_{i=1}^{n-1}q_{y_i,y_{i+1}}}\exp\big(\theta\sum_{i=1}^{n} f(y_i)\big)\Big) \\
&= \frac{1}{n}\log\Big(\sum_{y_1} K_{q,\theta}(1)(y_1)\frac{p(y_1)^2}{q(y_1)}\exp\big(f(y_1)\big)\Big) \\
&= \frac{1}{n}\log\Big(\sum_{y_1} K_{q,\theta}^{(n-1)}(1)(y_1)\frac{p(y_1)^2}{q(y_1)}\exp\big(f(y_1)\big)\Big) \\
&\to_{n\to\infty} \log\big(\gamma_q(\theta)\big) \\
&= c_q(\theta).
\end{aligned}$$

Recall that $k\,Var(\hat{\rho}_n) = F_n - \rho^2$, and we have that the rate for F_n is given by

$$\lim_{n\to\infty}\frac{1}{n}\log F_n = -R(T) = -\sup_{\theta}[\theta T - \log(\gamma_q(\theta))].$$

Recall the linear operator T_θ, $\{T_\theta : \mathcal{G}\to\mathcal{G}\}$,

$$T_\theta(g)(x) = \sum_{y=0}^{k-1}\exp\big(\theta f(y)\big)g(y)p_{xy},$$

which has a largest eigenvalue $\lambda(\theta)$ and associated (right) eigenvector ψ_θ.

We also know that

$$\begin{aligned}
\lim_{n\to\infty}\frac{1}{n}\log\rho_n &= -\sup_{\theta}[\theta T - \log\lambda(\theta)] \\
&= -[\theta_T T - \log\lambda(\theta_T)] \\
&= -I(T),
\end{aligned}$$

where θ_T satisfies the equation $T = \lambda'(\theta_T)/\lambda(\theta_T)$.

Suppose we choose the simulation transition probabilities $\{q_{ij}\}$ as

$$q_{o,xy} = p_{xy}\exp\big(\theta_T f(y)\big)\frac{\psi_{\theta_T}(y)}{\psi_{\theta_T}(x)\lambda(\theta_T)}. \tag{5.5}$$

In this case the variance operator takes the form

$$\begin{aligned}
K_{q_o,\theta}(g)(x) &= \sum_{y=0}^{k-1}\exp\big(\theta f(y)\big)g(y)\frac{p_{xy}^2}{q_{xy}} \\
&= \lambda(\theta_T)\psi_{\theta_T}(x)\sum_{y=0}^{k-1}\exp\big((\theta-\theta_T)f(y)\big)\frac{g(y)p_{xy}}{\psi_{\theta_T}(y)}.
\end{aligned}$$

Note this operator has positive eigenvector $\psi_{\theta_T}(y)\psi_{\theta-\theta_T}(y)$ with resulting eigenvalue $\gamma_{q_o}(\theta) = \lambda(\theta_T)\lambda(\theta - \theta_T)$. By the Perron–Frobenius Theorem this turns out to be the largest eigenvalue of the operator. The variance rate for the estimator is then

$$\begin{aligned}
R_{q_o}(T) &= \sup_\theta[\theta T - \log(\gamma_{q_o}(\theta))] \\
&= \sup_\theta[\theta T - \log\lambda(\theta_T) - \log\lambda(\theta - \theta_T)] \\
&= \sup_\theta[\theta_T T - \log\lambda(\theta_T) + (\theta - \theta_T)T - \log\lambda(\theta - \theta_T)] \\
&= \sup_\theta[I(T) + (\theta - \theta_T)T - \log\lambda(\theta - \theta_T)] \\
&= 2I(T).
\end{aligned}$$

We have thus proved the first part of the following theorem

Theorem 5.2.3. *In the setting of this subsection, the Markov chain with transition probabilities q_o given by (5.5) is an efficient simulation distribution. Furthermore, in the class of all Markov simulation distributions, this q_o is the unique efficient choice.*

Proof (Uniqueness). We prove that q_o is unique in the class of all Markov simulation distributions. We already are assuming that the Markov chain with transition probabilities $\{p_{ij}\}$ is irreducible and aperiodic. Hence, some power of the transition matrix will have all positive entries. This turns out to be a nice property that will simplify some of the arguments given below. Instead of working with some power of the transition matrix, we thus (without loss of generality) just assume that the transition matrix itself has this property; that is, $p_{xy} > 0$ for all x, y.

Suppose we have another efficient set of efficient transition probabilities $\{s_{ij}\}$. Then $R_s(T) = 2I(T)$. By the same convexity arguments of the previous subsection, we must then have $\gamma_s(2\theta_T) = \lambda(\theta_T)^2$. Suppose that $K_{s,2\theta_T}$ has largest eigenvalue $\gamma_s(2\theta_T)$ and associated right eigenvector $r_s(x)$. Define

$$\eta = \max_x \frac{\psi_{\theta_T}(x)^2}{r_s(x)} = \frac{\psi_{\theta_T}(x^*)^2}{r_s(x^*)}.$$

Then, applying Jensen's inequality,

$$\begin{aligned}
\gamma_s(2\theta_T) &= \frac{1}{r_s(x)} \sum_{y=0}^{k-1} \exp(2\theta_T y) r_s(y) \frac{p_{xy}^2}{s_{xy}} \\
&\geq \frac{1}{r_s(x)} \sum_{y=0}^{k-1} \exp(2\theta_T y) r_s(y) \frac{\psi_{\theta_T}(y)^2}{\eta r_s(y)} \frac{p_{xy}^2}{s_{xy}} \\
&= \frac{1}{\eta r_s(x)} \sum_{y=0}^{k-1} \exp(2\theta_T y) \psi_{\theta_T}(y)^2 \frac{p_{xy}^2}{s_{xy}^2} s_{xy} \\
&\geq \frac{1}{\eta r_s(x)} \left(\sum_{y=0}^{k-1} \exp(\theta_T y) \psi_{\theta_T}(y) \frac{p_{xy}}{s_{xy}} s_{xy} \right)^2 \\
&= \frac{1}{\eta r_s(x)} \left(\lambda(\theta_T) \psi_{\theta_T}(x) \right)^2 .
\end{aligned}$$

Hence if we choose $x = x^*$, we obtain $\gamma_s(2\theta_T) \geq \lambda(\theta_T)^2$, with equality if and only if

$$\exp(\theta_T y) \psi_{\theta_T}(y) \frac{p_{x^* y}}{s_{x^* y}}$$

is a constant for every y; that is, $s_{x^* y} = q_{o,x^* y}$ for all y.

This still isn't sufficient to show the uniqueness of q_o as the efficient Markovian simulation distribution. We need to show that $s_{xy} = q_{o,xy}$ for *all* x and y. Define

$$\tilde{S}_\epsilon = \{y : r_s(y) < \frac{1+\epsilon}{\eta} \psi_{\theta_T}(y)^2\}.$$

Note that $\lim_{\epsilon \to 0} \tilde{S}_\epsilon = \tilde{S} = \{y : r_s(y) = \psi_{\theta_T}(y)^2/\eta\}$. Now a key inequality is

$$r_s(y) \geq \frac{\psi_{\theta_T}(y)^2}{\eta} + \frac{\epsilon}{\eta} \psi_{\theta_T}(y)^2 1_{\{\tilde{S}_\epsilon^c\}}(y).$$

Hence,

$$\begin{aligned}
\gamma_s(2\theta_T) &= \frac{1}{r_s(x)} \sum_{y=0}^{k-1} \exp(2\theta_T y) r_s(y) \frac{p_{xy}^2}{s_{xy}} \\
&\geq \frac{1}{\eta r_s(x)} \sum_{y=0}^{k-1} \exp(2\theta_T y) \psi_{\theta_T}(y)^2 \frac{p_{xy}^2}{s_{xy}} \\
&\quad + \frac{\epsilon}{\eta r_s(x)} \sum_{y \in \tilde{S}_\epsilon^c} \exp(2\theta_T y) \psi_{\theta_T}(y)^2 \frac{p_{xy}^2}{s_{xy}} .
\end{aligned}$$

Now if we choose $x = x^*$ in the above, we obtain

$$\begin{aligned}
&\gamma_s(2\theta_T) \\
&\geq \gamma_s(2\theta_T) + \frac{\epsilon}{\eta r_s(x^*)} \sum_{y \in \tilde{S}_\epsilon^c} \exp(2\theta_T y) \psi_{\theta_T}(y)^2 \frac{p_{x^* y}^2}{s_{x^* y}} \\
&= \gamma_s(2\theta_T) \\
&\quad + \frac{\epsilon}{\eta r_s(x^*)} \sum_{y \in \tilde{S}_\epsilon^c} \exp(2\theta_T y) \psi_{\theta_T}(y)^2 p_{x^* y} \frac{\exp(-\theta_T y) \lambda(\theta_T) \psi_{\theta_T}(x^*)}{\psi_{\theta_T}(y)} \\
&= \gamma_s(2\theta_T) + \frac{\epsilon \psi_{\theta_T}(x) \lambda(\theta_T)}{\eta r_s(x^*)} \sum_{y \in \tilde{S}_\epsilon^c} \exp(\theta_T y) \psi_{\theta_T}(y) p_{x^* y}.
\end{aligned}$$

Now due to the strict positivity of the eigenvectors and of p_{xy}, we have that this last summand is strictly positive for all y. Hence, in order to avoid a contradiction, we must have that $\tilde{S}_\epsilon^c = \emptyset$ for all $\epsilon > 0$. Hence $\lim_{\epsilon \to 0} \tilde{S}_\epsilon = \tilde{S} = \{0, 1, 2, \ldots, k-1\}$. In other words, $r_s(x) = \psi_{\theta_T}(x)^2/\eta$ for all x. Now arguing as we did after (5.6), we immediately obtain that $s_{xy} = q_{o,xy}$ for all x, y. □

Example 5.2.9. Suppose we have a two-state $\{0, 1\}$ Markov chain that stays where it is with probability p and flips state with probability $1 - p = q$. Thus it has transition matrix

$$P = \begin{pmatrix} p & q \\ q & p \end{pmatrix}.$$

We are interested in simulating the probability that $\rho_n = P\big(\sum_{i=1}^n f(X_i) > nT\big)$, where $f(x) = 1 - x$. We know from Example 3.4.1 that

$$\lambda(\theta) = \frac{p\big(\exp(\theta) + 1\big) + \sqrt{p^2\big(1 + \exp(\theta)\big)^2 + 4\exp(\theta)(1 - 2p)}}{2},$$

which is the largest eigenvalue of the matrix T_θ:

$$T_\theta = \begin{pmatrix} p\exp(\theta) & q \\ q\exp(\theta) & p \end{pmatrix}.$$

The associated (right) eigenvector is easily found to be

$$\psi_\theta = \begin{pmatrix} 1 \\ \frac{q\exp(\theta)}{\lambda(\theta) - p} \end{pmatrix}.$$

We can't solve the equation $T = \lambda'(\theta_T)/\lambda(\theta_T)$ in closed form. Numerically for case $T = .75, p = .25$ we find that $\theta_T = 2.63$. We now have all the terms needed to solve for the efficient transition probabilities $\{q_{o,xy}\}$.

Example 5.2.10. Consider a finite state Markov chain $\{X_i\}$ whose state space is $\{0, 1, 2, \ldots, K-1\}$. We describe the probabilistic dynamics as follows: suppose at time n, $X_n = j$. We flip a coin with success probability p. If we have a success then we choose an independent sample Z_{n+1} from a probability mass function v on the K values of the state space $\{0, 1, 2, \ldots, K-1\}$. We then set $X_{n+1} = Z_{n+1}$. If the coin flip is not successful, $X_{n+1} = X_n = j$. We wish to simulate in order to estimate $\rho_n = P\big(\sum_{i=1}^{n} f(X_i) > nT\big)$, where f is some arbitrary function mapping from $N^+ \to \mathcal{R}$.

The transition probabilities of the chain are given by

$$p_{ij} = P\big(X_{n+1} = j \mid X_n = i\big) = \begin{cases} v(i)p + (1-p) & i = j \\ v(j)p & \text{otherwise,} \end{cases}$$

where $i, j = 0, 1, \ldots, K$.

T_θ (as an operator) is defined as

$$T_\theta(g)(x) = \sum_{y=0}^{k-1} \exp\big(\theta f(y)\big) g(y) p_{xy},$$

with the associated matrix

$$T_\theta = \Big(\exp\big(\theta f(j)\big)\big(v(j)p + \delta_{ij}(1-p)\big)\Big).$$

In order to find $\lambda(\theta)$, we must find the largest eigenvalue of the matrix T_θ. Let ψ_θ be the eigenvector associated with the largest eigenvalue of T_θ. We must then have

$$\left[\sum_{j=0}^{K-1} \exp\big(\theta f(j)\big) v(j) \psi_\theta(j) p\right] + \psi_\theta(i) \exp\big(\theta f(i)\big)(1-p) = \lambda(\theta)\psi_\theta(i).$$

In order to calculate ρ_n efficiently, another Markov chain $\{Y_n\}$ on the same state space with transition probabilities $\{q_{ij}\}$ is simulated.

We first compute θ_T iteratively. First set $\theta = 0$; find the largest eigenvalue $\lambda(\theta)$ of T_θ and associated eigenvector ψ_θ; check if $a = \lambda'(\theta_T)/\lambda(\theta_T)$ or not; if yes, return θ_T; if not, set $\theta = \theta + \triangle\theta$ and repeat.

Once we find θ_T and its associated eigenvector ψ_{θ_T}, we can compute the efficient simulation transition probabilities as

$$q_{xy} = p_{xy} \exp\big(\theta_T f(y)\big) \frac{\psi_{\theta_T}(y)}{\psi_{\theta_T} \lambda(\theta_T)}.$$

We have gone as far as we can with the closed form analysis. Let's choose some particular chain and continue.

Let $K = 3, f(i) = i,\ i = 0, 1, 2\ T = 1.5,$ and $v = [0.5\ 0.4\ 0.1]$.

We have

$$P = \begin{pmatrix} 0.80 & 0.16 & 0.04 \\ 0.20 & 0.76 & 0.04 \\ 0.20 & 0.16 & 0.64 \end{pmatrix} .$$

We compute that $\theta_T = 0.3440$ and the efficient Markov chain simulation distribution is

$$Q = \begin{pmatrix} 0.565 & 0.2284 & 0.2066 \\ 0.0986 & 0.7572 & 0.1442 \\ 0.0384 & 0.0621 & 0.8994 \end{pmatrix} .$$

In Fig. 5.12 we plot the value of $\sum_{i=1}^{n} Y_i^{(j)}/n$ for each simulation run j (for a value of $n = 100$). One can see that its mean value is centered at $a = 1.5$, a graphical demonstration of the dominating point property here in the functional of a Markov chain setting.

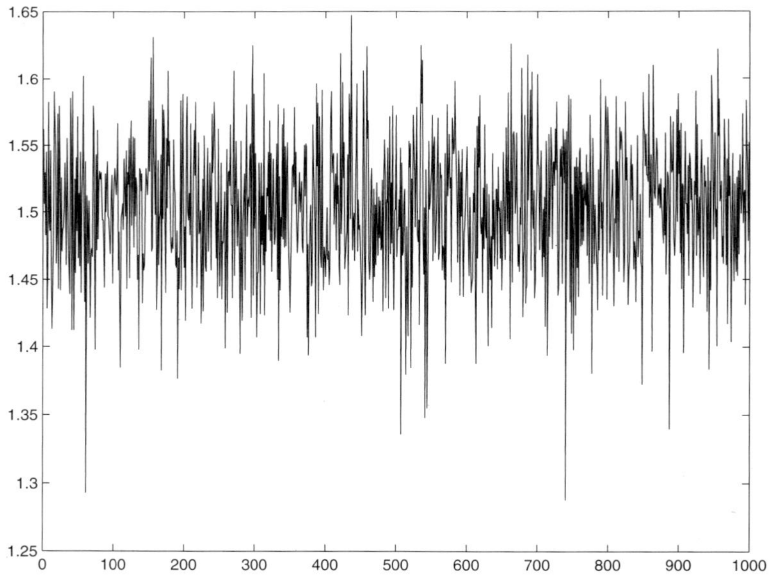

Fig. 5.12. Plot of $\sum_{i=1}^{100} Y_i^{(j)}/100$ for each simulation run j.

In Fig. 5.13, we show the estimate $\hat{\rho}_{100}$ as a function of the number of simulation runs. $K^* = 372510$ is the number of simulation runs needed to give us a 1% error with 95% confidence.

Example 5.2.11 (A Radiometer Simulation). Consider the digital communication system model given in Fig. 5.14. The receiver (a so-called radiometer)

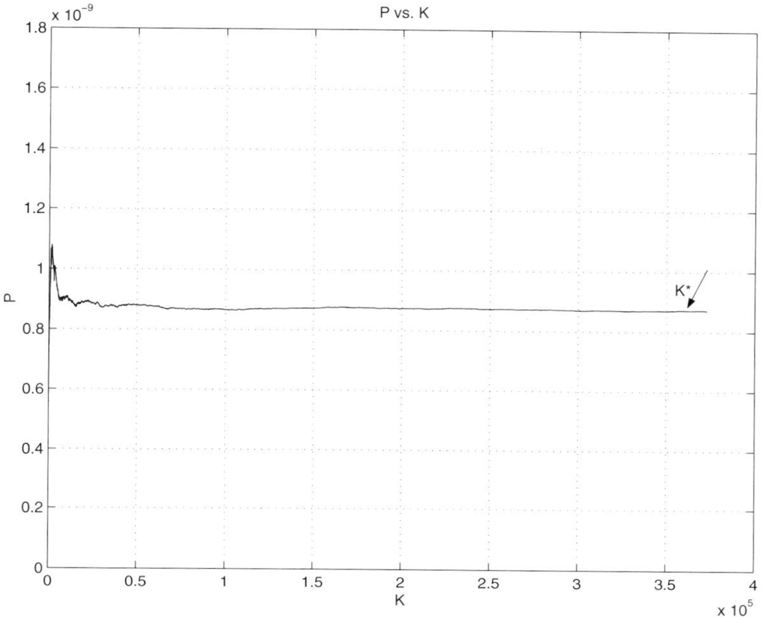

Fig. 5.13. Our estimate ρ_{100} plotted as a function of the number of simulation runs.

consists of a square law nonlinearity followed by a summer (a discrete time integrate and dump). Intersymbol interference is modeled by a first order Markov chain; that is, the channel input is a sequence of i.i.d. mean zero, variance σ_p^2 Gaussian random variables $\{N_k\}$, whereas the input to the receiver is $\{X_k\}$, where $X_k = \alpha_p X_{k-1} + N_k$ and $0 \leq \alpha_p < 1$. For the "initial condition," take $X_0 = N_0/\sqrt{1-\alpha_p^2}$, which makes $\{X_k\}$ a stationary Markov sequence. We wish to simulate in order to estimate,

$$\rho_n = P\Big(\sum_{i=1}^{n} X_i^2 > nT\Big) = P\big(A_T\big).$$

Instead of simulating directly the ISI sequence $\{X_k\}$, we consider simulating other Markovian sequences $\{\tilde{X}_k\}$, and employ the estimator

$$\hat{\rho}_n = \frac{1}{k}\sum_{j=1}^{k} 1_{\{\sum_{i=1}^{n} \tilde{X}_i^{(j)} > nT\}} \frac{p_X(\tilde{X}_1, \tilde{X}_2, \ldots, \tilde{X}_n)}{p_{\tilde{X}}(\tilde{X}_1, \tilde{X}_2, \ldots, \tilde{X}_n)},$$

where p_X is the joint probability density of the original $X_1, X_2, \ldots, X_N$ and $p_{\tilde{X}}$ is the joint probability density of the $\tilde{X}_1, \tilde{X}_2, \ldots, \tilde{X}_n$ random variables. The importance sampling *weight function* w is

$$w(x_1, x_2, \ldots, x_n) = \frac{p_X(x_1, x_2, \ldots, x_n)}{p_{\tilde{X}}(x_1, x_2, \ldots, x_n)}.$$

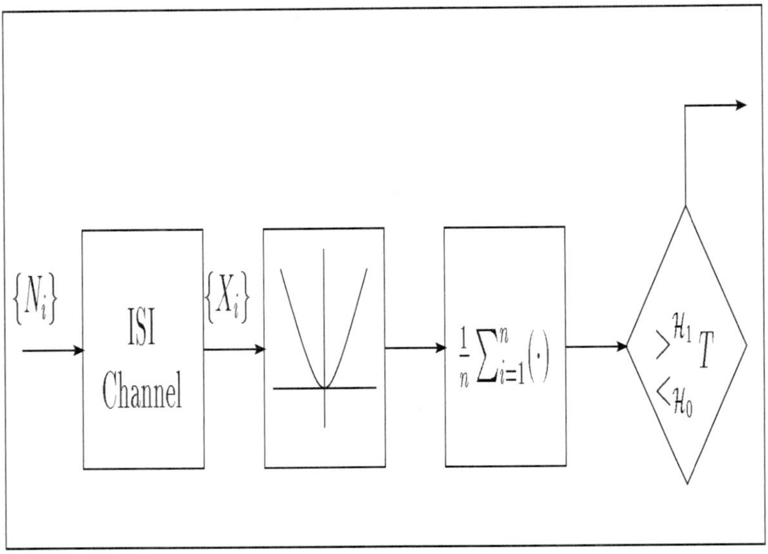

Fig. 5.14. Radiometer.

There is a variety of choices for biasing distributions that have been proposed in the literature. The mean bias method seeks to increase the error "hit" probability by adding a mean value ψ to the input i.i.d. random variables $\{N_i\}$. One can see that due to the symmetry of the square law device, we have a situation with multiple minimum rate points. For example, if we mean bias the input $\{N_i\}$ random variables with a certain vector b, then the negative of that vector $-b$ will give exactly the same performance. We will discuss this problem of multiple minimum rate points later on in this chapter, but what it means here is that any simple mean bias method is doomed to failure. Indeed a careful analysis can show that the optimal mean bias vector is exactly zero.

The variance scaling method seeks to increase the "hit" probability by multiplying the variance of the i.i.d. inputs $\{N_i\}$ by ψ^2. We wish to compute the weight function for this strategy. Denote the Gaussian probability density function as follows,

$$\phi(x, m, \sigma^2) = \frac{\exp\left(-\frac{(x-m)^2}{2\sigma^2}\right)}{\sqrt{2\pi\sigma^2}}.$$

Now

$$\begin{aligned} p_{\tilde{X}}(x_1, x_2, \ldots, x_n) &= p_{\tilde{X}}(x_n|x_1, x_2, \ldots, x_{n-1}) p_{\tilde{X}}(x_1, x_2, \ldots, x_{n-1}) \\ &= \phi(x_n, \alpha_p x_{n-1}, \psi^2 \sigma_p^2) p_{\tilde{X}}(x_1, x_2, \ldots, x_{n-1}) \\ &= \prod_{i=1}^{n-1} \phi(x_{i+1}, \alpha_p x_i, \psi^2 \sigma_p^2) p_{\tilde{X}}(x_1). \end{aligned}$$

Now

$$\tilde{X}_1 = \alpha_p \frac{\tilde{N}_0}{\sqrt{1-\alpha_p^2}} + \tilde{N}_1$$

which is normal with mean zero and variance $\sigma_p^2\psi^2/(1-\alpha_p^2)$. The p_X function can be gotten from $p_{\tilde{X}}$ by the simple artifice of setting $\psi = 1$. Hence, performing a little algebra, we obtain

$$\begin{aligned} &w(x_1, x_2, \ldots, x_n) \\ &= \psi^n \exp\Big(-\frac{1}{2\sigma_p^2}(1-\frac{1}{\psi^2})\big((1-\alpha_p^2)x_1^2 + \sum_{i=1}^{n-1}(x_{i+1}-\alpha_p x_i)^2\big)\Big). \end{aligned}$$

We now need to find the variance rate for this scaled variance methodology. The transition density for the variance scaling method is

$$q_v(y|x) = \frac{1}{\sqrt{2\pi}\psi\sigma_p} \exp(-\frac{(y-\alpha_p x)^2}{2\psi^2\sigma_p^2}).$$

The variance rate operator of interest is thus

$$\begin{aligned} K_\theta(f)(x) &= \int \exp(\theta y^2) f(y) \frac{p(y|x)^2}{q_v(y|x)} dy \\ &= \int \exp(\theta y^2) f(y) \frac{\psi}{\sqrt{2\pi}\sigma_p} \exp(-\frac{1}{\sigma_p^2}(1-\frac{1}{2\psi^2})(y-\alpha_p x)^2). \end{aligned}$$

It is not difficult to verify that the (positive) eigenfunction of this operator is

$$r_{v,\theta}(y) = \exp(a_v y^2),$$

where

$$a_v = \frac{1}{2\hat{\sigma}_p^2}[1-\alpha_p^2-\hat{\sigma}_p^2\theta - \sqrt{(1-\alpha_p^2-\hat{\sigma}_p^2\theta)^2 - 4\alpha_p^2\hat{\sigma_p}^2\theta}]$$

$$\hat{\sigma_p}^2 = \sigma_p^2(1-\frac{1}{2\psi^2})$$

and with associated (greatest) eigenvalue

$$\lambda_v(\theta) = \frac{\psi}{\sqrt{2(1-\frac{1}{2\psi^2}) - 2\sigma_p^2(\theta + a_v)}}.$$

$R_v(T)$ is found by maximizing $\theta T - \log \lambda_v(\theta)$. This can't be done in closed form. For $T = 4, \alpha_p = .25, \sigma_p^2 = 1$, one can numerically show that the fastest variance rate is obtained by choosing $\psi \approx 1.664$, and $R_v(4) \approx .964$.

To investigate the efficient exponential shift choice in this setting, we need to compute the Markov chain large deviation operator,

$$\begin{aligned} T_\theta(f)(x) &= \int \exp(\theta y^2) f(y) p(y|x) dy \\ &= \int \exp(\theta y^2) f(y) \frac{1}{\sqrt{2\pi}\sigma_p} \exp(-\frac{(y-\alpha_p x)^2}{2\sigma_p^2}) dy. \end{aligned}$$

The positive eigenfunction for this operator is

$$r_\theta(y) = \exp(ay^2),$$

where

$$a = \frac{1}{4\sigma_p^2}[1 - \alpha_p^2 - 2\sigma_p^2\theta - \sqrt{(1-\alpha_p^2 - 2\sigma_p^2\theta)^2 - 8\alpha_p^2\sigma_p^2\theta}]$$

with associated eigenvalue

$$\lambda(\theta) = \frac{1}{\sqrt{1 - 2\sigma_p^2(a+\theta)}}.$$

The efficient simulation distribution is

$$\begin{aligned} q(y|x) &= \frac{\exp(\theta_T y^2)}{\lambda(\theta_T)} \frac{r_{\theta_T}(x)}{r_{\theta_T}(y)} p(y|x) \\ &= \frac{1}{\sqrt{2\pi}\sigma_e} \exp(-\frac{(y-\alpha_e x)^2}{2\sigma_e^2}), \end{aligned}$$

where

$$\begin{aligned} \sigma_e^2 &= \frac{\sigma_p^2}{1 - 2\sigma_p^2(\theta_T + a)} \\ \alpha_e &= \frac{\alpha_p}{1 - 2\sigma_p^2(\theta_T + a)}. \end{aligned}$$

Again θ_T is the solution to $T = \lambda'(\theta_T)/\lambda(\theta_T)$. In other words, the efficient simulation estimation should be carried out under a new Markov distribution where $\tilde{X}_k = \alpha_e \tilde{X}_{k-1} + \tilde{N}_k$ and the $\{\tilde{N}_k\}$ are an i.i.d. mean zero Gaussian sequence with variance σ_e^2.

We can compute $R(T) = 2I(T)$ from the above (numerically) to find that, for example, if $T = 4, \alpha_p = .25, \sigma_p^2 = 1$, $R(4) \approx 1.109$, which is indeed an improvement over the best variance scaling strategy.

Example 5.2.12. We consider an optical communications system that uses a direct detection integrate and dump receiver. This receiver employs an avalanche photodiode to convert the received optical signal into electrical current. The incident light first produces primary electrons which are then multiplied by a random avalanche gain to produce a number of secondary electrons. The number of primary electrons is a Poisson random variable that depends on the light intensity of the optical signal. We then amplify this electrical signal which will add a Gaussian amplifier thermal noise component. We can model the test statistic as

$$D = \sum_{k=1}^{N} X_k + G,$$

where under hypothesis H_0, N is Poisson with parameter λ_0, and under hypothesis H_1, N is Poisson with parameter λ_1 ($\lambda_0 < \lambda_1$) and G is Gaussian with mean zero and variance σ_g^2. Our goal is to estimate the bit error probabilities $\rho_0 = P\big(D \geq \gamma | H_1 \text{ is true}\big) = P_0\big(D \geq \gamma\big)$ and $\rho_1 = P\big(D \leq \gamma | H_0 \text{ is true}\big) = P_1\big(D \leq \gamma\big)$. We analyze this system for the large intensity limit. We let $\lambda_0 = c_0\gamma$, $\lambda_1 = c_1\gamma$, and $\sigma_g^2 = c_g\gamma$, and consider the limit as $\gamma \to \infty$.

The $\{X_k\}$ are taken to be i.i.d. with moment generating function $M_x(\cdot)$.

Let us consider the moment generating function sequence of D (as a function of γ). To be definite let's consider the problem of estimating ρ_1 first; that is, the probability of error given that H_0 is true. Under hypothesis H_0,

$$\begin{aligned}
\mathbb{E}[\exp(\theta D)] &= \mathbb{E}[\exp\left(\theta(\sum_{k=1}^{N} X_k + G)\right)] \\
&= \mathbb{E}[\exp\left(\theta\sum_{k=1}^{N} X_k\right)]\mathbb{E}[\exp(\theta G)] \\
&= \mathbb{E}[\mathbb{E}[\exp\left(\theta\sum_{k=1}^{N} X_k\right)|N]]\exp(\frac{\theta^2 c_g\gamma}{2}) \\
&= \mathbb{E}[M_x(\theta)^N]\exp(\frac{\theta^2 c_g\gamma}{2}) \\
&= \exp\big(\lambda_0(M_x(\theta) - 1)\big)\exp(\frac{\theta^2 c_g\gamma}{2}) \\
&= \exp\big(c_0\gamma(M_x(\theta) - 1)\big)\exp(\frac{\theta^2 c_g\gamma}{2}).
\end{aligned}$$

Hence,

$$\lim_{\gamma\to\infty}\frac{1}{\gamma}\log\mathbb{E}[\exp(\theta D)] = \frac{\theta^2 c_g}{2} + c_0(M_x(\theta) - 1) = \phi(\theta).$$

Therefore, from the Gärtner–Ellis Theorem,

$$\begin{aligned}
\lim_{\gamma\to\infty} \rho_0 &= I(1) \\
&= \sup_\theta[\theta - \phi(\theta)] \\
&= \theta^* - \phi(\theta^*) \\
&= \theta^* - \frac{1}{2}(\theta^*)^2 c_g - c_0(M_x(\theta^*) - 1),
\end{aligned}$$

where θ^* is the solution to

$$1 = \theta^* c_g + c_0 M_x'(\theta^*).$$

An output estimator in this setting makes little sense. We propose to investigate an input estimator strategy. We replace each of the relevant random variables separately, $G, N, \{X_i\}$, (with distributions p_g, p_n, p_x) with the simulation variables $Z, M, \{Y_i\}$ with respective distributions q_z, q_m, q_y. Our estimator is thus of the form

$$\rho_0 = \frac{1}{k}\sum_{j=1}^{k} 1_{\{\sum_{i=1}^{M^{(j)}} Y_i^{(j)} + Z^{(j)} \geq \gamma\}} \frac{p_g(Z^{(j)})p_n(M)}{q_z(Z^{(j)})q_m(M)} \prod_{i=1}^{M} \frac{p_x(Y_i^{(j)})}{q_y(Y_i^{(j)})}.$$

Our problem now is how to go about choosing these simulation distributions. An obvious first choice is to choose exponential shifts. In that case, the Poisson random variable N with intensity $c_0\gamma$ (under H_0) N would be another Poisson with a different intensity, say $w\gamma$. Of course the Gaussian G would also be just another Gaussian with a mean shift and the same variance. We suppose that its shift parameter is θ_g; that is, $q_z(z') = p_g(z')\exp(\theta_g z')/M_g(\theta_g)$. Similarly $q_y(y') = p_x(y')\exp(\theta_x y')/M_x(\theta_x)$.

Let's consider computing the variance rate function. We first need to compute

$$\begin{aligned}
c_\gamma(\theta) &= \frac{1}{\gamma}\log\sum_{m'=0}^{\infty}\int \exp(\theta(\sum_{i=1}^{m'} y_i' + z'))\frac{(p_g(z')p_n(m'))^2}{q_z(z')q_m(m')} \\
&\quad \times \prod_{i=1}^{m'}\frac{(p_x(y_i'))^2}{q_y(y_i')} dz' dy_1' y_2' \cdots y_{m'}' \\
&= \frac{1}{\gamma}\log\sum_{m'=0}^{\infty}\int \exp(\theta(\sum_{i=1}^{m'} y_i' + z'))\exp(-\theta_x\sum_{i=1}^{m'} y_i')M_x(\theta_x)^{m'} \\
&\quad \times \exp(-\theta_g z')M_g(\theta_g)\frac{\exp(-(2c_0\gamma - w\gamma))}{m'!}\left(\frac{c_0^2}{w}\right) \\
&\quad \times p_y(z')\prod_{i=1}^{m'} p_x(y_i')dz' dy_1' \cdots dy_{m'}' \\
&= \frac{1}{\gamma}\log M_g(\theta_g)M_g(\theta - \theta_g) - \frac{1}{\gamma}(2c_0\gamma - w\gamma)
\end{aligned}$$

$$
\begin{aligned}
&+\frac{1}{\gamma}\log\sum_{m'=0}^{\infty}\frac{[M_x(\theta_x)M_x(\theta-\theta_x)^{\frac{(c_0\gamma)^2}{w\gamma}}]^{m'}}{m'!}\\
&=\frac{\theta_g^2 c_g}{2}+\frac{(\theta-\theta_g)^2 c_g}{2}-(2c_0-w)+\frac{c_0^2}{w}M_x(\theta_x)M_x(\theta-\theta_x)\\
&=c(\theta).
\end{aligned}
$$

Hence, the variance rate function R evaluated at 1 is given by

$$
\begin{aligned}
R(1)&=\sup_\theta[\theta-c(\theta)]\\
&=\sup_\theta[\theta-\frac{\theta_g^2 c_g}{2}-\frac{(\theta-\theta_g)^2 c_g}{2}\\
&\quad+(2c_0-w)-\frac{c_0^2}{w}M_x(\theta_x)M_x(\theta-\theta_x)].
\end{aligned}
$$

We know that $R(1)\le 2I(1)$. If we can find values of the parameters θ_g,θ_x,w, so that for one value of θ the bracketed expression equals $2I(1)$, we would have that $R(1)=2I(1)$ and those parameter values give us an efficient simulation. Thus we embark upon trying to maximize the above bracketed expression for each θ.

Consider maximizing the expression in the brackets over w for *fixed* θ,θ_x,θ_g. We easily obtain $w=c_0\sqrt{M_x(\theta_x)M_x(\theta-\theta_x)}$. Substituting this expression for w, we obtain for the bracketed expression

$$
\sup_\theta[\theta-\frac{\theta_g^2 c_g}{2}-\frac{(\theta-\theta_g)^2 c_g}{2}+2c_0(1-\sqrt{M_x(\theta_x)M_x(\theta-\theta_x)})].
$$

Again, maximize this expression over θ_g, holding θ,θ_x constant. We obtain $\theta_g=\theta/2$, which upon substituting gives

$$
\theta-\frac{\theta^2 c_g}{4}+2c_0(1-\sqrt{M_x(\theta_x)M_x(\theta-\theta_x)}).
$$

Setting the derivative with respect to θ_x to zero in this expression obtains $M_x'(\theta_x)M_x(\theta-\theta_x)=M_x(\theta_x)M_x'(\theta-\theta_x)$ which has a solution of $\theta_x=\theta/2$. Substituting this expression then obtains

$$
\theta-\frac{\theta^2 c_g}{4}+2c_0(1-M_x(\frac{\theta}{2})).
$$

However, if we choose $\theta=2\theta^*$, the above expression equals $2I(1)$. Hence the choices of $\theta_g=\theta^*,\theta_x=\theta^*,w=c_0M_x(\theta^*)$ give an efficient simulation.

To go any further, we need to suppose some sort of distribution for the random gains $\{X_i\}$. Suppose $p_x(z)=\exp(-z),\ z\ge 0$. Then $M_x(\theta)=1/(1-\theta)$. The equation for θ^* then becomes

$$
c_g\theta^3-(1+2c_g)\theta^2+(c_g+2)\theta+(c_0-1)=0.
$$

We now let $\gamma = 150$, $\lambda_0 = 30$, and $\sigma_g^2 = 50$. Solving the third-order polynomial above, we find $\theta^* = .5092$. In Fig. 5.15, we plot a representative simulation output. The 5% error with 95% confidence estimate of the type I error is 4.7042×10^{-19}, where k^* was found to be 23,221.

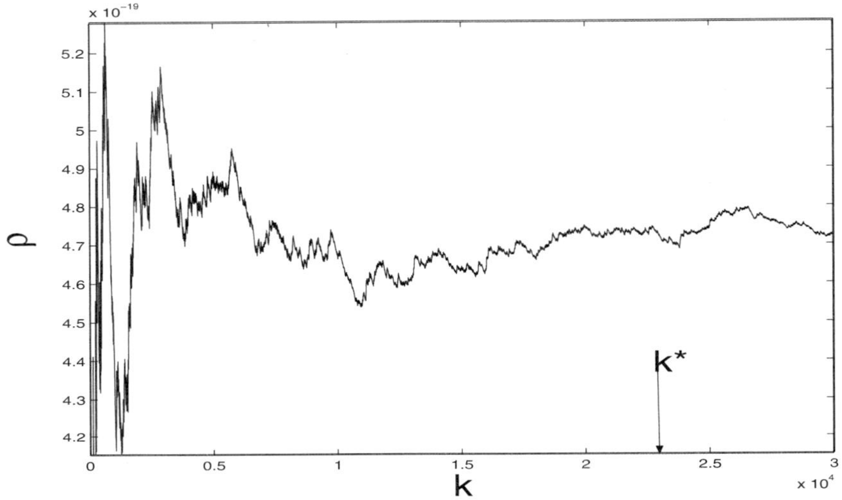

Fig. 5.15. Plot of the estimate $\hat{\rho}$ as a function of k, the number of simulation runs. k^* is the number of simulation runs that gives us 5% error with 95% confidence. $\hat{\rho}(k^*) = 4.7042 \times 10^{-19}$.

5.2.2 Sets Coverable with Finitely Many HyperPlanes

Unfortunately, dominating points don't always exist. In general, we would expect that all minimum rate points are "asymptotically important." (Surprisingly enough we will discover that even points that are NOT minimum rate points can be asymptotically important!) An efficient family of simulation distributions should account for all these points. Of course our simple exponential shift can only account for one. It turns out that we can eliminate this problem by merely considering simulation distributions that are convex combinations (or mixtures) of exponential shifts.

For our set E, we suppose that there exist $m < \infty$ points $\nu_1, \nu_2, \ldots, \nu_m$ such that

1) For each ν_i, there exists a unique $\theta_{\nu_i} \in \mathcal{R}^d$ such that $\nabla\phi(\theta_{\nu_i}) = \nu_i$;
2) $I(\nu_i) \geq I(E)$ for each $i = 1, \ldots, m$; and
3) $E \subset \cup_{i=1}^m \mathcal{H}(\nu_i)$,

where we recall our definition of a half space $\mathcal{H}(\nu) = \{x : \langle\theta_\nu, x - \nu\rangle \geq 0\}$. We note that not all these points need be minimum rate points as in for example, the case in Fig. 5.5.

Our main result is the following theorem

Theorem 5.2.4. *Let E be as above. Let $p_1, p_2, \dots, p_m$ be any probability vector with all elements strictly positive. Then the sequence of simulation distributions $\{Q_n\}$ given by*

$$dQ_n(z) = \left[\sum_{i=1}^{m} \exp\bigl(n[\langle\theta_{\nu_i}, z\rangle - \phi_n(\theta_{\nu_i})]\bigr)p_i\right] dP_n(z)$$

is efficient.

Proof. We proceed by directly upper bounding F_n for this family of distributions.

$$\begin{aligned} F_n &= \int_E \left[\frac{dP_n}{dQ_n}(z)\right]^2 dQ_n(z) \\ &\le \sum_{i=1}^{m} \int_{\mathcal{H}(\nu_i)} \left[\frac{dP_n}{dQ_n}(z)\right]^2 dQ_n(z) \\ &\le \sum_{i=1}^{m} \int_{\mathcal{H}(\nu_i)} \left[\sum_{i=1}^{m} \exp\bigl(n[\langle\theta_{\nu_i}, z\rangle - \phi_n(\theta_{\nu_i})]\bigr)p_i\right]^{-2} dQ_n(z) \\ &\le \sum_{i=1}^{m} p_i^{-2} \int_{\mathcal{H}(\nu_i)} \exp\bigl(-2n[\langle\theta_{\nu_i}, z\rangle - \phi_n(\theta_{\nu_i})]\bigr) dQ_n(z) \\ &= \sum_{i=1}^{m} p_i^{-2} \exp\bigl(-2n[\langle\theta_{\nu_i}, \nu_i\rangle - \phi_n(\theta_{\nu_i})]\bigr) \\ &\quad \int_{\mathcal{H}(\nu_i)} \exp\bigl(-2n[\langle\theta_{\nu_i}, z - \nu_i\rangle]\bigr) dQ_n(z). \end{aligned}$$

On the set $\mathcal{H}(\nu_i)$, $\langle\theta_{\nu_i}, z - \nu_i\rangle \ge 0$, and hence the integrals in the last line are all less than or equal to one. Hence

$$F_n \le \sum_{i=1}^{m} p_i^{-2} \exp\bigl(-2n[\langle\theta_{\nu_i}, \nu_i\rangle - \phi_n(\theta_{\nu_i})]\bigr).$$

For each index i,

$$\lim_{n\to\infty} \langle\theta_{\nu_i}, \nu_i\rangle - \phi_n(\theta_{\nu_i}) = \langle\theta_{\nu_i}, \nu_i\rangle - \phi(\theta_{\nu_i}) = I(\nu_i).$$

Therefore,

$$\limsup_{n\to\infty} \frac{1}{n} \log F_n \le -2 \min_{i=1,\dots,m} I(\nu_i) \le -2I(E).$$

Hence, this family of simulation distributions is efficient. □

Consider the situation depicted in Fig. 5.5, where ν_2 is a unique minimum rate point, but an additional point ν_1 is required to cover E with half spaces. The above result says that Q_n is an efficient family with $m = 2$. If we only use ν_2, we might think that the resulting simulation distribution is efficient. It would seem that the part of E not covered by $\mathcal{H}(\nu_2)$ is so insignificant that it should be neglectable by the simulation distribution. Surprisingly enough, this is not true! Consider the following example.

Example 5.2.13. Choose $a > 1, m > 0$ and suppose $E = (-\infty, -am] \cup [m, \infty)$. We are interested in the probability $\rho_n = P(\sum_{i=1}^n X_i \in n \cdot E)$, where $\{X_i\}$ are i.i.d. standard Gaussian random variables. The unique minimum rate point is $\nu_2 = m$ but it is *not* a dominating point since E contains that $(-\infty, -am]$ portion. Suppose we ignore that fact and simulate with the sequence $\{Y_i\}$ which are i.i.d. mean m, variance one, random variables. Let us compute the variance rate function R for this setting.

$$
\begin{aligned}
c(\theta) &= \log \int \frac{p(x)^2}{q(x)} \exp(\theta x)dx \\
&= \log \int \frac{\exp(-x^2)}{2\pi} \frac{\sqrt{2\pi}}{\exp(-\frac{(x-m)^2}{2})} \exp(\theta x)dx \\
&= \frac{\theta^2}{2} - m\theta + m^2.
\end{aligned}
$$

The variance rate function

$$
\begin{aligned}
R(x) &= \sup_\theta [\theta x - (\frac{\theta^2}{2} - m\theta + m^2)] \\
&= \frac{(x+m)^2}{2} - m^2.
\end{aligned}
$$

We can see that for values of x near $-m$, $R(x)$ is negative. We can be a little more precise than this. Doing some simple algebra, if $a \in (1, 1+\sqrt{2})$, then

$$
\inf_{x \in E} R(x) = R(-am) = m^2(\frac{(1-a)^2}{2} - 1) < 0.
$$

This means that the variance of the importance sampling estimator blows up exponentially fast in n. Thus we cannot neglect portions of sets *even* if those portions do not contain minimum rate points.

In Fig. 5.16 we plot the importance sampling estimate as a function of the number of simulation runs for the case of $n = 4, m = 1, a = 1.1$. During this run of 100,000 there were only three hits of the simulation variables on the $(-\infty, -am]$ component of the set E. As one can see in the figure, this causes the three large jumps in the simulation output. As a comparison, in Fig. 5.18 we plot the simple relative frequency direct Monte Carlo estimate. It behaves quite well and has converged (after 100,000 runs to .0361 which is quite close to the true value $P(Z > 2) + P(Z < -1.1 \times 2) \approx .0367$.

Perhaps even more disturbing is Fig. 5.17, which plots our estimate of the number of simulation runs needed for a 5% relative error with 95% confidence for the inefficient importance sampling estimate. The value of $k^\star(30{,}000) = 3{,}779$. We might have stopped our simulation at 30,000 runs and with enormous confidence have announced the value of .0227 (the value of the importance sampling estimator at 30,000 runs).

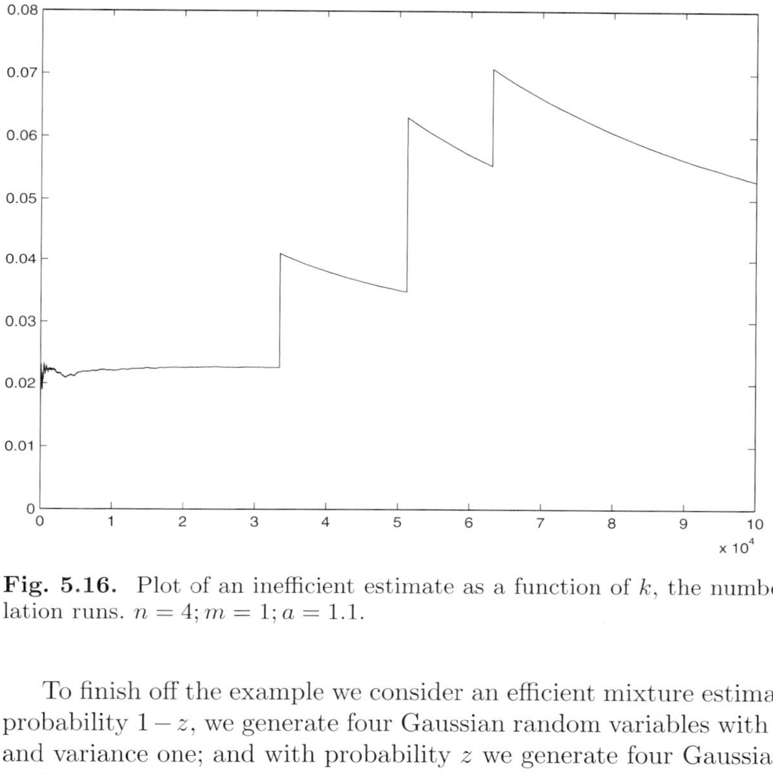

Fig. 5.16. Plot of an inefficient estimate as a function of k, the number of simulation runs. $n = 4; m = 1; a = 1.1$.

To finish off the example we consider an efficient mixture estimator. With probability $1 - z$, we generate four Gaussian random variables with mean one and variance one; and with probability z we generate four Gaussian variates with mean minus one, and variance one. In Fig. 5.19, we plot the performance of this estimator (choosing $z = .5$) for 100,000 runs. As expected it behaves quite well. The estimator is insensitive to the choice of mixture parameter. In Fig. 5.20 we plot the computed number of runs needed for 5% error and 95% confidence as a function of z. It is quite flat except for z near zero or one.

Example 5.2.14. We consider the problem of evaluating the probability of symbol error in eight-phase Phase Shift Keying (8-PSK). The digital signaling set is

$$s_i(t) = \sqrt{2E}\cos(2\pi f_0 t + \phi_i) \qquad 0 \le t \le 1, \qquad i = 0, 1, \ldots, 8,$$

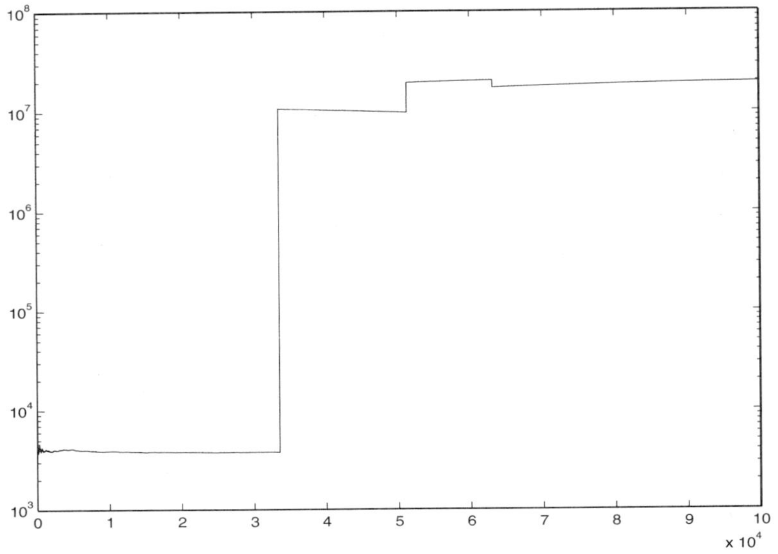

Fig. 5.17. For the inefficient estimator, a plot of the estimate of the number of simulation runs needed to achieve a 5% error with 95% confidence as a function of k, the number of simulation runs. $n = 4; m = 1; a = 1.1$.

where $\phi_i = 2\pi i/8$, $i = 0, 1, \ldots, 8$. E is the energy in each transmitted symbol. We receive this signal in thermal noise, giving a received signal model of

$$r(t) = s_i(t) + n(t) \qquad 0 \leq t \leq 1,$$

where $n(t)$ is a white Gaussian noise of spectral height $N_0/2$. Our receiver computes the statistic

$$R = (\int_0^1 r(t)\cos(2\pi f_0 t), \int_0^1 r(t)\sin(2\pi f_0 t)),$$

and then makes the minimum probability of error decision (assuming that all the symbols are equally likely to be sent). Suppose that $E/N_0 = 50$. The objective is to come up with a simulation scheme to determine the probability of error and to simulate the same.

Due to the symmetry of the symbol set, the symbol error will be the same regardless of which one is actually transmitted. Hence, without loss of generality, we just compute the probability of symbol error given that signal $s_0(t)$ is the one transmitted. Thus the received signal vector is

$$R = (R_1, R_2) = (\sqrt{2E} + N_1, N_2),$$

where (N_1, N_2) is a zero mean Gaussian random vector with covariance matrix

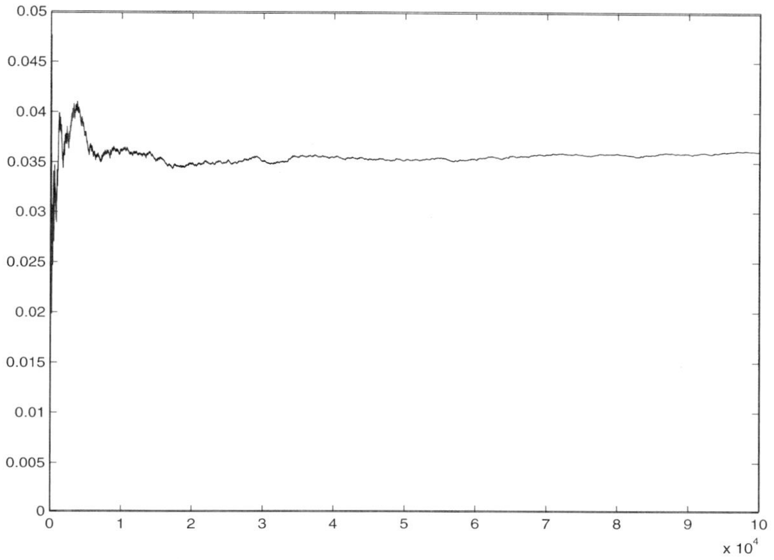

Fig. 5.18. Plot of a direct Monte Carlo estimate as a function of k , the number of simulation runs. $n = 4; m = 1; a = 1.1$.

$$K = \begin{bmatrix} N_0/4 & 0 \\ 0 & N_0/4 \end{bmatrix}.$$

An error occurs when the phase angle of R, defined as $\tan^{-1}(R_2/R_1)$ is greater than $\pi/8$ or less than $-\pi/8$ (see Fig. 5.21). We see that, in this case, the set of interest does not have a dominating point. In this setting we will have two important points. Because the rate function of the Gaussian is equivalent to Euclidean distance, we need only find the two points of the error set closest to the point $S_0 = (\sqrt{E/2}, 0)$. By simple geometry these two points are determined to be $\nu_1 = (\sqrt{(E/2)}\cos^2(\pi/8), \sqrt{(E/2)}\cos(\pi/8)\sin(\pi/8))$ and $\nu_2 = (\sqrt{(E/2)}\cos^2(\pi/8), -\sqrt{(E/2)}\cos(\pi/8)\sin(\pi/8))$.

An efficient simulation distribution would be to use a symmetric mixture of two Gaussian distributions, one with mean ν_1 and one with mean ν_2. Thus our proposed important sampling estimate is

$$\hat{\rho} = \frac{1}{k}\sum_{j=0}^{k} 1_{\{\tan^{-1}(Q_2^{(j)}/Q_1^{(j)}) > \frac{\pi}{8} \text{ OR } \tan^{-1}(Q_2^{(j)}/Q_1^{(j)}) < -\frac{\pi}{8}\}} \frac{p_r(Q^{(j)})}{q_r(Q^{(j)})}, \tag{5.6}$$

where $\{Q^{(j)} = (Q_1^{(j)} Q_2^{(j)})\}$ are generated from the simulation distribution q_r. Denote the two-dimensional Gaussian density with mean m and covariance K by $\Phi(m, K)$. Then

$$q_r = \frac{\Phi(\nu_1, K) + \Phi(\nu_2, K)}{2}$$

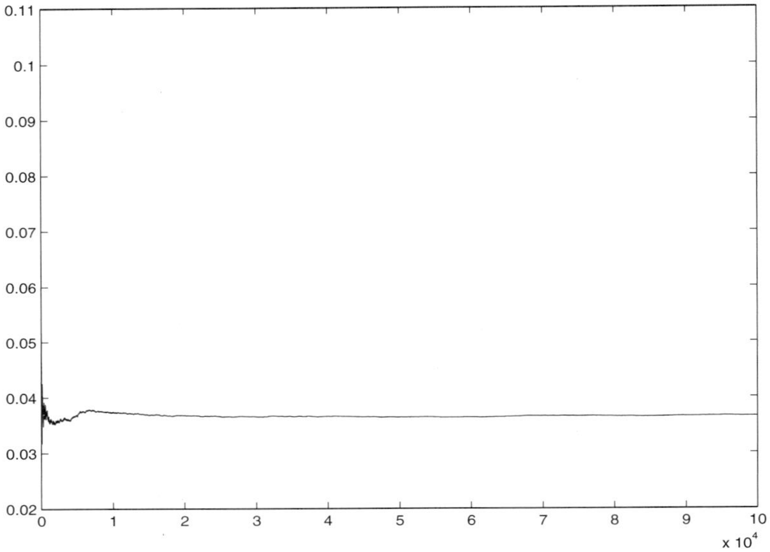

Fig. 5.19. Plot of an efficient estimate (with mixture parameter .5) as a function of k , the number of simulation runs. $n = 4; m = 1; a = 1.1$.

and p_r, the original distribution, is just $\Phi(S_0, K)$. In Fig. 5.22, we see a typical simulation.

5.2.3 Sets Not Coverable with Finitely Many Hyper-Planes

A problem with the above constructions is that the set of minimum rate points may not be finite. Assume we have our usual setup with the standard assumptions A1 to A4 holding.

Let $E \subset \mathcal{R}^d$ be an arbitrary closed set. The large deviation theory covering theorem given in Appendix B states that for any $\varepsilon > 0$, we can find a finite set of points in $\mathcal{R}^d$, $\nu_1, \nu_2, \ldots, \nu_m$ such that $E \subset \cup_{i=1}^m \mathcal{H}(\nu_i)$ and $I(\nu_i) \geq I(E) - \varepsilon$. Hence there exists a doubly indexed sequence of points $\nu_{1j}, \nu_{2j}, \ldots, \nu_{m_j j}$, $j = 1, 2, \ldots$ such that for each j: $A \subset \cup_{i=1}^{m_j} H(\nu_{ij})$ and $I(\nu_{ij}) \geq I(A) - 1/j$. Let $\{p_{ij}\}$ $i = 1, 2, \ldots, m_j$, $j = 1, 2, \ldots$ be any doubly indexed sequence of strictly positive numbers that sums to one. Define the family of simulation distributions

$$dQ_n = \left[\sum_{j=1}^{\infty} \sum_{i=1}^{m_j} p_{ij} \exp\bigl(n[\langle \theta_{\nu_{ij}}, z \rangle - \phi_n(\theta_{\nu_{ij}})]\bigr) \right] dP_n(z).$$

Our claim is that this choice of family of biasing distributions is efficient.

We can exactly mimic the derivation of the previous subsection to see that for every integer j,

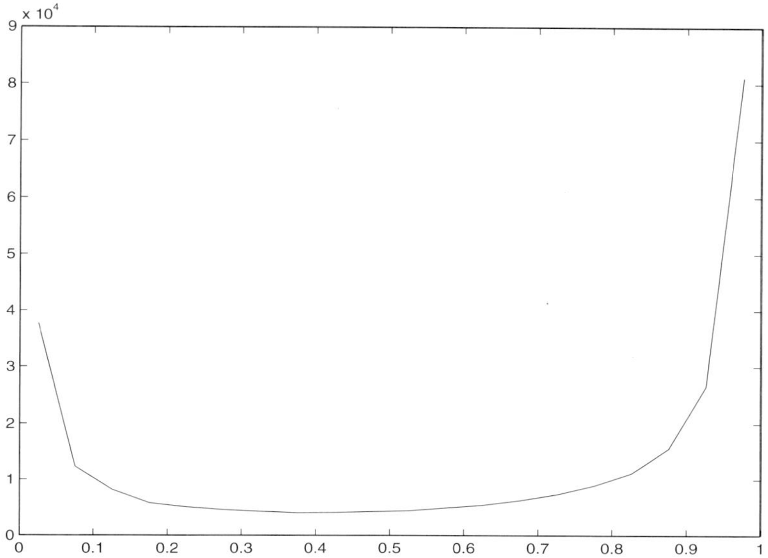

Fig. 5.20. Plot of the estimate of the number of simulation runs needed to achieve a 5% error with 95% confidence as as a function of the mixture probability. $n = 4; m = 1; a = 1.1$.

$$F_n \le \sum_{i=1}^{m_j} p_{ij}^{-2} \exp\bigl(-2n[\langle \theta_{\nu_{ij}}, \nu_{ij} \rangle - \phi_n(\theta_{\nu_{ij}})]\bigr),$$

and hence that

$$\limsup_{n\to\infty} \frac{1}{n} \log F_n \le -2 \min_{i=1,\dots,m} I(\nu_{ij}) \le -2(I(E) - \frac{1}{j}).$$

Since j can be taken to be arbitrarily large, we must have

$$\limsup_{n\to\infty} F_n \le -2I(E),$$

and hence the family is efficient.

Unfortunately, this construction is not very satisfying in that it doesn't explicitly tell us how to choose the points $\{\nu_{ij}\}$. We return to this theme when we study universal simulation distributions later in the text. At that point we are able to give a more explicit representation of how to choose the biasing distribution sequence in the setting of infinite numbers of "important" points in the set.

5.3 Notes and Comments

This chapter is by far the most important of the book. Embedded in the myriad arcane formulae there is a philosophy of rare event simulation. When

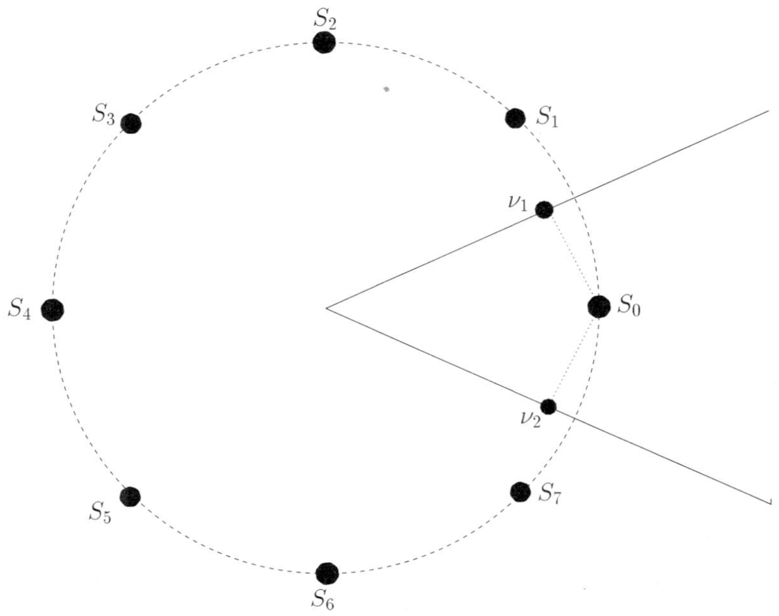

Fig. 5.21. A plot of the signal set, the error region, and the two important points ν_1, ν_2.

confronted with simulating the probability of a rare event, we first attempt to embed that probability as but one member of a sequence of problems that are being controlled by a large deviation theorem. For example, we might desire to estimate

$$\rho = P(\frac{1}{10}\sum_{i=1}^{10} X_i > T).$$

Here we have a very obvious choice of embedding, where we consider the sequence of problems

$$\rho_n = P(\frac{1}{n}\sum_{i=1}^{n} X_i > T) \qquad n = 1, 2, \ldots$$

whose large deviation theory behavior is given by Cramér's Theorem. For this *sequence* of problems, we can define a *sequence* of simulation distributions that is efficient. We then choose from that sequence of simulation distributions the distribution corresponding to $n = 10$ to use in our computer simulation. As with any procedure based upon asymptotics it can from time to time give unsatisfactory results. For example, if instead of choosing the tenth member of the sequence, our problem corresponded to choosing the first member, the simulation designer may not be comforted much by the claimed asymptotic

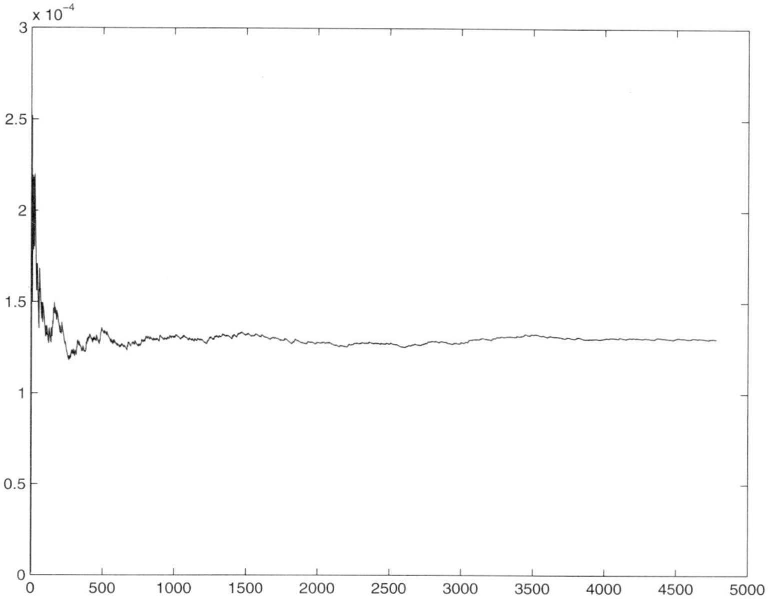

Fig. 5.22. A plot of the estimator $\hat{\rho}(k)$ as a function of the number of simulation runs k ($E/N_0 = 50$). The simulation was terminated at the computed value of $k^* = 4767$ (for a 5 % error with .90 probability criterion). The estimate of the probability is $1.3011e - 04$.

optimality of the procedure. Simulation is and will continue to be an art. The author believes that the pursuit of efficient simulation distributions will provide simulation practitioners with a valuable set of new tools to pursue that art. As with any tool, however, its usefulness is directly proportional to the skill of its wielder.

Theorem 5.1.1 has been more or less known in various forms since the early 1990s. An early version is found in [71]. Our definition of efficiency is also given for the first time there although the concept had been elucidated previously (more or less) in other places. Theorem 5.2.4 is also presented in [71]. The material on dominating points, minimum rate points, etc. is crucial to the understanding of the large deviation theory of importance sampling. There is a heuristic that says that the twisted (or tilted) distributions used in the (lower bound) proof of the large deviation theorem will make good (efficient) simulation distributions. This heuristic is basically true as long as there is a dominating point. In Example 5.2.13, we saw that it is necessary to cover *all* of the set with hyperplanes. Points and portions of the set that are not minimum rate points can contribute to the spectacular failure of importance sampling schemes designed without taking them into account. This phenomenon was first clearly identified in [33]. In succeeding chapters we will elaborate on this necessity in much greater detail which gives rise to the so-called "universal" simulation distributions.

6. Variance Rate Theory of Conditional Importance Sampling Estimators

If at first you don't succeed, try, try again. Then quit. There's no point in being a damn fool about it. *W. C. Fields*

This is a one line proof...if we start sufficiently far to the left. *Anon.*

6.1 The Variance Rate of Conditional Importance Sampling Estimators

We consider again our usual framework: for every integer n, let $Z_{p,n}$ be a random variable taking values in some complete separable metric space $\mathcal{S}_n$. Let P_n be the probability measure induced by $Z_{p,n}$ on $\mathcal{S}_n$. We suppose that for each random variable $Z_{p,n}$ we have access to some information about it contained in an *information* random variable $Z_{pi,n}$ taking values in some other probability space $\mathcal{I}_n$. Let $P_{i,n}$ denote the probability measure induced by $Z_{pi,n}$ on $\mathcal{I}_n$. Let f_n be an $\mathcal{R}^d$-valued measurable function on the space $\mathcal{S}_n$; that is, $f_n : \mathcal{S}_n \to \mathcal{R}^d$, and we suppose we are interested in $\rho_n = P\big(f_n(Z_{p,n}) \in nE\big)$, for some Borel set $E \subset \mathcal{R}^d$.

We also make the standard assumptions A1, A2, and A3 for the sequence of convex functions $\phi_n(\theta) = \frac{1}{n} \log \mathbb{E}[\exp(\langle\theta, f_n(Z_{p,n})\rangle)]$. In particular, $\lim_n \phi_n(\theta) = \phi(\theta)$, and $\lim_n \frac{1}{n} \log \rho_n = -I(E)$.

Instead of directly simulating the information random variable $Z_{pi,n}$, we choose to simulate with another $\mathcal{I}_n$-valued random variable $Z_{qi,n}$ with associated probability measure on $\mathcal{I}_n$, $Q_{i,n}$. For $z \in \mathcal{I}_n$, let

$$g(z) = P\big(f_n(Z_{p,n}) \in nE | Z_{pi,n} = z\big).$$

The conditional importance sampling estimator is

$$\hat{\rho}_{g,n} = \frac{1}{k} \sum_{j=1}^{k} g(Z_{qi,n}^{(j)}) \frac{dP_{i,n}}{dQ_{i,n}}(Z_{qi,n}^{(j)}).$$

Define

$$\phi_n(\theta, z) = \frac{1}{n} \log \int \exp\big(\langle\theta, f_n(z_{p,n})\rangle\big) dP(z_{p,n} | Z_{pi,n} = z).$$

Also let us define,

$$c_{g,n}(\theta) = \frac{1}{n}\log \int \exp\big(2n\phi_n(\theta, z)\big)\frac{dP_{i,n}}{dQ_{i,n}}(z)dP_{i,n}(z).$$

We now define the *conditional importance sampling rate function* as

$$R_g(x) = \sup_\theta 2\langle \theta, x\rangle - c_g(\theta).$$

We define the following set of assumptions

Assumption C1. $c_g(\theta) = \lim_n c_{g,n}(\theta)$ exists for all $\theta \in \mathcal{R}^d$, where we allow ∞ both as a limit value and as an element of the sequence $\{c_{g,n}(\theta)\}$.

Assumption C2. The origin belongs to the interior of the domain of c_g, (i.e., $0 \in \overset{\circ}{D}_{c_g}$) and c_g itself is a lower semi-continuous, convex function.

It turns out that in order to prove the lower bound, we need a stronger version of C1 plus some additional assumptions. First let us define, for all θ and η in $\mathcal{R}^d$,

$$c^*_{g,n}(\theta, \eta) = \int \exp\big(n[\phi_n(\theta + \eta, z) + \phi_n(\eta, z)]\big)\frac{dP_{i,n}}{dQ_{i,n}}(z)dP_{i,n}(z).$$

Assumption C1*. $c^*_g(\theta, \eta) = \lim_n c^*_{g,n}(\theta, \eta)$ exists for all $\theta, \eta \in \mathcal{R}^d$, where we allow ∞ both as a limit value and as an element of the sequence $\{c^*_{g,n}(\theta, \eta)\}$.

Assumption C3. We assume $c^*_g(\theta, \eta)$ and $c_g(\theta, \eta)$ are essentially smooth as functions of θ for all $\eta \in \mathcal{R}^d$.

There is a variety of assumptions that we can make in order to bring the derivative with respect to θ inside the defining integrals for $c_{g,n}(\theta)$ and $c^*_g(\theta, \eta)$. If this were possible, then we would have

$$\begin{aligned}
&\frac{\partial}{\partial\theta}c_{g,n}(\theta)\,\big|_{\theta=\theta_\psi}\\
&= \frac{1}{n}\frac{\int \exp\big(2n\phi_n(\theta_\psi, z)\big)2n\phi_n'(\theta_\psi, z)\frac{dP_{i,n}}{dQ_{i,n}}(z)dP_{i,n}(z)}{\int \exp\big(2n\phi_n(\theta_\psi, z)\big)\frac{dP_{i,n}}{dQ_{i,n}}(z)dP_{i,n}(z)}\\
&= 2\exp\big(-nc_{g,n}(\theta_\psi)\big)\int \exp\big(2n\phi_n(\theta_\psi, z)\big)\phi_n'(\theta_\psi, z)\frac{dP_{i,n}}{dQ_{i,n}}(z)dP_{i,n}(z).
\end{aligned}$$

Operating on the other function, we would have

$$\begin{aligned}
&\frac{\partial}{\partial\theta} c^*_{g,n}(\theta,\theta_\psi)\,|_{\theta=0} \\
&= \frac{1}{n}\frac{\int \exp\big(2n\phi_n(\theta_\psi,z)\big)n\phi_n'(\theta_\psi,z)\frac{dP_{i,n}}{dQ_{i,n}}(z)dP_{i,n}(z)}{\int \exp\big(2n\phi_n(\theta_\psi,z)\big)\frac{dP_{i,n}}{dQ_{i,n}}(z)dP_{i,n}(z)} \\
&= \exp\big(-nc_{g,n}(\theta_\psi)\big)\int \exp\big(2n\phi_n(\theta_\psi,z)\big)\phi_n'(\theta_\psi,z)\frac{dP_{i,n}}{dQ_{i,n}}(z)dP_{i,n}(z) \\
&= \frac{1}{2}c_{g,n}'(\theta_\psi).
\end{aligned}$$

Remarkably (see Theorem A.0.6), if convex functions converge, then so do their derivatives. Thus taking limits in the above relationships we find $c_g^{*\prime}(0,\theta_\psi) = c_g'(\theta_\psi)$.

Hence, we capture all these arguments in the following assumption.

Assumption C4. For all $\psi \in \mathcal{R}^d$,

$$c_g^{*\prime}(0,\theta_\psi) = \frac{1}{2}c_g'(\theta_\psi).$$

Theorem 6.1.1. *Let E be any Borel set such that $\mathring{E} \neq \emptyset$, $\bar{E} = \bar{\mathring{E}}$, and $0 < R_g(E) < \infty$. Assume C1 and C2, hold; then*

$$\lim_{n\to\infty} \frac{1}{n}\log(F_n) \leq -R_g(E).$$

If furthermore C1, C3, and C4 hold, then*

$$\lim_{n\to\infty} \frac{1}{n}\log(F_n) \geq -R_g(E).$$

Proof. Due to Assumptions C1 and C2, from the covering lemma of Appendix B, we have that there exist (for each $\varepsilon > 0$) $m < \infty$ points $\nu_1, \nu_2, \ldots, \nu_m$ such that

1) For each ν_i, there exists a unique $\theta_{\nu_i} \in \mathcal{R}^d$ such that $\nabla c_g(\theta_{\nu_i}) = \nu_i$;
2) $R_g(\nu_i) \geq R_g(E) - \varepsilon$ for each $i = 1, \ldots, m$;
3) $E \subset \cup_{i=1}^m \mathcal{H}(\nu_i)$,

where we recall our definition of a half space $\mathcal{H}(\nu) = \{x : \langle\theta_\nu, x-\nu\rangle \geq 0\}$.

Also note that

$$\begin{aligned}
g(z) &= P\big(f_n(Z_{p,n}) \in nE | Z_{pi,n} = z\big) \\
&= \int 1_{\{f_n(z_{p,n})\in nE\}} dP(z_{p,n}|Z_{pi,n} = z) \\
&\leq \sum_{i=1}^m \int 1_{\{f_n(z_{p,n})\in n\mathcal{H}(\nu_i)\}} dP(z_{p,n}|Z_{pi,n} = z)
\end{aligned}$$

$$\leq \sum_{i=1}^{m} \int \exp\big(n\langle\theta_{\nu_i}, \frac{f_n(z_{p,n})}{n} - \nu_i\rangle\big) dP(z_{p,n}|Z_{pi,n} = z)$$
$$= \sum_{i=1}^{m} \exp\big(n\phi_n(\theta_{\nu_i}, z)\big) \exp(-n\langle\theta_{\nu_i}, \nu_i\rangle)$$

$$F_n = \int g(z_{qi,n})^2 \big(\frac{dP_{i,n}}{dQ_{i,n}}(z_{qi,n})\big)^2 dQ_{i,n}(z_{qi,n})$$
$$\leq \int \left(\sum_{i=1}^{m} \exp\big(n\phi_n(\theta_{\nu_i}, z)\big) \exp(-n\langle\theta_{\nu_i}, \nu_i\rangle) \right)^2 \frac{dP_{i,n}}{dQ_{i,n}}(z) dP_{i,n}(z)$$
$$= \int m^2 \left(\frac{1}{m} \sum_{i=1}^{m} \exp\big(n\phi_n(\theta_{\nu_i}, z)\big) \exp(-n\langle\theta_{\nu_i}, \nu_i\rangle) \right)^2 \frac{dP_{i,n}}{dQ_{i,n}}(z) dP_{i,n}(z)$$
$$\leq \int m^2 \frac{1}{m} \sum_{i=1}^{m} \exp\big(2n\phi_n(\theta_{\nu_i}, z)\big) \exp(-2n\langle\theta_{\nu_i}, \nu_i\rangle) \frac{dP_{i,n}}{dQ_{i,n}}(z) dP_{i,n}(z)$$
$$= m \sum_{i=1}^{m} \exp(-2n\langle\theta_{\nu_i}, \nu_i\rangle) \exp\big(nc_{g,n}(\theta_{\nu_i})\big).$$

Hence

$$\limsup_n \frac{1}{n} \log F_n \leq \sup_i -[2\langle\theta_{\nu_i}, \nu_i\rangle - c_g(\theta_{\nu_i})]$$
$$\leq -R_g(E) + \varepsilon.$$

Since $\varepsilon > 0$ is arbitrary, the upper bound part of the theorem is proved.

We now prove the lower bound. Let $\psi \in \overset{\circ}{E}$. Fix $\varepsilon > 0$. Define

$$G_n(\psi) = \{z' : \|\frac{f_n(z')}{n} - \psi\| < \varepsilon\}.$$

Therefore for $z' \in G_n$,

$$|\langle\theta_\psi, \frac{f_n(z')}{n} - \psi\rangle| \leq \|\theta_\psi\|\varepsilon$$
$$|\langle\theta_\psi, f_n(z') - n\psi\rangle| \leq n\|\theta_\psi\|\varepsilon.$$

Hence we may lower bound the conditional probability $g(z)$ as follows.

$$g(z) = P\big(f_n(Z_{p,n}) \in nE|Z_{pi,n} = z\big)$$
$$\geq P\big(f_n(Z_{p,n}) \in G_n|Z_{pi,n} = z\big)$$
$$= \int_{G_n} dP(z_{p,n}|Z_{pi,n} = z)$$
$$\geq \exp(-n\|\theta_\psi\|\varepsilon) \int_{G_n} \exp(\langle\theta_\psi, f_n(z') - n\psi\rangle) dP(z_{p,n}|Z_{pi,n} = z).$$

We now use this result to lower bound F_n as follows.

$$
\begin{aligned}
&F_n \\
&= \int g(z)^2 \Big(\frac{dP_{i,n}}{dQ_{i,n}}(z)\Big)^2 dQ_{i,n}(z) \\
&\geq \exp(-2n\|\theta_\psi\|\varepsilon)\exp(-2n\langle\theta_\psi,\psi\rangle) \\
&\quad \times \int \Big(\int_{G_n} \exp\big(\langle\theta_\psi, f_n(z')\rangle\big) dP(z_{p,n}|Z_{pi,n}=z)\Big)^2 \frac{dP_{i,n}}{dQ_{i,n}}(z) dP_{i,n}(z) \\
&= \exp(-2n\|\theta_\psi\|\varepsilon)\exp(-2n\langle\theta_\psi,\psi\rangle) \\
&\quad \times \int \Big(\int_{\mathcal{R}^d} \exp\big(\langle\theta_\psi, f_n(z')\rangle\big) dP(z_{p,n}|Z_{pi,n}=z) \\
&\quad - \int_{G_n^c} \exp\big(\langle\theta_\psi, f_n(z')\rangle\big) dP(z_{p,n}|Z_{pi,n}=z)\Big)^2 \frac{dP_{i,n}}{dQ_{i,n}}(z) dP_{i,n}(z) \\
&= \exp(-2n\|\theta_\psi\|\varepsilon)\exp(-2n\langle\theta_\psi,\psi\rangle) \int \Big(\exp\big(n\phi_n(\theta_\psi,z)\big) \\
&\quad - \int_{G_n^c} \exp\big(\langle\theta_\psi, f_n(z')\rangle\big) dP(z_{p,n}|Z_{pi,n}=z)\Big)^2 \frac{dP_{i,n}}{dQ_{i,n}}(z) dP_{i,n}(z) \\
&\geq \exp(-2n\|\theta_\psi\|\varepsilon)\exp(-2n\langle\theta_\psi,\psi\rangle) \int \Big(\exp\big(2n\phi_n(\theta_\psi,z)\big) - 2\exp\big(n\phi_n(\theta_\psi,z)\big) \\
&\quad \int_{G_n^c} \exp\big(\langle\theta_\psi, f_n(z')\rangle\big) dP(z_{p,n}|Z_{pi,n}=z)\Big) \frac{dP_{i,n}}{dQ_{i,n}}(z) dP_{i,n}(z) \\
&= \exp(-2n\|\theta_\psi\|\varepsilon)\exp(-2n\langle\theta_\psi,\psi\rangle) \Big[\exp\big(nc_{g,n}(\theta_\psi)\big) - \int \Big(2\exp\big(n\phi_n(\theta_\psi,z)\big) \\
&\quad \int_{G_n^c} \exp\big(\langle\theta_\psi, f_n(z')\rangle\big) dP(z_{p,n}|Z_{pi,n}=z)\Big) \frac{dP_{i,n}}{dQ_{i,n}}(z) dP_{i,n}(z)\Big] \\
&= \exp(-2n\|\theta_\psi\|\varepsilon)\exp(-2n\langle\theta_\psi,\psi\rangle)\exp\big(nc_{g,n}(\theta_\psi)\big) \\
&\quad \times \Big[1 - \int \Big(2\exp\big(n\phi_n(\theta_\psi,z)\big) \int_{G_n^c} \exp\big(\langle\theta_\psi, f_n(z')\rangle\big) \\
&\quad \times dP(z_{p,n}|Z_{pi,n}=z)\Big) \frac{dP_{i,n}}{dQ_{i,n}}(z) dP_{i,n}(z) \exp\big(-nc_{g,n}(\theta_\psi)\big)\Big].
\end{aligned}
\tag{6.1}
$$

We now show that the integral over the complement of the set G_n becomes negligible for large n. To do this we use Theorem 3.2.1 on a certain sequence of probability measures. Consider the θ_ψ shifted conditional probability measure

$$\exp\big(\langle\theta_\psi, f_n(z_p)\rangle\big)\exp\big(-n\phi_n(\theta_\psi,z)\big) dP(z_p|Z_{pi,n}=z),$$

with a probability measure on the $Z_{pi,n}$ random variable given by

$$\exp\big(-nc_{g,n}(\theta_\psi)\big)\exp\big(2n\phi_n(\theta_\psi,z)\big) \frac{dP_{i,n}}{dQ_{i,n}}(z) dP_{i,n}(z).$$

Together these measures define a sequence of probability measures for the $Z_{p,n}$ random variable, which we denote as $N_{Q,n}$. Let us consider the moment generating function sequence,

$$\begin{aligned}
&\mathbb{E}_{N_{Q,n}}[\exp(\langle\theta, f_n(Z_p)\rangle)] \\
&= \int\int \exp(\langle\theta, f_n(z_p)\rangle)\exp(\langle\theta_\psi, f_n(z_p)\rangle)\exp(-n\phi_n(\theta_\psi, z)) \\
&\quad \times dP(z_p|Z_{pi,n}=z)\exp(-nc_{g,n}(\theta_\psi))\exp(2n\phi_n(\theta_\psi, z))\frac{dP_{i,n}}{dQ_{i,n}}(z)dP_{i,n}(z) \\
&= \exp(-nc_{g,n}(\theta_\psi)) \\
&\quad \times \int \exp(n[\phi_n(\theta+\theta_\psi, z)+\phi_n(\theta_\psi, z)])\frac{dP_{i,n}}{dQ_{i,n}}(z)dP_{i,n}(z) \\
&= \exp(-nc_{g,n}(\theta_\psi))\exp(nc^*_{g,n}(\theta,\theta_\psi)).
\end{aligned}$$

Hence,

$$\lim_{n\to\infty}\frac{1}{n}\log\mathbb{E}_{N_{Q,n}}[\exp(\langle\theta, f_n(Z_p)\rangle)] = c^*_g(\theta,\theta_\psi) - c_g(\theta_\psi).$$

Due to Assumption C4, we have

$$\begin{aligned}
\frac{\partial}{\partial\theta}c^*_g(\theta,\theta_\psi) - c_g(\theta_\psi)\,|_{\theta=0} &= \frac{\partial}{\partial\theta}c^*_g(\theta,\theta_\psi)\,|_{\theta=0} \\
&= c^{*\prime}_g(0,\theta_\psi) \\
&= \frac{1}{2}c'_g(\theta_\psi) \\
&= \frac{1}{2}[2\psi] \\
&= \psi.
\end{aligned}$$

Thus from Theorem 3.2.5 we have that

$$\begin{aligned}
&N_{Q,n}(G^c_n) \\
&\quad = \quad \int \exp(n\phi_n(\theta_\psi, z))\int_{G^c_n}\exp(\langle\theta_\psi, f_n(z')\rangle)dP(z_{p,n}|Z_{pi,n}=z) \\
&\qquad\quad \times \frac{dP_{i,n}}{dQ_{i,n}}(z)dP_{i,n}(z)\exp(-nc_{g,n}(\theta_\psi)) \\
&\to_{n\to\infty} 0.
\end{aligned}$$

In fact from the proof of Theorem 3.2.5, we know that this convergence is exponentially fast with a strictly positive rate constant.

Thus from (6.1), we have

$$\begin{aligned}
\lim_{n\to\infty}\frac{1}{n}\log F_n &\geq -2\|\theta_\psi\|\varepsilon - 2\langle\theta_\psi,\psi\rangle + c_g(\theta_\psi) \\
&= -2\|\theta_\psi\|\varepsilon - R_g(\psi).
\end{aligned}$$

Since $\varepsilon > 0$ is arbitrary and ψ can be varied over $\overset{\circ}{E}$, we have that

$$\lim_{n\to\infty} \frac{1}{n} \log F_n \geq -R_g(E).$$

□

6.2 Efficient Conditional Importance Sampling Estimators

We consider again the same framework and notation as in the previous section. Our principal assumption here is that the set E can be covered by finitely many hyperplanes; in particular, we suppose that there exist $m < \infty$ points $\nu_1, \nu_2, \ldots, \nu_m$ such that

1) For each ν_i, there exists a unique $\theta_{\nu_i} \in \mathcal{R}^d$ such that $\nabla\phi(\theta_{\nu_i}) = \nu_i$.
2) $I(\nu_i) \geq I(E)$ for each $i = 1, \ldots, m$.
3) $E \subset \cup_{i=1}^m \mathcal{H}(\nu_i)$,

where we recall our definition of a half-space $\mathcal{H}(\nu) = \{x : \langle\theta_\nu, x - \nu\rangle \geq 0\}$.

We also assume that all of the standard assumptions needed for the Gärtner–Ellis Theorem hold: in particular if $n\phi_n(\theta) = \log \mathbb{E}[\exp(\langle\theta, f_n(Z_{p,n})\rangle)]$ and $\lim_n \phi_n(\theta) = \phi(\theta)$, then $\lim_n \frac{1}{n}\log\rho_n = -I(E)$.

Instead of directly simulating the information random variable $Z_{pi,n}$, we choose to simulate with another $\mathcal{I}_n$-valued random variable $Z_{qi,n}$ with associated probability measure on $\mathcal{I}_n$, $Q_{i,n}$. For $z \in \mathcal{I}_n$, let

$$g(z) = P\big(f_n(Z_{p,n}) \in nE | Z_{pi,n} = z\big).$$

The conditional importance sampling estimator is

$$\hat{\rho}_{g,n} = \frac{1}{k}\sum_{j=1}^{k} g(Z_{qi,n}^{(j)})\frac{dP_{i,n}}{dQ_{i,n}}(Z_{qi,n}^{(j)}).$$

Define again

$$\phi_n(\theta, z) = \frac{1}{n}\log\int \exp\big(\langle\theta, f_n(z_{p,n})\rangle\big)dP(z_{p,n}|Z_{pi,n} = z).$$

Note that

$$\begin{aligned}\mathbb{E}_{P_{pi,n}}[\exp\big(n\phi_n(\theta, Z_{pi,n})\big)] &= \mathbb{E}[\exp\big(\langle\theta, f_n(Z_{p,n})\rangle\big)] \\ &= \exp\big(n\phi_n(\theta)\big). \end{aligned} \tag{6.2}$$

Also note that

$$
\begin{aligned}
g(z) &= P\big(f_n(Z_{p,n}) \in nE | Z_{pi,n} = z\big) \\
&= \int 1_{\{f_n(z_{p,n}) \in nE\}} dP(z_{p,n} | Z_{pi,n} = z) \\
&\leq \sum_{i=1}^{m} \int 1_{\{f_n(z_{p,n}) \in n\mathcal{H}(\nu_i)\}} dP(z_{p,n} | Z_{pi,n} = z) \\
&\leq \sum_{i=1}^{m} \int \exp\big(n\langle \theta_{\nu_i}, \frac{f_n(z_{p,n})}{n} - \nu_i \rangle\big) dP(z_{p,n} | Z_{pi,n} = z) \\
&= \sum_{i=1}^{m} \exp\big(n\phi_n(\theta_{\nu_i}, z)\big) \exp\big(-n\langle \theta_{\nu_i}, \nu_i \rangle\big).
\end{aligned}
$$

As always $k\,Var(\hat{\rho}_{g,n}) = F_n - \rho_n^2$, where

$$
F_n = \int g(z_{qi,n})^2 \left(\frac{dP_{i,n}}{dQ_{i,n}}(z_{qi,n}) \right)^2 dQ_{i,n}(z_{qi,n}).
$$

Theorem 6.2.1. *With the above assumptions, the sequence of probability distributions given by*

$$
dQ_{i,n}(z) = \left(\sum_{i=1}^{m} p_i \exp\big(n\phi_n(\theta_{\nu_i}, z)\big) \exp\big(-n\phi_n(\theta_{\nu_i})\big) \right) dP_{i,n}(z),
$$

is efficient, where $(p_1, p_2, \dots, p_m)$ is a probability vector with strictly positive components.

Proof. Note that this is a valid probability distribution due to (6.2). Also let $p^* = \min_{1 \leq i \leq m} p_i$.

Then for this choice of biasing distribution,

$$
\begin{aligned}
F_n &\leq \int \frac{\big(\sum_{i=1}^{m} \exp\big(n\phi_n(\theta_{\nu_i}, z_{qi,n})\big) \exp\big(-n\langle \theta_{\nu_i}, \nu_i \rangle\big)\big)^2}{\big(\sum_{i=1}^{m} p_i \exp\big(n\phi_n(\theta_{\nu_i}, z_{qi,n})\big) \exp\big(-n\phi_n(\theta_{\nu_i})\big)\big)} dP_{i,n}(z_{qi,n}) \\
&\leq \int \frac{\big(\sum_{i=1}^{m} \exp\big(n\phi_n(\theta_{\nu_i}, z_{qi,n})\big) \exp\big(-n\langle \theta_{\nu_i}, \nu_i \rangle\big)\big)^2}{p^* \big(\sum_{i=1}^{m} \exp\big(n\phi_n(\theta_{\nu_i}, z_{qi,n})\big) \exp\big(-n\phi_n(\theta_{\nu_i})\big)\big)} dP_{i,n}(z_{qi,n}).
\end{aligned}
$$

Fix $\varepsilon > 0$. There exists an n_0 such that for all $n \geq n_0$ we must have $|\phi_n(\theta_{\nu_i}) - \phi(\theta_{\nu_i})| < \varepsilon$ for $i = 1, 2, \dots, m$. Thus since $I(\nu_i) = \langle \theta_{\nu_i}, \nu_i \rangle - \phi(\theta_{\nu_i})$, we have $\phi_n(\theta_{\nu_i}) \leq I(\nu_i) - \langle \theta_{\nu_i}, \nu_i \rangle + \varepsilon$. Hence, for $n \geq n_0$,

$$
\begin{aligned}
&F_n \\
&\leq \int \frac{\big(\sum_{i=1}^{m} \exp\big(n\phi_n(\theta_{\nu_i}, z_{qi,n})\big) \exp\big(-n\langle \theta_{\nu_i}, \nu_i \rangle\big)\big)^2}{p^* \big(\sum_{i=1}^{m} \exp\big(n\phi_n(\theta_{\nu_i}, z_{qi,n})\big) \exp\big(-n\langle \theta_{\nu_i}, \nu_i \rangle + nI(\nu_i) - n\varepsilon\big)\big)} \\
&\quad \times dP_{i,n}(z_{qi,n}) \\
&\leq \int \frac{\big(\sum_{i=1}^{m} \exp\big(n\phi_n(\theta_{\nu_i}, z_{qi,n})\big) \exp\big(-n\langle \theta_{\nu_i}, \nu_i \rangle\big)\big)^2}{p^* \big(\sum_{i=1}^{m} \exp\big(n\phi_n(\theta_{\nu_i}, z_{qi,n})\big) \exp\big(-n\langle \theta_{\nu_i}, \nu_i \rangle + nI(E) - n\varepsilon\big)\big)}
\end{aligned}
$$

$$
\begin{aligned}
&\times dP_{i,n}(z_{qi,n}) \\
&\leq \frac{1}{p^*}\exp\big(-n(I(E)-\varepsilon)\big) \\
&\quad \times \int \sum_{i=1}^{m} \exp\big(n\phi_n(\theta_{\nu_i}, z_{qi,n})\big)\exp\big(-n\langle\theta_{\nu_i},\nu_i\rangle\big)dP_{i,n}(z_{qi,n}) \\
&\leq \frac{1}{p^*}\exp\big(-n(I(E)-\varepsilon)\big)\left[\sum_{i=1}^{m}\exp\big(n\phi_n(\theta_{\nu_i})\big)\exp(-n\langle\theta_{\nu_i},\nu_i\rangle)\right] \\
&\leq \frac{1}{p^*}\exp\big(-nI(E)\big)\big[m\exp\big(-n(I(E)-\varepsilon)\big)\big] \\
&\leq \frac{m}{p^*}\exp\big(-2n(I(E)-\varepsilon)\big).
\end{aligned}
$$

Since $\varepsilon > 0$ is arbitrary, this simulation distribution sequence is efficient. □

6.2.1 Conditioning Estimators for I.I.D. Sums

Consider the i.i.d. sum of $\mathcal{R}^d$-valued random variables setting. Let E be a set as in the theorem statement with m important points $\nu_1, \nu_2, \ldots, \nu_m$. Suppose we are interested in

$$
\rho_n = P\big(\sum_{i=1}^{n} Y_i + X_i \in nE\big),
$$

where $\{X_i\}$ is an i.i.d. sequence of $\mathcal{R}^d$-valued random variables independent of $\{Y_i\}$ another i.i.d. sequence of $\mathcal{R}^d$-valued random variables. We take

$$
\begin{aligned}
Z_{p,n} &= (X_1, Y_1, X_2, Y_2, \ldots, X_n, Y_n) \\
Z_{pi,n} &= (X_1, X_2, \ldots, X_n) \\
f(Z_{p,n}) &= \sum_{i=1}^{n} Y_i + X_i.
\end{aligned}
$$

Thus we can compute

$$
\begin{aligned}
&\exp\big(n\phi_n(\theta, z)\big) \\
&= \int \exp\big(\langle\theta, f_n(z_{p,n})\rangle\big)dP(z_{p,n}|Z_{pi,n} = z) \\
&= \int \exp\big(\langle\theta, \sum_{i=1}^{n}(x_i + y_i)\rangle\big) \\
&\quad \times dP(x_1, y_1, x_2, y_2, \ldots, x_n, y_n | X_1 = z_1, X_2 = z_2, \ldots X_n = z_n) \\
&= \mathbb{E}[\exp\big(\langle\theta, \sum_{i=1}^{n} z_i + Y_i\rangle\big)]
\end{aligned}
$$

$$= \exp\big(\langle\theta, \sum_{i=1}^{n} z_i\rangle\big) M_Y(\theta)^n.$$

Also, we have

$$\exp\big(n\phi_n(\theta)\big) = \mathbb{E}[\exp\big(\langle\theta, f_n(Z_{p,n})\rangle\big)] = M_Y(\theta)^n M_x(\theta)^n.$$

Hence our efficient simulation distribution becomes

$$dQ_{i,n}(z) = \left(\sum_{i=1}^{m} p_i \exp\big(\langle\theta_{\nu_i}, \sum_{j=1}^{n} z_j\rangle\big) M_x(\theta_{\nu_i})^{-n}\right) dP_{i,n}(z).$$

Since the $\{X_i\}$ are independent, this means that $P_{i,n}$ is a product measure which implies from the above equation that $Q_{i,n}$ is a mixture of product measures.

$$dQ_{i,n}(z = (z_1, z_2, \ldots, z_n))$$
$$= \sum_{i=1}^{m} p_i \left(\prod_{j=1}^{n} \exp(\langle\theta_{\nu_i}, z_j\rangle) M_x(\theta_{\nu_i})^{-1} dP_{i,1}(z_j)\right).$$

This is exactly the efficient exponential shift of the $\{X_i\}$ random variables that we would use if we were interested in simulating

$$\rho'_n = P\big(\sum_{i=1}^{n} X_i \in nE\big).$$

Remark 6.2.1. This startling property of the efficient conditional estimator is noted in the scalar i.i.d. sum case (with $m = 1$) in [78, 79].

Example 6.2.1. Suppose that $Z_{p,n} = \sum_{i=1}^{n} W_i + X_i$, where the sequence $\{W_i\}$ is an i.i.d. $\mathcal{R}$-valued standard normal sequence and is independent of $\{X_i\}$ which are $\mathcal{R}$-valued i.i.d. exponential with parameter one. We suppose that we are interested in

$$\rho_n = P\big(Z_{p,n} > nT\big),$$

where $T > 1$. The large deviation rate of the probability is simple to compute. The rate function is given by

$$I(T) = \sup_\theta[\theta T - \log M_W(\theta) M_X(\theta)]$$
$$= \sup_\theta[\theta T - \log \frac{\exp\big(\frac{\theta^2}{2}\big)}{1-\theta}]$$
$$= \theta_T T - \frac{\theta_T^2}{2} + \log(1 - \theta_T),$$

where $\theta_T = (T+1-\sqrt{T^2-2T+5})/2$. We decide to use a conditional importance sampling procedure with the choice of the information random variables as $Z_{i,n} = (X_1, X_2, \ldots, X_n)$. Note that

$$\begin{aligned}
P\big(Z_{p,n} > nT|Z_{i,n}\big) &= P\big(\sum_{i=1}^{n} W_i > nTj - \sum_{i=1}^{n} X_i\big) \\
&= \Phi^c\left(\sqrt{n}T - \frac{1}{\sqrt{n}}\sum_{i=1}^{n} X_i\right),
\end{aligned}$$

where $\Phi^c(x) = P\big(W_1 > x\big)$ is the complement of the standard normal cumulative distribution function. Conditioned on knowledge of $Z_{i,n}$, $Z_{p,n}$ is normal with mean $\sum_{X_i}$ and variance n. Thus its moment generating function is

$$\exp\big(n\phi_n(\theta, Z_{pi,n})\big) = \exp\left(\theta\sum_{i=1}^{n} X_i + \frac{\theta^2 n}{2}\right).$$

Therefore an efficient choice for biasing the information random variables is

$$\begin{aligned}
dQ_{i,n}(x) &= \exp\big(n\phi_n(\theta_T, x)\big)\exp\big(-n\phi_n(\theta_T)\big)dP_{i,n}(x) \\
&= \exp\left(\theta\sum_{i=1}^{n} x_i + \frac{\theta^2 n}{2}\right)\exp\big(-n\phi_n(\theta_T)\big)dP_{i,n}(x) \\
&= \exp\left(\theta\sum_{i=1}^{n} x_i + \frac{\theta^2 n}{2}\right)\exp\big(-n\phi_n(\theta_T)\big)\prod_{i=1}^{n}\exp(-x_i) \\
&= C\prod_{i=1}^{n}\exp\big(-(1-\theta_T)x_i\big),
\end{aligned}$$

where C denotes the necessary constant so that the above product of exponentials is a valid probability density. Performing that simple computation reveals that $C = (1-\theta_T)^n$. Thus an efficient choice is to bias the information random variables to be i.i.d. exponential with parameter $1-\theta_T$.

Example 6.2.2. A simplified model of an optical wavelength division multiplexing (WDM) network is given in [66]. The reader should consult this reference for an understanding of how the model comes about. We will just concentrate on the simulation problem in this example.

The photo-current generated by a photodiode is approximated by

$$\begin{aligned}
i_d = {} & \frac{a_1 E^2}{2} + \sum_{m=2}^{M}\sqrt{\epsilon}a_1 E^2 \times \cos(\phi_m) \\
& + \sum_{m,n=2,m>n}^{M} \epsilon E^2 \times \cos(\phi_m - \phi_n) + (M-1)\epsilon\frac{E^2}{2} + n_G,
\end{aligned}$$

where $a_1 \in \{0, 1\}$ is the information bit, E is the pulse amplitude, and n_G is the Gaussian zero mean, variance σ_G^2, receiver thermal noise which is independent of the signal and the crosstalk components. The "worst case" crosstalk effects of the $M-1$ interfering channels are represented by the terms involving M. The parameter ϵ controls the crosstalk intensity. The $\{\phi_m\}$ are modeled as independent uniform $[0, 2\pi]$ phase angles.

We are interested in the probability of error when $a_1 = 1$. To proceed with our analysis, we assume that the third and fourth terms in the equation for i_d can be neglected (small ϵ). Thus we have

$$i_d = \frac{a_1 E^2}{2} + \sum_{m=2}^{M} \sqrt{\epsilon} a_1 E^2 \cos(\phi_m) + n_G. \tag{6.3}$$

When the decision threshold τ is set at one half the ON-signal output current (symmetric setting, i.e., $\tau = \frac{E^2}{4}$) and under hypothesis that $a_1 = 1$, then

$$\begin{aligned}
\rho & \\
&= P\big(Error|a_1 = 1\big) \\
&= P\big(i_d < \tau|a_1 = 1\big) \\
&= P\left(\frac{a_1 E^2}{2} + \sum_{m=2}^{M} \sqrt{\epsilon} a_1 E^2 \cos(\phi_m) + n_G < \frac{E^2}{4} \middle| a_1 = 1\right) \\
&= P\left(\sum_{m=2}^{M} \sqrt{\epsilon} E^2 \cos(\phi_m) + n_G < -\frac{E^2}{4}\right) \\
&= P\left(\sum_{m=2}^{M} (-\cos(\phi_m)) + \frac{(-n_G)}{\sqrt{\epsilon}E^2} \frac{1}{4\sqrt{\epsilon}}\right) \\
&= P\left(\sum_{m=2}^{M} \cos\phi_m + n_{G'} > \frac{1}{4\sqrt{\epsilon}}\right),
\end{aligned}$$

where $\{\phi_m\}$ are uniformly distributed on $[0, 2\pi]$ and $n_{G'} \sim N(0, \sigma_G^2/\epsilon E^4)$. We need to embed this probability into a sequence of probabilities that satisfy a large deviation principle. Consider the sequence of probabilities

$$\gamma_N = P\big(\frac{1}{N} \sum_{m=1}^{N} (\cos(\phi_m) + n_{G'',m}) > T\big).$$

Let us take $n_{G'',m}$ to be i.i.d. Gaussian zero mean, variance $\sigma_G^2/\big(\epsilon E^4(M-1)\big)$, and $T = 1/\big(4\sqrt{\epsilon}(M-1)\big)$. This is a large deviation sequence in N and furthermore when $N = M - 1$, we match up with our original problem; that is, $\gamma_{M-1} = \rho$.

The large deviations of γ_N are easily worked out.

$$\begin{aligned}\phi_N(\theta) &= \frac{1}{N}\log \mathbb{E}[\exp\left(\theta \sum_{m=1}^{N} (\cos\phi_m + n_{G'',m})\right)]\\ &= \log\bigl(M_c(\theta)M_g(\theta)\bigr),\end{aligned}$$

where $M_c(\theta)$ and $M_g(\theta)$ are the moment generating functions for the cosines and the Gaussians, respectively, and are obtained as

$$\begin{aligned}M_c(\theta) &= \int_0^{2\pi} \frac{1}{2\pi}\exp(\theta\cos x)dx = I_0(\theta)\\ M_g(\theta) &= \int_{-\infty}^{+\infty} \exp(\theta y)\frac{1}{\sqrt{2\pi\sigma^{\ 2}}}\exp\bigl(-\frac{1}{2}\frac{y^2}{\sigma^2}\bigr)dy = \exp\bigl(\frac{1}{2}\theta^2\sigma^2\bigr),\end{aligned}$$

where $I_0(\theta)$ is modified Bessel function of the first kind. Thus

$$\lim_{N\to\infty} \phi_N(\theta) = \phi(\theta) = \log\bigl(M_c(\theta)M_g(\theta)\bigr).$$

The large deviation rate of the probability can be computed as

$$\begin{aligned}I(T) &= \sup_\theta[\theta T - \phi(\theta)]\\ &= \sup_\theta[\theta T - \log M_c(\theta)M_g(\theta)]\\ &= \sup_\theta[\theta T - \log\left(I_0(\theta)\exp\bigl(\frac{1}{2}\theta^2\sigma^2\bigr)\right)]\\ &= \theta_T T - \log I_0(\theta_T) - \frac{1}{2}\theta_T^2\sigma^2.\end{aligned}$$

To determine θ_T, we set

$$\begin{aligned}&\frac{\partial}{\partial\theta}[\theta T - \log\left(I_0(\theta)\exp\bigl(\frac{1}{2}\theta^2\sigma^2\bigr)\right)] = 0\\ &\qquad\qquad \Rightarrow T - \frac{I_0'(\theta)}{I_0(\theta)} - \theta\sigma^2 = 0.\end{aligned}$$

Noting that

$$I_n'(x) = \frac{n}{x}I_n(x) + I_{n+1}(x)$$

we have

$$T - \frac{I_1(\theta)}{I_0(\theta)} - \theta\sigma^2\Bigr|_{\theta=\theta_T} = 0.$$

This equation can be solved to obtain θ_T.

To find the efficient conditional importance sampling distributions, we first compute

$$\begin{aligned}
&\exp\big(N\phi_N(\theta, z)\big) \\
&= \mathbb{E}_{p,n}\left[\exp\big(\theta f_n(z_{p,n})\big)\,\middle|\, Z_{pi,n} = z\right] \\
&= \mathbb{E}\left[\exp\left(\theta \sum_{i=1}^{N} (\cos\phi_i + n_i)\right)\middle|\, \phi_i = z_i, i = 1, 2, \ldots, N\right] \\
&= \exp\left(\theta \sum_{i=1}^{N} \cos z_i + \frac{N\theta\sigma^2}{2}\right).
\end{aligned}$$

Hence an asymptotically efficient biasing p.d.f. for the information random variables $Z_{pi,N} = \{\phi_i\}$ is

$$\begin{aligned}
&q_{i,N}(x) \\
&= \exp\big(N\phi_N(\theta_T, x)\big)\exp\big(-N\phi_N(\theta_T)\big) \cdot \big(1_{\{x_i \in [0,2\pi]\}}\big)^N \\
&= \exp\left(\theta_T \sum \cos x_i + \frac{N\theta_T^2\sigma^2}{2}\right) \frac{\exp\left(\frac{-N\theta_T^2\sigma^2}{2}\right)}{\big(I_0(\theta_T)\big)^N} \left(\frac{1}{2\pi} \cdot 1_{\{x_i \in [0,2\pi]\}}\right)^N \\
&= \prod_{i=1}^{N} \left(\frac{\exp(\theta_T \cos x_i)}{2\pi I_0(\theta_T)} 1_{\{x_i \in [0,2\pi]\}}\right).
\end{aligned}$$

Since the $\{\phi_i\}$ are i.i.d. random variables, the biasing probability density functions are also. Thus the probability density function for a single biased random variable ϕ_i can be written as

$$q_{\phi_i}(x) = \frac{\exp(\theta_T \cos x)}{2\pi I_0(\theta_T)} 1_{\{x \in [0,2\pi]\}}.$$

Hence the conditional importance sampling estimator of ρ is

$$\begin{aligned}
&\hat{\rho}_g(k) \\
&= \frac{1}{k} \sum_{j=1}^{k} Q\left(\frac{NT - \sum_{i=1}^{N} \cos x_i^{(j)}}{\sqrt{N\sigma^2}}\right) \\
&\quad \times \prod_{i=1}^{N} \frac{\frac{1}{2\pi} \cdot 1_{\{x_i^{(j)} \in [0,2\pi]\}}}{\frac{\exp(\theta_T \cos x_i^{(j)})}{2\pi I_0(\theta_T)} 1_{\{x_i^{(j)} \in [0,2\pi]\}}} \\
&= \frac{1}{k} \sum_{j=1}^{k} \frac{1}{2} \mathit{erfc}\left(\frac{NT - \sum_{i=1}^{N} \cos x_i^{(j)}}{\sqrt{2N\sigma^2}}\right) \\
&\quad \times \big(I_0(\theta_T)\big)^N \exp\left(-\theta_T \sum_{i=1}^{N} \cos x_i^{(j)}\right).
\end{aligned}$$

To simulate, we choose the following system parameters,

$$M = 4$$
$$N = M - 1 = 3$$
$$\frac{E^4}{\sigma_G^2} = 1024.$$

In the simulation, the crosstalk-to-signal ratio XSR varies from -55 dB to -15 dB in increments of 1 dB.

For a particular XSR, we obtain

$$\epsilon = 10^{XSR/10}$$
$$\sigma^2 = \frac{1}{\epsilon N \frac{E^4}{\sigma_G^2}}$$
$$T = \frac{1}{4\sqrt{\epsilon N}}.$$

Once these values are obtained, we compute θ_T above using the MATLAB symbolic toolbox. We then implement the conditional estimator in MATLAB.

A plot of $\hat{\rho}_g(k)$ versus k for XSR $= -50$ dB is shown in Fig. 6.1. A plot of $\hat{\rho}_g$ versus XSR is shown in Fig. 6.2.

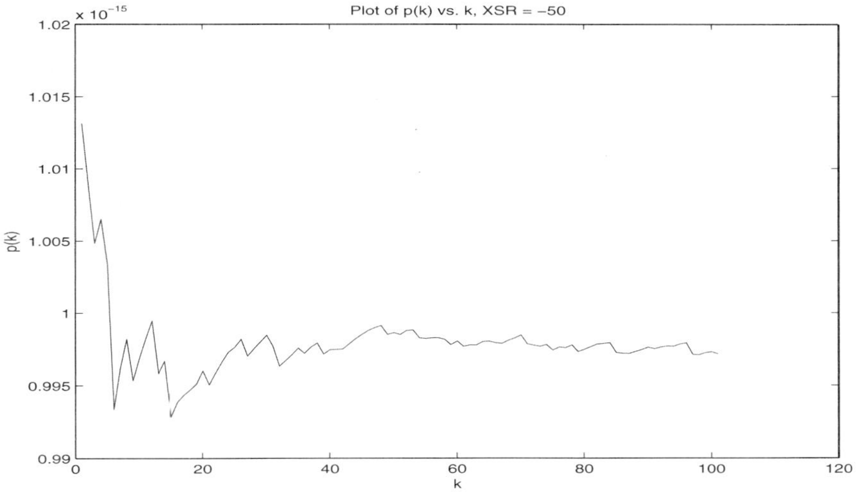

Fig. 6.1. Plot of $\hat{\rho}(k)$ vs. k.

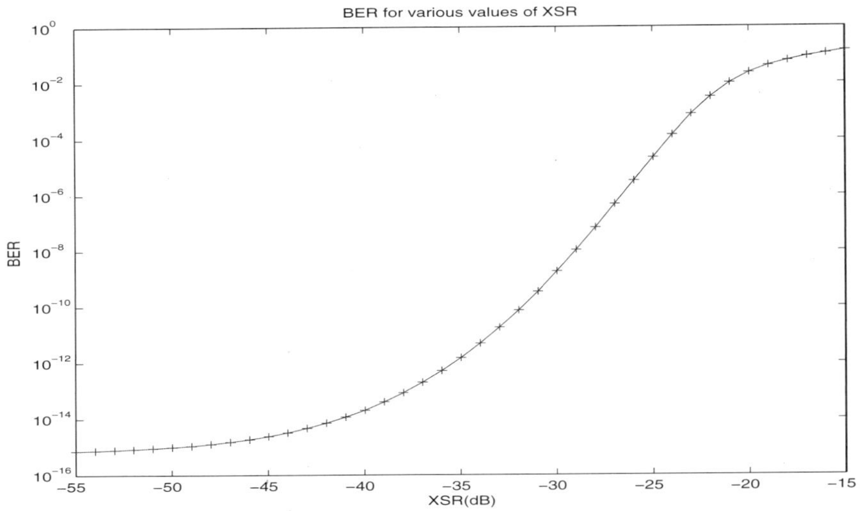

Fig. 6.2. Plot of bit error rate vs. XSR.

6.3 Notes and Comments

The class of conditional importance sampling estimators was introduced in [78] in the context of the simulation of i.i.d. sums and called there the g-method.

In the previous chapter we expounded the large deviation theory philosophy of how to intelligently choose importance sampling simulation distributions for rare event simulation problems. Our argument was that for highly reliable systems, we should view our simulation problem from the point of view of trying to find efficient biasing distributions. The reason for following this philosophy is that by first embedding our problem as but one of a parametric sequence of problems, we can concern ourselves with maximizing the estimator variance *rate* to zero instead of minimizing the actual estimator variance itself. The mathematics of maximizing the variance rate is often far simpler than trying to minimize the actual variance over some class of simulation distributions. This is intuitive since our large deviation framework is only trying to maximize a rate parameter instead of the actual variance. Trying to minimize analytically the estimator variance directly almost always leads to a very complicated functional minimization problem. Of course, when this minimization can be carried out, it is very desirable to do so. In most practical situations, it really can't be done.

It is now apparent that we have a more or less encompassing theory that seems to explain quite well exactly what is occurring in this search for good biasing distributions. The theory and definitely the practice of large deviation theory techniques is still very much in its infancy. This chapter is about expanding this large deviation framework to include the conditional importance sampling estimators. These estimators are always guaranteed to perform better than conventional estimators and should be used whenever possible.

7. The Large Deviations of Bias Point Selection

Success is not a good criterion if there is no significant penalty for being wrong. Banging on pots always keeps the sun from being eaten by the moon during eclipses. *Anon.*

For non-deterministic system read 'Inhabited by pixies'. *Anon.*

7.1 The Variance Rate of Input and Output Estimators

Suppose the inputs to the system are i.i.d. random variables $\{X_i\}$, which have the scalar density function $p(\cdot)$. We are interested in $\rho = P(\sum_{j=1}^{n} X_i > na)$. We simulate with i.i.d. $\{S_i\}$ whose individual density functions are $q(\cdot)$. In the light of Theorem 5.1.1, let us compute the variance rates for the input and output estimators, respectively.

For the output estimator we have $\mathcal{S}_n = \mathcal{R}\ \ \forall n$. Also $d = 1$ and $f_n(x) = x\ \ \forall n$. $dP_n/dQ_n = p*p*\cdots*p/q*q*\cdots*q$, where $*$ denotes the convolution operator and there are n convolutions each in the numerator and denominator respectively. For the output estimator we define

$$c_{n,o}(\theta) = \frac{1}{n}\log\left(\int \frac{(p*\cdots*p)(z)^2}{(q*\cdots*q)(z)}\exp(\theta z)dz\right)$$

and $c_o(\theta) = \lim_{n\to\infty} c_{n,o}(\theta)$.

For the input estimator we have $\mathcal{S}_n = \mathcal{R}^n\ \forall n$. Also $d = 1$ and $f_n(x_1, x_2, \ldots, x_n) = \sum_{i=1}^{n} x_i\ \forall n$. $(dP_n/dQ_n)(x_1, x_2, \ldots, x_n) = \prod_{i=1}^{n}(p(x_i)/q(x_i))$. For the input estimator we define

$$\begin{aligned} c_{n,i}(\theta) &= \frac{1}{n}\log(\int\prod_{i=1}^{n}\frac{p(x_i)^2}{q(x_i)}dx_1dx_2\cdots dx_n \\ &= \log(\int\frac{p^2(x)}{q(x)}\exp(\theta x))dx \end{aligned}$$

and hence $c_i(\theta) = c_{1,i}(\theta)$.

The variance rate of the input estimator is already known from our calculations in the previous chapters and is given by

$$\lim_{n\to\infty}\frac{1}{n}\log(Var(\hat{\rho}_i)) = -\sup_{\theta}[\theta x - c_i(\theta)].$$

Example 7.1.1. For some $1/2 < a < 1$, we wish to estimate via simulation the probability

$$P(\sum_{k=1}^{n} B_k \geq na),$$

where the $\{B_k\}$ are i.i.d. symmetric Bernoulli random variables. An output simulation would have us simulate the binomial random variable with some other distribution. The exponential shift of a binomial is another binomial. Rather than trying to generate binomials with large n parameter (which is slow) we decide (somewhat arbitrarily) to generate Poisson variates with mean parameter na. Thus

$$p_n(k) = b(k; n, 1/2)$$

and we simulate with

$$q_n(k) = p(k; na).$$

In the following derivation we make use of the following equality for approximating the binomial distributions with a Poisson [28, pg. 172, Eq. 10.3]

$$p(k;\lambda)\exp(k\lambda/n) > b(k;n,p) > p(k;\lambda)\exp(-k^2/(n-k) - \lambda^2/(n-\lambda)),$$

where $\lambda = np$. Thus we can show that

$$\exp\big(nc_{n,o}(\theta)\big) = \exp(na)4^{-n}\sum_{k=0}^{n}\frac{\exp(\theta k)}{(na)^k}\frac{(n!)^2}{(k!)[(n-k)!]^2}$$

has the following asymptotic logarithmic limit behavior,

$$\lim_{n\to\infty} c_{n,o}(\theta) = c_o(\theta) = a - \log(4) + \log(1 + \frac{\exp(\theta)}{a}).$$

The corresponding input bias scheme for this problem is to simulate the symmetric Bernoulli random variables with Poisson random variables of mean a. For this scheme, $c_i(\theta) = c_1(\theta)$ and thus

$$\begin{aligned} c_i(\theta) &= \log(\sum_{k=0}^{1}\frac{p_1(k)^2}{q_1(k)}) \\ &= a - \log(4) + \log(1 + \frac{\exp(\theta)}{a}). \end{aligned}$$

In other words the $c_i(\theta) = c_o(\theta)$, and thus the variance rates are equal!

So far, we have seen for the case of bias distributions of the form of exponential shifts (see Remark 4.3.2) and in the above example that the variance rate for the two schemes has been equal. This is not always the case, as the following example shows.

Example 7.1.2. Suppose we wish to estimate via simulation the probability

$$P(\sum_{k=1}^{n} X_k \geq n),$$

where the $\{X_k\}$ are i.i.d. exponential random variables with parameter $\lambda > 1$ (thus $\mathbb{E}[X_1] = 1/\lambda$). An output simulation would have us simulate the sum of exponential random variables (which has an Erlang (or Gamma) distribution) with some other distribution. In this example we choose the bias distribution to be that of a sum of standard Gaussian squared random variables (which has a $\mathcal{X}^2$ distribution). In other words, we select

$$p_n(x) = \frac{\lambda^n x^{n-1} \exp(-\lambda x)}{\Gamma(n)}$$

and we simulate with

$$q_n(x) = \frac{x^{(n/2)-1} \exp(-\frac{x}{2})}{2^{n/2}\Gamma(\frac{n}{2})}$$

Therefore the variance rate is dependent on

$$\begin{aligned}
&\exp\big(nc_{n,o}(\theta)\big) \\
&= \frac{\lambda^{2n} 2^{n/2}\Gamma(n/2)}{\Gamma(n)^2} \int_0^\infty x^{2n-(n/2)-1} \exp\big(-(2\lambda - \theta - \frac{1}{2})x\big)dx \\
&= \frac{\lambda^{2n} 2^{n/2}\Gamma(\frac{n}{2})}{\Gamma(n)^2} \frac{\Gamma(\frac{3n}{2})}{(2\lambda - \theta - \frac{1}{2})^{3n/2}}
\end{aligned}$$

Note that by Stirling's formula, we have

$$\Gamma(z) \approx \exp(-z) z^{z-(1/2)} \sqrt{2\pi}.$$

So $\frac{\Gamma(\frac{n}{2})\Gamma(\frac{3n}{2})}{\Gamma(n)^2} \approx (\frac{1}{3})^{n/2}(\frac{3}{2})^{2n}\frac{2}{\sqrt{(3)}}$ and therefore, as $n \to \infty$,

$$\exp\big(nc_{n,o}(\theta)\big) \approx \frac{(\lambda^{2n})(2^{n/2})}{(2\lambda - \theta - \frac{1}{2})^{3n/2}} (\frac{2}{\sqrt{3}})(\frac{1}{3})^{n/2}(\frac{3}{2})^{2n}.$$

Hence,

$$\lim_{n\to\infty} c_{n,o}(\theta) = c_o(\theta) = \log(\frac{\lambda^2(\sqrt{2})}{(2\lambda - \theta - \frac{1}{2})^{3/2}} \frac{3\sqrt{3}}{4}).$$

For the input bias scheme we have to simulate each of the exponential random variables in the sum of interest with a Gaussian square random variable. Computing the variance rate for this scheme is straightforward. Note that the limit of the moment generating sequence for the input bias scheme $c_i(\theta)$ is just equal to $c_{n,o}(\theta)$ for the value $n = 1$! Therefore

$$c_i(\theta) = c_{1,o}(\theta) = \log(\frac{\lambda^2(\sqrt{2})}{(2\lambda - \theta - \frac{1}{2})^{3/2}} \frac{\pi}{2}).$$

Hence the rates are indeed slightly different. Remember that any difference in the rate will eventually translate into huge differences in the actual estimator variance as n grows large.

Example 7.1.3. A common noncoherent communication receiver is the energy detector. In the simplest binary setting, we must choose between two hypotheses: that of $H_1 = \{\text{signal one present}\}$ and $H_0 = \{\text{signal zero present}\}$. We suppose that signal one has more power than signal zero. When signal zero is being transmitted (i.e., H_0 is true), we suppose that during the signaling time period of n samples, we receive the m-periodic signal sequence $\{s_i\}$ in the presence of an additive white Gaussian noise $\{N_i\}$ (mean zero, variance one). Denote the average power of the sequence as $P_s = (\sum_{i=1}^m s_i^2)/m$ and the d.c. value as $m_s = (\sum_{i=1}^m s_i)/m$. Also, we assume, for simplicity, that n is a multiple of m. Denote the received sequence as $\{R_i\}$. Of course the purpose of the communications receiver is to determine which signal is present. An energy detector would compute the following test statistic,

$$\sum_{i=1}^n R_i^2 \mathop{\gtrless}_{H_0}^{H_1} Tn,$$

where the symbol $\mathop{\gtrless}_{H_0}^{H_1}$ signifies that we choose hypothesis H_1 (H_0) if the left-hand (right-hand) side of the equation is greater than the other side.

To compute the false alarm probability (the miss probability is computed similarly), we need to compute

$$P(\sum_{i=1}^n R_i^2 > Tn | H_0 \text{ is true})$$
$$= P(\sum_{i=1}^n (s_i + N_i)^2 > Tn).$$

The exponential rate with which this probability goes to zero can be computed easily from Theorem 3.2.1. Note that the moment generating function of a general term in the sum is given by

$$M_i(\theta) = \mathbb{E}[\exp(\theta(s_i + N_i)^2)]$$
$$= \frac{1}{\sqrt{1 - 2\theta}} \exp(\frac{\theta^2 s_i^2}{\frac{1}{2} - \theta} + \theta s_i^2).$$

Thus

$$\phi(\theta) = \lim_{n\to\infty} \frac{1}{n} \sum_{i=1}^{n} \log M_i(\theta)$$
$$= P_s \frac{\theta}{1-2\theta} - \frac{1}{2}\log(1-2\theta).$$

The rate of the probability going to zero is thus

$$I(T) = \sup_{\theta}[\theta T - \phi(\theta)]$$
$$= \frac{2T-\alpha}{4} - P_s \frac{2T-\alpha}{2\alpha} + \frac{1}{2}\log\frac{\alpha}{2T},$$

where $\alpha = 1 + \sqrt{1+4TP_s}$.

Suppose we choose to simulate with an input strategy where we compute the sum of independent random variables $\sum_{i=1}^{n}(Y_{i,\nu}+s_i)^2$ where $Y_{i,\nu}$ is the (functional) exponential shift of N_i; that is, the probability density of Y_i is given by

$$q_{i,\nu}(x) = \frac{\exp\big(\nu(x+s_i)^2\big)\exp\big(-\frac{x^2}{2}\big)}{M_i(\nu)\sqrt{2\pi}}.$$

Completing the square, we find that $q_{i,\nu}$ is a Gaussian probability density function with mean $2\nu s_i/(1-2\nu)$ and variance $1/(1-2\nu)$. If we choose $\nu = \nu_o = (2T-\alpha)/4T$, the resulting simulation distributions $\{q_{i,\nu_o}\}$ are Gaussian with means $s_i(2T-\alpha)/\alpha$ and common variance $2T/\alpha$. We can easily compute $R(T)$ for this choice to find that $R(T) = 2I(T)$ and thus this is an *efficient* simulation strategy. Since the input strategy is efficient, the output strategy must also be efficient (by Theorem 4.3.1).

We recall that if $\{U_i\}$ are i.i.d. standard normal random variables (denote the standard normal density as p) and $\{\delta_i\}$ are constants, then $\sum_{i=1}^{n}(U_i + \delta_i)^2$ has a noncentral chi-square distribution with n degrees of freedom and noncentrality parameter $\lambda = \sum_{i=1}^{n}\delta_i^2$. An explicit expression for the density is given by [42, Eq. 28.3.5 (note typographical error)],

$$f_n(\lambda, x) = \frac{1}{2}\big(\frac{x}{\lambda}\big)^{(1/4)(n-2)} I_{(1/2)(n-2)}(\sqrt{\lambda x})\exp\big(-\frac{1}{2}(\lambda+x)\big).$$

Thus the distribution of $\sum_{i=1}^{n}(Y_{i,\nu_o}+s_i)^2$ has probability density

$$f_n(\lambda_q, \frac{\alpha}{2T}x)\frac{\alpha}{2T},$$

where

$$\lambda_q = \frac{2\alpha}{T}\sum_{i=1}^{n} s_i^2.$$

Thus, we can conclude, by Theorem 4.3.1, that the output simulation strategy with this as the biasing probability density is efficient ($R(T) = 2I(T)$). (The authors note that to try to show this directly from the definition of efficiency appears to be very difficult.)

Now let us consider a simpler input simulation strategy. We simulate $\sum_{i=1}^{n}(Y_i'+s_i)^2$ where $\{Y_i'\}$ are i.i.d. Gaussian with mean m_q and variance σ_q^2. This is a simpler biasing distribution since the mean is held constant and does not vary with the index i. Let us choose m_q, σ_q^2 so that the associated output simulation distribution also has density $f_n(\lambda_q, \frac{\alpha}{2T}x)\frac{\alpha}{2T}$. Thus the associated output simulation is efficient. (The important point to note here is that even if the output simulation is efficient, an associated input simulation may not be. That is indeed the case here with this simpler input strategy.) We can choose our parameters so that the output simulation is efficient if we choose $\sigma_q^2 = 2T/\alpha$ and

$$m_q = -m_s + \sqrt{m_s - (1 - \frac{2T}{\alpha}P_s)}.$$

We denote the normal density with these choices for mean and variance as q'. We still need to compute $R(T)$ for this input strategy. We thus compute

$$\begin{aligned} c(\theta) &= \lim_{n\to\infty} \frac{1}{n} \log \int \exp\bigl(\theta(x+s_i)^2\bigr)\frac{p(x)^2}{q'(x)}dx \\ &= \frac{1}{2}\log(\frac{\sigma_q^2}{2}) - \frac{1}{2}\log(1-\theta-\frac{1}{2\sigma_q^2}) + \theta P_s + \frac{m_q^2}{2\sigma_q^2} \\ &\quad + \frac{\theta P_s}{1-\theta-\frac{1}{2\sigma_q^2}} - \frac{\theta m_s m_q}{\sigma_q^2(1-\theta-\frac{1}{2\sigma_q^2})} \\ &\quad + \frac{m_q^2}{4\sigma_q^4(1-\theta-\frac{1}{2\sigma_q^2})}. \end{aligned}$$

Now we must compute $R(T) = \sup_\theta[\theta T - c(\theta)] = \theta_T T - c(\theta_T)$. The optimizing value of θ is given by

$$\begin{aligned} \theta_T &= 1 - \frac{1}{2\sigma_q^2} - \frac{1}{4T} \\ &\quad + \frac{\sqrt{\frac{1}{4} + 4T[P_s(1-\frac{1}{2\sigma_q^2})^2 - \frac{m_s m_q}{\sigma_q^2}(1-\frac{1}{2\sigma_q^2}) + \frac{m_q^2}{4\sigma_q^4}]}}{2T}. \end{aligned}$$

We can compute the variance rate for the constant mean estimator and compare it to an efficient estimator for a variety of signal sequences. In Fig. 7.1 we see that for values of T near $1+P_s$, the constant mean estimator works quite well for sawtooth waveforms but has increasingly bad performance as T gets large. In Fig. 7.2 we see that for T values near $1 + P_s$, we actually have a *negative* variance rate, indicating that we are doing far worse than just a direct Monte Carlo simulation. This isn't too surprising given that for

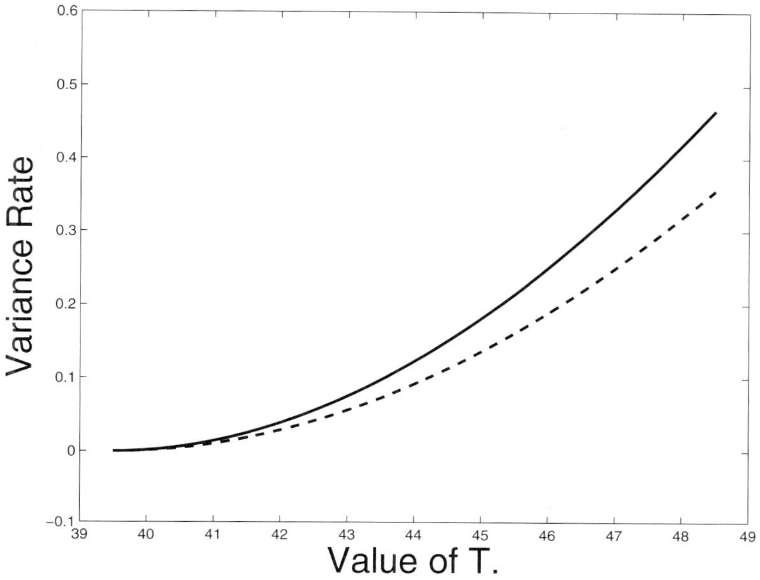

Fig. 7.1. The variance rate plotted as a function of T for the constant mean estimator (dashed line) compared to that of the efficient estimator (solid line) for a sawtooth input.

this sinusoidal signal, one would think that a constant mean estimator should have mean near zero (which our choice doesn't have).

In Fig. 7.3, we plot a typical simulation of the two input estimators and the associated output estimator as a function of the number of simulation runs. We choose $T = 48.5$ and $s(j) = j,\ j = 1, \ldots, m,\ m = 10$. The two efficient estimators are giving (after 7000 runs) a value of 8.85×10^{-5} which has an error of at most 5% with 95% confidence. To get this same level of accuracy and confidence with the constant mean estimator, we estimate that 1.8×10^7 simulation runs would be required.

7.2 Notes and Comments

A lot of times when one is designing a simulation methodology, we forget that we almost always have several options available to us with regard to the bias point. If we just take the distributions of the input random variables from the random number generators and product them to form the weight function, this is an input formulation. This is almost always the easiest way to proceed. If we can do some analysis up to some intermediate point of the systems, and then bias the derived distributions at that intermediate point, then we have an intermediate formulation.

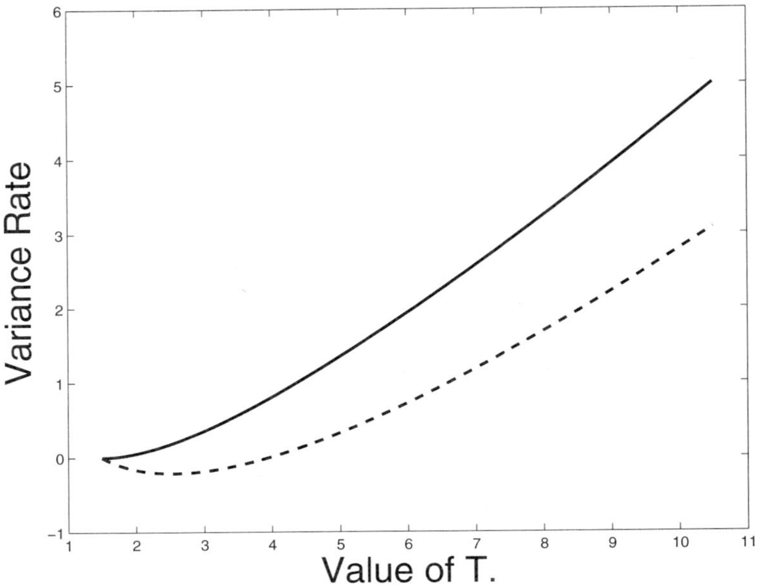

Fig. 7.2. The variance rate plotted as a function of T for the constant mean estimator (dashed line) compared to that of the efficient estimator (solid line) for a sinusoidal input.

If we can carry out the analysis all the way to the end, we have an output formulation, which sometimes means that we have an analytical solution and don't really need to simulate in the first place. One really shouldn't turn up one's nose too much at the study of output formulations. Firstly, it isn't really quite true that knowledge of the distribution gives us an analytical answer. We still have to integrate that distribution over a complicated, possibly multidimensional, set. In multidimensional settings, very often the best way to integrate is to use a Monte Carlo importance sampling technique. In a later chapter, we will show that even if we don't know analytically the output distribution, we can still construct the associated estimators using the so-called "blind" techniques.

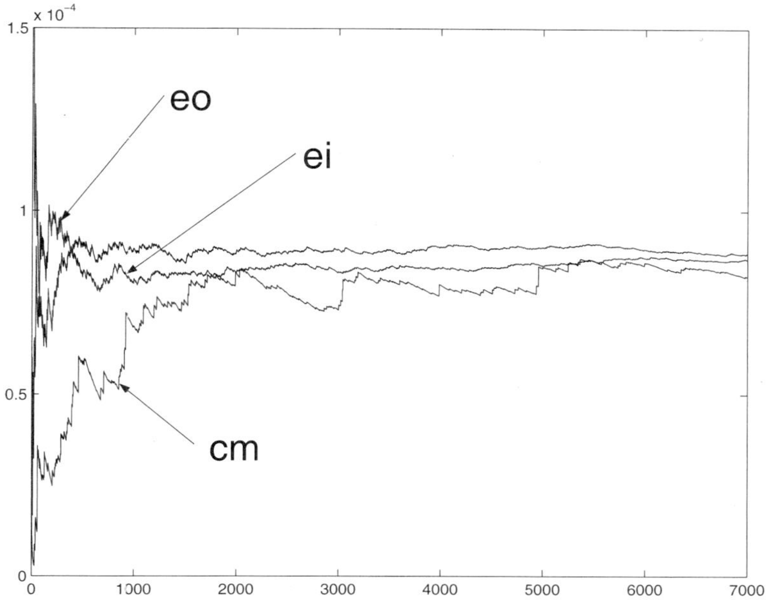

Fig. 7.3. Estimator values as a function of the number of simulation runs. The trajectories of the efficient output estimator (eo), the efficient input estimator (ei), and the inefficient constant mean input estimator (cm), are indicated by labeled arrows.

8. Chernoff's Bound and Asymptotic Expansions

Though the mills of God grind slowly yet,
they grind exceeding small. *F. F. von Logau*

The large deviation theory of Chapter 3 tells us what the asymptotic *rates* to zero are of various sequences of rare event probabilities. The very reason for this book is that in almost all applications, this knowledge is not sufficient. We actually need to know what the probability is (or at least have a good estimate of it) in order to complete an engineering system analysis or design. The philosophy of this book is that we are going to simulate in order to obtain that number.

In order to evaluate the performance of various simulation methodologies, we need to have a bit more information about the behavior of the large deviation probabilities than just their rate. In the first section, we consider calculating the "fine structure" of the rare event probabilities in the one-dimensional setting. We then use the ideas developed in this section to make some general statements about what we might expect in the more general $\mathcal{R}^d$ situation.

8.1 $\mathcal{R}$-Valued Random Variables

Lemma 8.1.1. *Suppose we have a sequence of probability measures μ_n converging weakly to a probability measure μ and a sequence of $\mathcal{R}$-valued functions f_n converging uniformly to f, a bounded continuous function. Then*

$$\int f_n d\mu_n \to \int f d\mu.$$

Proof. By the uniform convergence and the finiteness of the probability measures we have that $\int |f_n(x) - f(x)| d\mu_n(x) \to 0$. But this implies that $|\int f_n(x) d\mu_n(x) - \int f(x) d\mu_n(x)| \to 0$. Since $f(x)$ is bounded and continuous and $\mu_n \to \mu$ weakly, we have by the definition of weak convergence (see, e.g., [8]) $\int f(x) d\mu_n(x) \to \int f(x) d\mu$. This finishes the proof of the lemma. □

We consider the usual Gärtner–Ellis framework. Let Y_n be a sequence of $\mathcal{R}$-valued random variables and define $M_n(\theta) = \mathbb{E}[\exp(\theta Y_n)]$ and $\phi_n(\theta) = (1/n)\log M_n(\theta)$. We assume the standard assumptions hold for the sequence of convex functions $\{\phi_n\}$.

If we assume A1 to A3 and that there exists a $\theta_a \in C \cap (0,\infty)$ such that $\phi'(\theta_a)$ exists and $\phi'(\theta_a) = a$, then by Theorem 3.2.1 we have

$$\lim_{n\to\infty} \frac{1}{n} \log P\big(\frac{Y_n}{n} \geq a\big) = -\{a\theta_a - \phi(\theta_a)\}.$$

By convexity, notice that the assumption that $\theta_a > 0$ is equivalent to $\phi'(0) < a$. In the i.i.d. sum case, this is just $\mathbb{E}[X_i] = \phi'(0) < a$ which by the law of large numbers is necessary for $P\big(\frac{Y_n}{n} \geq a\big) \to 0$.

We hereafter consider only the case $a = 0$, which by the transformation $Y_n' - an = Y_n$ is no loss of generality and somewhat simplifies the notation in the following proofs. We assume $\phi'(0) < 0$ and $\phi(\theta_0) = 0$ for some $\theta_0 > 0$, and that $\phi''(\theta_0) = \sigma^2 > 0$. We are interested in precise estimates of the probability $P\big(Y_n/n \geq 0\big)$. Let $F_n(x)$ denote the distribution of Y_n. Define

$$dF_n^{(\theta)}(x) = \frac{\exp(\theta x)dF_n(x)}{\int \exp(\theta y)dF_n(y)} = \frac{\exp(\theta x)dF_n(x)}{M_n(\theta)}$$

as the "twisted" or "tilted" distribution. Let $Y_n^{(\theta)}$ be a random variable having $F_n^{(\theta)}$ as its probability distribution. Let $H_n(x)$ denote the distribution function of $Y_n^{(\theta_0)}/\sqrt{n}\sigma$.

Then

$$\begin{aligned} P\big(Y_n/n \geq 0\big) &= \int_0^\infty dF_n(x) \\ &= M_n(\theta_0)\int_0^\infty \exp(-\theta_0 x)dF_n^{(\theta_0)}(x) \\ &= M_n(\theta_0)\int_0^\infty \exp(-\sqrt{n}\theta_0 x\sigma)dH_n(x). \end{aligned} \tag{8.1}$$

We now need to consider the asymptotics of the integral

$$\int_0^\infty \exp(-\sqrt{n}\alpha x)dH_n(x) \tag{8.2}$$

(for $\alpha > 0$) which may be transformed by integration by parts to

$$\sqrt{n}\alpha \int_0^\infty \exp(-\sqrt{n}\alpha x)[H_n(x) - H_n(0)]dx. \tag{8.3}$$

Remark 8.1.1. In (8.2) we may upper bound $\exp(-\sqrt{n}\alpha x)$ by 1 to find that an upper bound to the integral is 1. This is the analogue to the Chernoff Bound for i.i.d. sums.

Let us define the sequence of complex functions

$$\Phi_n(\omega) = \frac{M_n(\frac{i\omega}{\sqrt{n}\sigma} + \theta_0)}{M_n(\theta_0)}.$$

8.1.1 The NonLattice Case

We may now state one of our principal lemmas for the nonlattice random variable case.

Lemma 8.1.2. *Suppose*

$$\Phi_n(\omega) \xrightarrow{\mathcal{L}_1} \exp(-\frac{1}{2}\omega^2).$$

Then

$$\sqrt{n}\int_0^\infty \exp(-\sqrt{n}\alpha x)dH_n(x) \longrightarrow \frac{1}{\sqrt{2\pi}\alpha}. \tag{8.4}$$

Proof. We note that $\Phi_n(\omega)$ is the characteristic function of $Y_n^{(\theta_0)}/\sqrt{n}\sigma$. The assumed $\mathcal{L}_1$ convergence implies that $\Phi_n(\omega)$ is $\mathcal{L}_1$ (for large enough n) and hence that $Y_n^{(\theta_0)}/\sqrt{n}\sigma$ has a density function $h_n(x)$ [20, pg. 155]. The Fourier inversion formula then implies

$$|h_n(x) - \frac{\exp(-\frac{x^2}{2})}{\sqrt{2\pi}}| \le \frac{1}{2\pi}\int_{-\infty}^{\infty} |\Phi_n(\omega) - \exp(-\frac{\omega^2}{2})|d\omega.$$

The $\mathcal{L}_1$ convergence of $\Phi_n(\omega)$ to $\exp(-\omega^2/2)$ immediately implies uniform convergence of $h_n(x)$ to $\exp(-x^2/2)/\sqrt{2\pi}$.

We now define the sequence of probability measures

$$\mu_n(dx) = \sqrt{n}\alpha \exp(-\sqrt{n}\alpha x)dx.$$

It is easy to check that μ_n is a probability measure for every n and that the sequence converges weakly to $\mu(dx) = \delta(x)dx$, point mass at the origin. Therefore in the integral we may replace $dH_n(x)$ by $h_n(x)dx$ and by invoking Lemma 8.1.1, we have completed the proof of the lemma. □

Theorem 8.1.1. *If* $\Phi_n(\omega) \xrightarrow{\mathcal{L}_1} \exp(-\omega^2/2)$ *then*

$$\lim_{n\to\infty} P(\frac{Y_n}{n} \ge 0)\sqrt{2\pi n}\sigma\theta_0/M_n(\theta_0) = 1.$$

Proof. The theorem follows immediately from (8.1) and the previous lemma. (Note that with the stated assumptions $\theta_0\sigma \neq 0$.) □

Corollary 8.1.1. *Assume $\Phi_n(\omega) \longrightarrow \exp(-\omega^2/2)$ where the convergence is pointwise. Suppose $\Phi_n(\omega)$ is $\mathcal{L}_1$ (for all n sufficiently large) or alternatively suppose $Y_n^{(\theta_0)}/\sqrt{n}\sigma$ has bounded density functions $h_n(\cdot)$ (for all n sufficiently large). Then*

$$\lim_{n\to\infty} P(\frac{Y_n}{n} \geq 0)\sqrt{2\pi n}\sigma\theta_0/M_n(\theta_0) = 1.$$

Proof. For the first alternative, note that characteristic functions are bounded. Pointwise convergence and the dominated convergence theorem then imply the $\mathcal{L}_1$ convergence needed as a condition of Theorem 8.1.1.

For the second alternative, note that if a probability density h_n is bounded, then its characteristic function Φ_n is $\mathcal{L}_1$. This completes the proof. □

Remark 8.1.2. It is known from the proof of Theorem 3.2.1 that with the stated assumptions, $Y_n^{(\theta_0)}/n \to 0$ almost surely. This fact and the fact that $M_n(\theta) \approx \phi(\theta)^n$ means that $Y_n^{(\theta_0)}/n$ is "behaving" as if it were the sample average of some random variables converging to a mean value. Hence the idea that by scaling with $1/\sqrt{n}$ instead of $1/n$, we may be able to have Central Limit Theorem type behavior. The proof of Lemma 8.4 hinges on the fact that $Y_n^{(\theta_0)}/\sqrt{n}\sigma$ converges in distribution to a standard normal.

Remark 8.1.3. As a consequence of the hypothesis of $\mathcal{L}_1$ convergence, we rule out "lattice type" random variables. From previous results for the i.i.d. [6] and Markov [61] cases, we know that these must be treated separately and have (in general) a limit dependent upon the lattice spacing. We consider this case in more detail in the following subsection.

Remark 8.1.4. It may be suspected that $\mathcal{L}_1$ convergence of the characteristic functions is a rather stringent condition. Perhaps so, but some further assumption on the moment generating sequence is required other than the three conditions used to invoke Theorem 3.2.1. For example, suppose $Y_n = \sum_{i=1}^n X_i$ where the $\{X_i\}$ are independent Gaussian random variables with variance one. Suppose that $\mathbb{E}[X_i] = m_i$ where the $\{m_i\}$ sequence is chosen so that $\sum_{i=1}^n m_i/n \to 0$ and $\sum_{i=1}^n m_i/\sqrt{n} \to \infty$. It is easy to check that $\lim_{n\to\infty} 1/n \log M_n(\theta) = \phi(\theta) = \theta^2/2$ which satisfies A1 to A3 and the other conditions for Theorem 3.2.1 hold. However, any fixed exponential shift merely corresponds to a fixed mean shift in this Gaussian setting. Hence since $\sum m_i/\sqrt{n} \to \infty$, $Y_n^{(\theta_0)}/\sqrt{n}\sigma$ does not converge in distribution for any θ_0.

8.1.2 The Lattice Case

We now consider the lattice random variable case. Suppose the $\{Y_n\}$ have support on the set $\{b+jd; j \in \mathbb{N}, d \in (0,\infty), 0 \leq b < d\}$. Let d be the largest number with the property. Then d is called the *span* or *lattice spacing.*

Lemma 8.1.3. *Suppose*

$$\Phi_n(\omega)1_{\{-\sqrt{n}\pi\sigma/d, \sqrt{n}\pi\sigma/d\}}(\omega) \xrightarrow{\mathcal{L}_1} \exp(-\frac{1}{2}\omega^2).$$

Then for $\alpha > 0$*,*

$$\sqrt{n}\int_{0^-}^{\infty} \exp(-\sqrt{n}\alpha x) dH_n(x) \longrightarrow \frac{d\exp(-\frac{\alpha b}{\sigma})}{\sqrt{2\pi}\sigma\Big(1-\exp(-\frac{\alpha d}{\sigma})\Big)}.$$

Proof. One should note that the $\{Y_n^{(\theta)}\}$ have support on the same lattice as the $\{Y_n\}$. The random variable $\Phi_n(\omega)$ is periodic with fundamental period $2\pi\sigma\sqrt{n}/d$. Now let us consider the expression in (8.3) multiplied by $\sqrt{n}$.

$$\begin{aligned}
& n\alpha\int_{0^-}^{\infty} \exp(-\sqrt{n}\alpha x)[H_n(x) - H_n(0^-)]dx \\
&= n\alpha \sum_{k=0}^{\infty} \int_{[kd+b]/\sqrt{n}\sigma}^{[(k+1)d+b]/\sqrt{n}\sigma} \exp(-\sqrt{n}\alpha x) dx \\
&\quad \times [H_n(\left[\frac{(k+1)d}{\sqrt{n}\sigma}\right]^-) - H_n(0^-)] \\
&= \sum_{k=0}^{\infty} \sqrt{n}[H_n(\left[\frac{(k+1)d+b}{\sqrt{n}\sigma}\right]^-) - H_n(0^-)] \\
&\quad \times [\exp(-\frac{\alpha}{\sigma}[kd+b]) - \exp(-\frac{\alpha}{\sigma}[(k+1)d+b])].
\end{aligned}$$

Define $t_n(x) = P\big(Y_n^{(\theta_0)}/(\sqrt{n}\sigma) = x\big)$. Then from properties of characteristic functions (see, e.g., [29, p. 511])

$$t_n(\frac{x}{\sqrt{n}})\frac{\sqrt{n}\sigma}{d} = \frac{1}{2\pi}\int_{-\sqrt{n}\pi\sigma/d}^{\sqrt{n}\pi\sigma/d} \Phi_n(\omega)\exp\big(-i\omega\frac{x}{\sqrt{n}}\big)d\omega.$$

Let $\backslash(\cdot)$ denote the density function of a zero mean, unit variance Gaussian random variable. Hence

$$\begin{aligned}
|t_n(\frac{x}{\sqrt{n}})\frac{\sqrt{n}\sigma}{d} - \backslash(\frac{x}{\sqrt{n}})| \leq & \int_{-\sqrt{n}\pi\sigma/d}^{\sqrt{n}\pi\sigma/d} |\Phi_n(\omega) - \exp\big(-\frac{1}{2}\omega^2\big)|d\omega \\
& + \int_{|\omega|>\sqrt{n}\pi/d} \exp\big(-\frac{1}{2}\omega^2\big)d\omega.
\end{aligned}$$

The assumed $\mathcal{L}_1$ convergence then implies

$$\lim_{n\to\infty} t_n(\frac{x}{\sqrt{n}})\frac{\sqrt{n}\sigma}{d} = \lim_{n\to\infty} \backslash(\frac{x}{\sqrt{n}}) = \backslash(0) = \frac{1}{\sqrt{2\pi}}$$

uniformly in x.

We then note that

$$\sqrt{n}\left[H_n\left(\left[\frac{[(k+1)d+b]}{\sqrt{n}\sigma}\right]^-\right)-H_n(0^-)\right] = \sum_{j=0}^{k}\sqrt{n}t_n\left(\frac{jd+b}{\sqrt{n}\sigma}\right) \xrightarrow[n\to\infty]{} \frac{d(k+1)}{\sigma\sqrt{2\pi}}.$$

The uniform convergence implies that we may interchange limit with sum in the following expression.

$$\begin{aligned}
&\lim_{n\to\infty} n\alpha\int_{0^-}^{\infty}\exp(-\sqrt{n}\alpha x)[H_n(x)-H_n(0^-)]dx\\
&= \lim_{n\to\infty}\sqrt{n}\sum_{k=0}^{\infty}[H_n(\left[\frac{[(k+1)d+b]}{\sqrt{n}\sigma}\right]^-)-H_n(0^-)]\\
&\quad\times[\exp(-\frac{\alpha}{\sigma}[kd+b])-\exp(-\frac{\alpha}{\sigma}[(k+1)d+b])]\\
&= \exp(-\frac{\alpha b}{\sigma})\sum_{k=0}^{\infty}\frac{d(k+1)}{\sigma\sqrt{2\pi}}[\exp(-\frac{\alpha}{\sigma}kd)-\exp(-\frac{\alpha}{\sigma}(k+1)d)]\\
&= \frac{d\exp(-\frac{\alpha b}{\sigma})}{\sqrt{2\pi}\sigma\left(1-\exp(-\frac{\alpha d}{\sigma})\right)}. \qquad (8.5)
\end{aligned}$$

This completes the proof. □

Theorem 8.1.2. *If* $\Phi_n(\omega)1_{\{-\sqrt{n}\pi\sigma/d,\sqrt{n}\pi\sigma/d\}}\xrightarrow{\mathcal{L}_1}\exp(-\frac{1}{2}\omega^2)$, *then (if* $\phi'(0)<0$*)*

$$\lim_{n\to\infty}P(\frac{Y_n}{n}\geq 0)\frac{\sqrt{2\pi n}\sigma(1-\exp(-\theta_0 d))}{d\exp(-\theta_0 b)M_n(\theta_0)}=1.$$

Proof. The corollary follows immediately from (8.1) and the preceding lemma. □

Remark 8.1.5. One can deduce from the above derivation that (for $b=0$),

$$\lim_{n\to\infty}P(\frac{Y_n}{n}>0)\frac{\sqrt{2\pi n}\sigma(1-\exp(-\theta_0 d))}{d\exp(-\theta_0 d)M_n(\theta_0)}=1.$$

Remark 8.1.6. Following an argument similar to that of [29, pg. 518], we can show that the above $\mathcal{L}_1$ convergence holds for sums of i.i.d. lattice type random variables. In this setting $Y_n^{(\theta_0)}$ appears as a sum of zero mean, unit variance, i.i.d. lattice type random variables. Let $\Phi(\cdot)$ be the characteristic function of one of the summands. Then

$$\Phi(\frac{\omega}{\sqrt{n}}) = 1 - \frac{\omega^2}{2n} + o(\frac{1}{n}). \tag{8.6}$$

It is easy to verify that $\Phi^n(\frac{\omega}{\sqrt{n}}) \to \exp(-\omega^2/2)$. Consider the integral

$$\int_{-\sqrt{n}\pi/d}^{\sqrt{n}\pi/d} |\Phi^n(\frac{\omega}{\sqrt{n}}) - \exp(-\frac{1}{2}\omega^2)|d\omega.$$

We split the integration over three regions, $|\omega| \leq a$, $a \leq |\omega| \leq \delta\sqrt{n}$, and $\delta\sqrt{n} \leq |\omega| \leq \sqrt{n}\pi/d$. The integral over the first region goes to zero by the convergence of $\Phi^n(\omega)$ and the dominated convergence theorem. Because of (8.6), it is possible to choose $\delta > 0$ so that $\Phi(\omega) \leq \exp(-\omega^2/4)$ for $|\omega| < \delta$, which implies that the integrand over the second region can be made negligible by making a large. In the third region, $\sup[\Phi(\omega/\sqrt{n}) : \delta\sqrt{n} \leq |\omega| \leq \pi\sqrt{n}/d] = \eta < 1$. The contribution of the third integral is then less than

$$\begin{aligned}
&\eta^{n-1} \int_{\sqrt{n}\delta}^{\sqrt{n}\pi/d} |\Phi(\frac{\omega}{\sqrt{n}})|d\omega + \int_{|\omega|>\delta\sqrt{n}} \exp(-\frac{1}{2}\omega^2)d\omega \\
&\leq \eta^{n-1}\sqrt{n}\pi/d + \int_{|\omega|>\delta\sqrt{n}} \exp(-\frac{1}{2}\omega^2)d\omega \\
&\to 0.
\end{aligned}$$

One should be cautioned that in the i.i.d. lattice variable case, our result holds only for $b = 0$. The general lattice case has an n-varying coefficient [6] or [30, pg. 193].

8.2 Examples for the $\mathcal{R}$-valued Case

Example 8.2.1 (Sum of Squares from a Gaussian Random Process). Let $\{N_i\}$ be a wide sense stationary Gaussian random process with continuous power spectrum $f(\omega), 0 \leq \omega \leq 2\pi$. We suppose $f(\omega) \geq \delta > 0$ for some $\delta \in \mathcal{R}^+$. We also require as a technical condition that $\lim_{\gamma\to 1/\|f\|_\infty} \int_0^{2\pi} \log(1-\gamma f(\omega))d\omega = \infty$. We are interested in the asymptotics of the "sum of squares" of this process; that is,

$$P(\frac{1}{n}\sum_{i=1}^{n} N_i^2 \geq \gamma),$$

where we suppose $\gamma > \mathbb{E}[N_1^2]$. Define $Y_n = \sum_{i=1}^{n}(N_i^2 - \gamma)$. Then

$$\begin{aligned}
M_n(\theta) &= \mathbb{E}[\exp(\theta Y_n)] \\
&= \exp(-\theta\gamma n)\prod_{i=1}^{n}\left(\frac{1}{1-2\theta\lambda_i^{(n)}}\right)^{1/2},
\end{aligned} \tag{8.7}$$

where $\{\lambda_i^{(n)}\}$ are the eigenvalues of the covariance matrix of $(N_1, N_2, \ldots, N_n)^T$. Therefore, for $\theta < 1/(2\,\|f\|_\infty)$,

$$\lim_{n\to\infty} \frac{1}{n} \log M_n(\theta) = -\theta\gamma - \frac{1}{4\pi}\int_0^{2\pi} \log(1 - 2\theta f(\omega))d\omega = \phi(\theta)$$

and $\phi(\theta) = \infty$ for θ outside this range. The convergence follows by the Toeplitz Distribution Theorem [35]. The technical condition implies that $\phi(\theta) \to \infty$ as $\theta \to 1/(2\,\|f\|_\infty)$. By convexity it is simple to verify that a solution to the equation $\phi'(\theta_0) = 0$ exists or

$$\gamma = \frac{1}{2\pi}\int_0^{2\pi} \frac{f(\omega)}{1 - 2\theta_0 f(\omega)} d\omega.$$

(If $f(\omega) = 1$, then $\gamma = 1/(1 - 2\theta_0)$ or $\theta_0 = (1 - 1/\gamma)/2$.) However, in general, this is a transcendental equation and such closed form solutions will not always be available. With some effort, we can check

$$\begin{aligned}
\Phi_n(\omega) &= \frac{M_n(\frac{i\omega}{\sqrt{n}} + \theta_0)}{M_n(\theta_0)} \\
&= \exp(-i\omega\gamma\sqrt{n}) \prod_{i=1}^{n} \left(1 - \frac{2i\omega\lambda_i^{(n)}/\sqrt{n}}{1 - 2\theta_0\lambda_i^{(n)}}\right)^{-1/2} \\
n &\overrightarrow{\to \infty} \exp(-\frac{\phi''(\theta_0)\omega^2}{2}),
\end{aligned} \tag{8.8}$$

where

$$\phi''(\theta) = \frac{1}{\pi}\int_0^{2\pi} \frac{f(\omega)^2}{(1 - 2\theta f(\omega))^2} d\omega.$$

$\Phi_n(\omega)$ is $\mathcal{L}_1$ for $n > 2$ certainly. The pointwise convergence, Remark 8.1.2, and Theorem 8.1.1 imply the following limit,

$$\lim_{n\to\infty} P(\frac{1}{n}\sum_{i=1}^{n} N_i^2 \geq \gamma)\sqrt{2\pi\phi''(\theta_0)n}/M_n(\theta_0) = 1.$$

Example 8.2.2 (Random Sums of I.I.D. Random Variables). Consider a random sum:

$$Y_\lambda = \sum_{i=1}^{N_\lambda} X_i,$$

where $\{X_i\}$ are i.i.d. mean $m < 0$ random variables with moment generation function $M_X(\cdot)$. Take $0 < p < 1$, λ a positive integer, and N_λ to be a "shifted" geometric random variable with $P(N_\lambda = n) = (1-p)p^{n-\lambda}(n \geq \lambda)$ and independent of the $\{X_i\}$. The characteristic function of Y_λ is

$$M_\lambda(i\omega) = \frac{(1-p)M_X(i\omega)^\lambda}{1 - M_X(i\omega)p}.$$

We then obtain (for values of θ such that $M_X(\theta) < 1/p$),

$$\phi(\theta) = \lim_{\lambda\to\infty} \frac{1}{\lambda} \log(M_\lambda(\theta)) = M_X(\theta)$$

and $\phi(\theta) = \infty$ otherwise. Hence $\phi'(\theta_0) = M'_X(\theta_0) = 0$ (as long as $M_X(\theta_0) < 1/p$) defines θ_0 and $\phi''(\theta_0) = M''_X(\theta_0) = \sigma^2$ defines σ. The characteristic function of interest is

$$\begin{aligned}
\Phi_\lambda(\omega) &= \frac{M_\lambda(\frac{i\omega}{\sqrt{\lambda}\sigma} + \theta_0)}{M_\lambda(\theta_0)} \\
&= \frac{M_X(\frac{i\omega}{\sqrt{\lambda}\sigma} + \theta_0)^\lambda (1 - M_X(\theta_0)p)}{M_X(\theta_0)^\lambda (1 - M_X(\frac{i\omega}{\sqrt{\lambda}\sigma} + \theta_0)p)} \\
&\overrightarrow{\lambda \to \infty} \exp(-\frac{1}{2}\omega^2).
\end{aligned}$$

By the arguments given in Remark 8.1.2, one can easily verify that if $M_X(\frac{i\omega}{\sigma} + \theta_0)/(1 - M_X(\frac{i\omega}{\sigma} + \theta_0)p)$ is $\mathcal{L}_1$, then the conditions of Theorem 8.1.1 hold. If X_1 is a lattice-type random variable with span d, then the conditions of Theorem 8.1.2 automatically hold.

We note that finding the asymptotics of $M_n(\theta_0)$ can itself be a nontrivial problem. Our philosophy in this chapter has been to assume that knowledge of the moment generating function sequence is complete. In many situations, this can be a nontrivial task, even though the logarithmic behavior is known.

8.3 $\mathcal{R}^d$-Valued Random Variables

We are going to be much more brief in this section. The finer estimates depend upon delicate estimates and arguments regarding the nature of the boundary of the set of interest near its minimal rate points. We consider only a special class of sets that should include most sets of applications interest. See the notes and comments section for further discussion and available literature for the multidimensional setting.

We consider the usual framework of Theorem 3.2.1. Let $\{Y_n\}$ be a sequence of $\mathcal{R}^d$-valued random variables and define for every n, $M_n(\theta) = \mathbb{E}[\exp(\langle\theta, Y_n\rangle)]$. As usual we define the limit $\lim_{n\to\infty} \frac{1}{n} \log M_n(\theta) = \phi(\theta)$ for all $\theta \in \mathcal{R}^d$. For $x \in \mathcal{R}^d$, $I(x) = \sup_\theta[\langle\theta, x\rangle - \phi(\theta)]$. We suppose that we are interested in the fine asymptotics as n grows large of

$$P(\frac{Y_n}{n} \in E).$$

We assume that the set E satisfies various conditions. E is convex (with nonempty interior) and has a unique dominating point ν_E that satisfies $\nabla\phi(\theta_E) = \nu_E$.

We also desire that $(Y_n^{(\theta_E)} - n\nu_E)/\sqrt{n}$ converge to a zero mean normal random vector. We actually make a stronger assumption than this in order to simplify the proof. The characteristic function of $(Y_n^{(\theta_E)} - n\nu_E)/\sqrt{n}$ is

$$\begin{aligned}
&\mathbb{E}[\exp\bigl(i\langle\omega, \frac{Y_n^{(\theta_E)} - n\nu_E}{\sqrt{n}}\rangle\bigr)] \\
&= \exp(-i\sqrt{n}\langle\omega, \nu_E\rangle)\mathbb{E}[\exp\bigl(i\langle\omega, \frac{Y_n^{(\theta_E)}}{\sqrt{n}}\rangle bigr)] \\
&= \exp(-i\sqrt{n}\langle\omega, \nu_E\rangle)\frac{M_n(\frac{i\omega}{\sqrt{n}} + \theta_E)}{M_n(\theta_E)}
\end{aligned}$$

We assume for all $\omega \in \mathcal{R}^d$,

$$\exp(i\sqrt{n}\langle\omega, \nu_E\rangle)\frac{M_n(\frac{i\omega}{\sqrt{n}} + \theta_E)}{M_n(\theta_E)} \xrightarrow{\mathcal{L}_1} \exp\bigl(-\frac{1}{2}\omega V\omega'\bigr),$$

where V is a symmetric positive definite (covariance) matrix. We further assume that for some vector K with all positive components,

$$\exp(\sqrt{n}\langle K, \nu_E\rangle)\frac{M_n(\frac{K}{\sqrt{n}} + \theta_E)}{M_n(\theta_E)} \to \exp\bigl(-\frac{1}{2}KVK'\bigr).$$

Define the (multidimensional) twisted distribution as

$$dF_n^{(\theta)}(x) = \frac{\exp(\langle\theta, x\rangle)dF_n(x)}{\int \exp(\langle\theta, y\rangle)dF_n(y)} = \frac{\exp(\langle\theta, x\rangle)dF_n(x)}{M_n(\theta)}.$$

Then,

$$\begin{aligned}
P(\frac{Y_n}{n} \in E) &= \int_{nE} dF_n(x) \\
&= M_n(\theta_E)\int_{nE} \exp(-\langle\theta_E, x\rangle)dF_n^{(\theta_E)}(x) \\
&= M_n(\theta_E)\exp(-n\langle\theta_E, \nu_E\rangle) \\
&\quad\times \int_{n(E-\nu_E)} \exp(-\langle\theta_E, x\rangle)dF_n^{(\theta_E)}(x + n\nu_E). \qquad (8.9)
\end{aligned}$$

To better estimate the behavior of the integral in (8.9), we make a change of variable. Let $y = Rx$, where R is the rotation that takes $\theta_E = (\theta_{E1}, \theta_{E2}, \dots, \theta_{Ed})$ into $R\theta_E = (\|\theta_E\|, 0, \dots, 0)$. Let $A = \{Rx : x \in E - \nu_E\}$. We note that A is convex, $0 \in \partial A$, and $A \subset \{x : x_1 \geq 0\}$ (the right half plane). Let $G_n(\cdot)$ denote the distribution function we obtain after performing the rotation

R on $F_n^{(\theta_E)}(\cdot + n\nu_E)$. (In other words $G_n(x) = P\big(Y_n^{(\theta_E)} \le R^{-1}x + n\nu_E\big)$.) Thus we may write the integral in (8.9) as

$$\int_{nA} \exp(-\|\theta_E\| y_1) dG_n(y). \tag{8.10}$$

To proceed further, we make some assumptions about the curvature of ∂E at ν_E. We suppose that ∂E has "small curvature" near ν_E (which is completely equivalent to A having small curvature near zero) in the sense that for some $\epsilon > 0$, $0 < r < \infty$, and $0 \le \gamma < 1/2$,

$$[A \cap H(0,\epsilon)] \supset [\Gamma(r,\gamma) \cap H(0,\epsilon)],$$

where

$$H(0,\epsilon) = \{x \in \mathcal{R}^d : 0 \le x_1 \le \epsilon\}$$

and

$$\Gamma(r,\gamma) = \{x \in \mathcal{R}^d : \sqrt{x_2^2 + x_3^2 + \cdots + x_d^2} \le r x_1^\gamma\}.$$

We wish to show that in this case the integral (8.10) is equal to $c[1 + o(n^{-\delta})]/\sqrt{n}$, $\delta > 0$ for some constant c. In other words, the integral goes to zero like $1/\sqrt{n}$. To see this observe that

$$0 \le \int_0^\infty \exp(-\|\theta_E\| y_1) dG_{n,1}(y_1) - \int_{nA} \exp(-\|\theta_E\| y_1) dG_n(y),$$

where $G_{n,1}$ is the distribution function of the first marginal of G_n,

$$\begin{aligned}
&\le \int_{n[\Gamma(r,\gamma)\cap H(0,\epsilon)]^c} \exp(-\|\theta_E\| y_1) dG_n(y) \\
&\le \int_{n[\Gamma(r,\gamma)\cap H(0,\epsilon)]^c} dG_n(y) \\
&\le G_n(n\Gamma(r,\gamma)^c) + G_n(nH(0,\epsilon)^c) \\
&= G_n(n\Gamma(r,\gamma)^c) + G_n(H(n\epsilon,\infty)).
\end{aligned} \tag{8.11}$$

We know that the first integral of the first line above goes to zero like $1/\sqrt{n}$ from our work done in the one-dimensional case of the previous section. The second integral is the one in which we're interested. If we can show that this difference goes to zero strictly faster than $1/\sqrt{n}$, then we will have that the integral (8.10) is equal to $c[1 + o(n^{-\delta})]/\sqrt{n}$, $\delta > 0$ as desired.

First, note that $G_n(H(n\epsilon,\infty)) = P\big(Y_n^{(\theta_E)} \in (R^{-1}H(n\epsilon,\infty)) + n\nu_E\big) = P\big(Y_n^{(\theta_E)}/n \in (R^{-1}H(\epsilon,\infty)) + \nu_E\big)$. We know by the standard large deviation theorems that this probability goes to zero at least exponentially fast. Hence we now concentrate ourselves on the first term of (8.11).

Note that

$$\Gamma = \{x \in \mathcal{R}^d : \sqrt{x_2^2 + x_3^2 + \cdots x_d^2} \le r x_1^\gamma\}$$
$$\Gamma^c = \{x \in \mathcal{R}^d : \sqrt{x_2^2 + x_3^2 + \cdots x_d^2} > r x_1^\gamma\}.$$

Note that $x \in n\Gamma^c$ if and only if $\frac{x}{n} \in \Gamma^c$ and hence

$$n\Gamma^c = \{x \in \mathcal{R}^d : \sqrt{(\frac{x_2}{n})^2 (\frac{x_3}{n})^2 + \cdots (\frac{x_d}{n})^2} > r(\frac{x_1}{n})^\gamma\}$$
$$= \{x \in \mathcal{R}^d : \sqrt{x_2^2 + x_3^2 + \cdots x_d^2} \le r n^{1-\gamma} x_1^\gamma\}.$$

Hence for $\psi > 0$,

$$n\Gamma^c \subset [\{x \in \mathcal{R}^d : \sqrt{x_2^2 + x_3^2 + \cdots + x_d^2} > r n^{1-\gamma-\gamma\psi}\}$$
$$\cup \{x \in \mathcal{R}^d : 0 \le x_1 < n^{-\psi}\}].$$

Now we choose ψ small enough so that $1 - \gamma - \gamma\psi = \eta > 1/2$. Therefore

$$G_n(n\Gamma^c) \le G_n(\{x \in \mathcal{R}^d : 0 \le x_1 < n^{-\psi}\})$$
$$+ G_n(\{x \in \mathcal{R}^d : \sqrt{x_2^2 + x_3^2 + \cdots + x_d^2} > r n^{1-\gamma-\gamma\psi}\}).$$

Now

$$G_n(\{x \in \mathcal{R}^d : 0 \le x_1 < n^{-\psi}\}) = G_{n,1}(0 \le x_1 \le n^{-\psi}).$$

Also we know that $G_n(x\sqrt{n}) = P(\frac{Y_n^{(\theta_E)} - n\nu_E}{\sqrt{n}} \le R^{-1}x)$ is converging to a Gaussian distribution. In fact due to the assumed $\mathcal{L}_1$ convergence of the characteristic functions, we have that $G_n(x\sqrt{n})$ has a density $g_n(x\sqrt{n})\sqrt{n}$ which is converging *uniformly* to a Gaussian density $\mathbf{n}(x)$. Thus

$$G_{n,1}(0 \le x_1 \le n^{-\psi}) = \int_0^{n^{-\psi}} dG_{n,1}(x)$$
$$= \int_0^{n^{-\psi}} g_{n,1}(x)dx$$
$$= \int_0^{n^{-\psi-\frac{1}{2}}} g_{n,1}(\sqrt{n}y)\sqrt{n}dy$$
$$\approx \mathbf{n}(0) n^{-(\psi+1/2)},$$

and thus $G_{n,1}(0 \le x_1 \le n^{-\psi})$ goes to zero strictly faster than $1/\sqrt{n}$.

Now let $Z_n = (Z_{n,1}, Z_{n,2}, \ldots, Z_{n,d})$ be a random variable that has distribution G_n. Then

$$G_n(\{x \in \mathcal{R}^d : \sqrt{x_2^2 + x_3^2 + \cdots + x_d^2} > r n^{1-\gamma-\gamma\psi}\})$$

$$
\begin{aligned}
&= P\Big(\sum_{i=2}^{d} Z_{n,i}^2 > r^2 n^{2\eta}\Big) \\
&\le \sum_{i=2}^{d} P\Big(Z_{n,i}^2 > \frac{r^2 n^{2\eta}}{d-1}\Big) \\
&= \sum_{i=2}^{d} P\Big(|Z_{n,i}| > \frac{r n^{\eta}}{\sqrt{d-1}}\Big).
\end{aligned}
$$

Note that the moment generating function for G_n, M_n' is calculated as

$$
\begin{aligned}
M_n'(\theta) &= \int \exp(\langle \theta, x\rangle) dG_n(x) \\
&= \int \exp(\langle \theta, x\rangle) dP(Y_n^{(\theta_E)} - n\nu_E \le R^{-1}x),
\end{aligned}
$$

perform a change of variable with $z = R^{-1}x + n\nu_E$

$$
\begin{aligned}
&= \int \exp(\langle \theta, R(z - n\nu_E)\rangle) dP(Y_n^{(\theta_E)} \le z) \\
&= \exp(-n\langle R^{-1}\theta, \nu_E\rangle) \int \exp(\langle R^{-1}\theta, z\rangle) dF_n^{(\theta_E)} \\
&= \exp(-n\langle R^{-1}\theta, \nu_E\rangle) \frac{M_n(R^{-1}\theta + \theta_E)}{M_n(\theta_E)}.
\end{aligned}
$$

Let $M_{n,i}$ be the moment generating function of $Z_{n,i}$. By assumption (for each i) there exists some $\epsilon > 0$ such that $M_{n,i}(\epsilon/\sqrt{n}) \to \mathcal{N}(\epsilon)$ where $\mathcal{N}$ is the moment generating function of some Gaussian random variable. Thus

$$
\begin{aligned}
P\big(|Z_{n,i}| > cn^{\eta}\big) &= \int 1_{\{x > cn^{\eta}\}} dG_{n,i}(x) \\
&\le \exp\big(\theta(x - cn^{\eta})\big) dG_{n,i}(x) \quad \theta > 0 \\
&= M_{n,i}'(\theta) \exp(-cn^{\eta}\theta),
\end{aligned}
$$

and letting $\theta = \epsilon/\sqrt{n}$ we obtain

$$
\begin{aligned}
&= M_{n,i}\Big(\frac{\epsilon}{\sqrt{n}}\Big) \exp\big(-cn^{\eta-(1/2)}\big) \\
&\le \mathcal{N}(\epsilon)(1+\epsilon) \exp\big(-cn^{\eta-(1/2)}\big)
\end{aligned}
$$

for all sufficiently large n. Thus the probability would go to zero faster than one over any polynomial in n.

Hence, we have shown from (8.9), and subject to all our conditions that

$$
\lim_{n\to\infty} P\Big(\frac{Y_n}{n} \in E\Big) \exp\big(nI(\nu_E)\big)\sqrt{n} = c \quad 0 < c < \infty.
$$

8.4 Variance Expansion of Importance Sampling Estimators

We again consider the setting of our original variance rate results of Chapter 5: for every integer n, let $Z_{p,n}$ be a random variable taking values in some complete separable metric space $\mathcal{S}_n$. Let P_n be the probability measure induced by $Z_{p,n}$ on $\mathcal{S}_n$. Instead of directly simulating $Z_{p,n}$, we choose to simulate with another $\mathcal{S}_n$-valued random variable $Z_{q,n}$ which in turn induces a probability measure on $\mathcal{S}_n$, Q_n. Let f_n be an $\mathcal{R}^d$-valued measurable function on the space $\mathcal{S}_n$; that is, $f_n : \mathcal{S}_n \to \mathcal{R}^d$.

To create the importance sampling estimators, we must assume that P_n is absolutely continuous with respect to Q_n for all n (and hence the Radon-Nikodym derivative dP_n/dQ_n exists). We suppose we are interested in $\rho_n = P\big(\frac{f_n(Z_{p,n})}{a_n} \in E\big)$, for some Borel set $E \subset \mathcal{R}^d$. For $\theta \in \mathcal{R}^d$, define

$$c_n(\theta) = \frac{1}{n}\log\Big(\int \frac{dP_n}{dQ_n}(z)\exp(\langle\theta, f_n(z)\rangle)dP_n(z)\Big)$$

and

$$c(\theta) = \lim_{n\to\infty} c_n(\theta),$$

where we assume that the limit exists and is finite for all $\theta \in \mathcal{R}^d$. The variance rate function is

$$R(x) = \sup_{\theta\in\mathcal{R}^d}\left[\langle\theta, x\rangle - c(\theta)\right] \qquad (\text{for } x \in \mathcal{R}^d).$$

We suppose that E has a dominating point ν_E, where $R(E) = \langle\theta_E, \nu_E\rangle - c(\theta_E)$.

Define a probability measure on $\mathcal{R}^d$ as

$$\mu_{Q,n}(B) = \int_{\{z\in\mathcal{S}_n : f_n(z)\in B\}} \frac{dP_n}{dQ_n}(z)\exp\big(\langle\theta_E, f_n(z)\rangle\big)\exp\big(-nc_n(\theta_E)\big)dP_n(z),$$

for all Borel sets $B \subset \mathcal{R}^d$. Its moment generating function

$$L_n(\theta) = \frac{\exp\big(nc_n(\theta+\theta_E)\big)}{\exp\big(nc_n(\theta_E)\big)}.$$

Just as in the previous section, we now suppose that this sequence of moment generating functions satisfies the same "Central Limit" assumptions as the moment generating functions of $\{Y_n^{(\theta_E)}\}$. We also suppose that the set E satisfies the same "small curvature" assumptions of the previous section. Then

$$
\begin{aligned}
F_n &= \int \frac{dP_n}{dQ_n}(z) 1_{\{z: \frac{f_n(z)}{n} \in E\}} dP_n(z) \\
&= \int_{\{z: \frac{f_n(z)}{n} \in E\}} \frac{dP_n}{dQ_n}(z) dP_n(z) \\
&= \exp\big(nc_n(\theta_E)\big) \int_{\{z: \frac{f_n(z)}{n} \in E\}} \exp\big(-\langle\theta_E, f_n(z)\rangle\big) \exp\big(-nc_n(\theta_E)\big) \\
&\quad \times \exp\big(-\langle\theta_E, f_n(z)\rangle\big) \frac{dP_n}{dQ_n}(z) dP_n(z) \\
&= \exp\big(nc_n(\theta_E)\big) \int_{nE} \exp(-n\langle\theta_E, y\rangle) d\mu_{Q,n}(y) \\
&= \exp\big(nc_n(\theta_E)\big) \int_{n(E-\nu_E)} \exp(-n\langle\theta_E, y\rangle) \\
&\quad \times \exp(-n\langle\theta_E, \nu_E\rangle) d\mu_{Q,n}(y + n\nu_E).
\end{aligned}
$$

We are now in exactly the same situation as in the previous chapter which allows us to assert,

$$
\lim_{n\to\infty} F_n \exp\big(nR(\nu_E)\big)\sqrt{n} = c' \qquad 0 < c' < \infty.
$$

Hence F_n is going to zero with our known exponential rate times a $1/\sqrt{n}$ term. Now $k\,Var(\hat{\rho}_n) = F_n - \rho^2$. From the previous section, we know that ρ^2 is going to zero with a known exponential rate times a $1/n$ term (due to the square!). Hence even if our estimator is efficient, the F_n term will *still* dominate and give us that our variance rate is the same as F_n. Hence, in almost all practical cases, studying the F_n term is sufficient to determine the variance rate.

8.5 Notes and Comments

Bounding or estimating the"fine structure" of rare event probabilities has a long history in communications theory. Chernoff's Bound is the most frequently resorted to communications theory technique for these sorts of situations. Chernoff's original paper [19] was concerned with the asymptotic discernibility of two i.i.d. sequences of random variables. He showed that the logarithmic rate of the probability of error to zero was given by the now-called Chernoff Entropy:

$$
\lim_{n\to\infty} \frac{1}{n} \log P\big(\text{error given first } n \text{ observations}\big) = \inf_{\alpha} \log(H_\alpha),
$$

where $H(\alpha) = \int (dp/dw)^\alpha (dq/dw)^{1-\alpha} dw$, where p, q are the distributions of one element of the sequence under the two hypotheses and w is any measure dominating both of them. This result is considered to be one of the first large deviation theory theorems. As a consequence of the proof (in fact by application of Markov's inequality), one can show

$$P\big(\text{error given first } n \text{ observations}\big) \le (\inf_{\alpha} H(\alpha))^n.$$

It is this last result that is typically known as Chernoff's Bound. Myriad other work followed. For example, Bahadur and Rao [6] showed that the following limit holds,

$$\lim_{n\to\infty} P\big(\text{error given first n observations}\big)\sqrt{n}/(\inf_{\alpha} H(\alpha))^n = c > 0.$$

Chernoff's result, with hindsight, can rightly be regarded as a particular application to the case of likelihood ratio tests of the large deviations theorem of Cramér. Cramér's Theorem dealt with tail probabilities of sums of i.i.d. random variables. (The likelihood ratio in the i.i.d. setting can be expressed as such a sum and hence establish Chernoff's result.)

The section on the $\mathcal{R}^d$-valued case follows closely and is inspired by Ney's path-breaking paper [63]. There the i.i.d. case is treated. We have merely made some minor modifications to update it to the Gärtner–Ellis setting. An even more careful and exhaustive treatment is due to Iltis [44].

9. Gaussian Systems

Question: Why did you use the Gaussian assumption?
Answer: Because it's the normal assumption! *S. Pasupathy*
¡Normal, pero un poquito acelera'o! *Los Van Van*

The design of information systems operating in and being corrupted by Gaussian (thermal) noise is a keystone subject area. The application areas are as numerous as are the uses of the radio spectrum. In this chapter we give a very general overview and explicitly present a theoretical framework from which a vast array of practical simulation problems can be viewed and solved.

9.1 Systems in Gaussian Noise

In this section we present a general simulation methodology to use for complex highly reliable stochastic systems operating in the presence of Gaussian noise or disturbance. Gaussian noise is the most common assumption used for the ambient thermal background in which most radio frequency communications systems have to operate.

We first recall some results from the large deviation theory of Gaussian random vectors. Let N be a d-dimensional Gaussian random vector, with mean m and covariance matrix K. The relevant rate function $I(\cdot)$ for this setting is

$$I(x) = \frac{1}{2}(x-m)^T K^{-1}(x-m).$$

We then know that (see, e.g., Example 3.2.3)

$$\limsup_{n\to\infty} \frac{1}{n} \log P\big(m + \frac{1}{\sqrt{n}}(N-m) \in F\big) \leq - \inf_{x\in F} I(x)$$

for each closed set $F \subset \mathcal{R}^d$ and

$$\liminf_{n\to\infty} \frac{1}{n} \log P\big(m + \frac{1}{\sqrt{n}}(N-m) \in O\big) \geq - \inf_{x\in O} I(x)$$

for each open set $O \subset \mathcal{R}^d$.

In applications, in the vast majority of cases, we can take our set of interest to be closed (or open) by just including (or not) the boundary of where a threshold decision is to be made without affecting the overall symbol probability of error. Hence we assume that for the sets of interest A in our applications that $\bar{A} = \bar{\mathring{A}}$, where $\bar{A}$ denotes the closure of the set A and $\mathring{A}$ denotes the interior of the set A. Hence

$$\lim_{n\to\infty} \frac{1}{n} \log\Big(P\big(m + \frac{1}{\sqrt{n}}(N - m) \in A\big)\Big) = -\inf_{x\in A} I(x) = -I(x_A).$$

A point x_A where the minimization of the rate function actually occurs is called a *minimum rate point* of the set A. Note that there may be many minimum rate points depending on the shape of set A. The set of points $B_a = \{x : I(x) = a\}$ is of course the boundary of a hyperellipsoid in $\mathcal{R}^d$. If $\inf_{x\in A} I(x) = a$, all of the points where the boundary of A intersects B_a are minimum rate points. The number of such points could be one or several or countably infinite or even uncountably infinite. Without some prior knowledge of the set, it can be very difficult to say beforehand how many minimum rate points exist. For example, if the set A is convex, there can only be one such minimum rate point.

Consider the system simulation problem depicted in Fig. 9.1. Here we suppose that a Gaussian noise vector N is input to a communication receiver along with an independent message symbol S. The receiver performs some operations on these inputs and outputs the function $h(N, S) \in \mathcal{R}^{d'}$. Depending on where in $\mathcal{R}^{d'}$ this d'-dimensional vector lies, we decode the symbol. In other words, the receiver partitions the space $\mathcal{R}^{d'}$ into disjoint sets $\{D_i\}$. If $h(N, S) \in D_i$, the receiver announces S_i as the transmitted symbol (or equivalently announces i, the subscript of the symbol). Note that without loss of generality (due to our very general system model) we can just assume that the noise N is zero mean. (If not, just modify the system model so as to add the mean value in as part of its operation.)

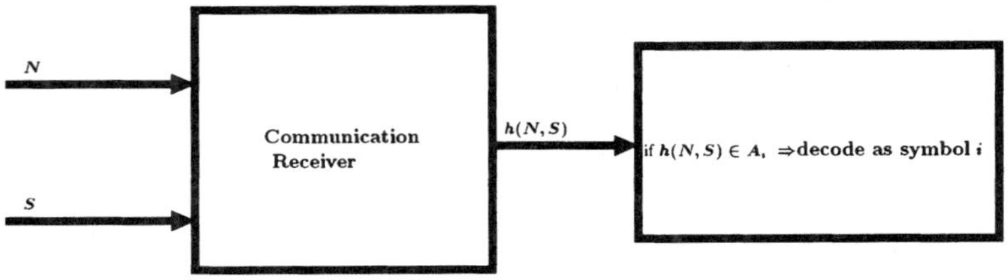

Fig. 9.1. A generic digital communication model.

An error occurs whenever symbol j is decoded as symbol i where $i \neq j$. To evaluate the average bit error rate, we need to estimate $p_{ji} = P\big(\text{announce symbol } j | \text{symbol } i \text{ transmitted}\big)$ for all pairs $i \neq j$. One of our underlying assumptions is that we are dealing with a highly reliable system and hence the probabilities $\{p_{ji}\}$ are very small. Hence we assume that our zero mean noise vector N can be written as

$$N = \frac{G}{\sqrt{n}}, \tag{9.1}$$

where G is some fixed zero mean Gaussian vector with covariance matrix K, multidimensional probability density $p(\cdot)$, and n is a large integer. (Note that the probability density of N is thus $p(\cdot\sqrt{n})\sqrt{n}^d$). At first sight, this inverse square root of n scaling may appear strange. To explain the scaling, note that Cramér's Theorem deals with the large deviations of an average of i.i.d. random variables such as $(\sum_1^n G_i)/n$. However, if the $\{G_i\}$ are Gaussian, then $(\sum_1^n G_i)/n$ has exactly the same distribution as $G_1/\sqrt{n}$. Thus $G_1/\sqrt{n}$ obeys exactly the same large deviation principle as does the sample average.

Given that the symbol transmitted is independent of the noise, it is true that there exists a set $A_{ji} \subset \mathcal{R}^d$ such that

$$\rho_{ji,n} = P\big(N \in A_{ji}\big).$$

The significance of the set A_{ji} is that, if the noise vector is in this set and if symbol i was transmitted, then the receiver will announce j, and thus create a certain type of symbol error.

Instead of directly simulating the noise vector G with density p and scaling it by dividing by $\sqrt{n}$, we use importance sampling and simulate another noise vector Y_n with density q_n. The form of our estimator is thus

$$\hat{\rho}_{ji,n} = \frac{1}{k}\sum_{l=1}^{k} 1_{\{Y_n^{(l)} \in A_{ji}\}} \frac{p(\sqrt{n}Y_n^{(l)})\sqrt{n}^d}{q_n(Y_n^{(l)})},$$

where $Y_n^{(l)}$ is the lth independent sample drawn from the q_n density.

The variance of this estimate goes to zero as $1/k$ as the number of sample runs k increases (since it is just an average of k independent random variables). As always, we wish to investigate the variance behavior as a function of n.

Since it is an unbiased estimate, we can write

$$k\,Var(\hat{\rho}_{ji,n}) = F_{q_n} - \rho_{ji,n}^2,$$

where

$$F_{q_n} = \mathbb{E}_{q_n}\left[1_{\{Y_n \in A_{ji}\}}\left(\frac{p(\sqrt{n}Y_n)\sqrt{n}^d}{q_n(Y_n)}\right)^2\right],$$

where we note that F_{q_n} is the just the mean square value of one of the summand terms in the estimator.

From our large deviation theorem, we know that

$$\lim_{n\to\infty} \frac{1}{n} \log \rho_{ji,n} = - \inf_{x\in A_{ji}} I(x) = -I(x_{A_{ji}}),$$

where

$$I(x) = \frac{1}{2} x^T K^{-1} x.$$

Hence $\rho_{ji,n}^2$ goes to zero exponentially in n with rate $2I(x_{A_{ji}})$. Since the variance is always nonnegative, we must have that the rate with which F_{q_n} goes to zero (if it does at all) must be less than or equal to $2I(x_{A_{ji}})$. If F_{q_n} goes to zero at exactly the rate $I(x_{A_{ji}})$, then of course the estimator is *efficient*.

In the next section we address the problem of finding efficient estimators in this Gaussian setting.

9.1.1 Efficient Estimators for Gaussian Disturbed Systems

Of course, in general there is no unique sequence of efficient estimators. If we impose enough conditions on what sort of bias distributions will be allowed, sometimes we can specify some uniquely optimal efficient estimator. We consider a class of efficient estimators based upon a generalization of the mean shifting strategy. The mean shift strategy says to choose the biasing distribution as a simple mean shift of the original distribution. We do something similar except that we employ "random" mean shifts, where the means are distributed over the level set of the rate function.

Recall that we wish to estimate $\rho_{ji,n}$ using the estimator $\hat{\rho}_{ji,n}$. The situation is much complicated by the question of the number of important points (minimum rate points and others). In the setting where there are only a finite number of important points $\nu_1, \nu_2, \ldots \nu_m$, we have given formulae for efficient simulation sequences. In the Gaussian setting, we have proved that the following choice is efficient,

$$q(x) = \sum_{l=1}^{m} p_l p((x - \nu_l)\sqrt{n})\sqrt{n}^d,$$

where $\{p_l\}$ is an arbitrary probability vector.

In the setting of an infinite number of important points, all we know is that efficient simulation sequences exist. We have no systematic way to generate them. The obvious conjecture is that if we replace $\{p_l\}$ by some arbitrary probability distribution with support exactly equaling the set of important points, we will have an efficient sequence of biasing distribution. Unfortunately we still cannot say whether even this particular choice is always efficient.

In the small Gaussian setting of this chapter we can indeed come up with a universal optimal choice. Consider an arbitrary closed set A and denote $\min_{\{x\in A\}} I(x) = I(A) = I$. We make no further specifications on the set. Hence the set could have an uncountable number of important points. We now give a proof of the following theorem.

Theorem 9.1.1. *The sequence of bias probability densities given by*

$$q_n(x) = p(x\sqrt{n})\sqrt{n}^d \exp(-nI)\frac{\Gamma(\frac{d}{2})\sqrt{2}^{(d/2)-1}}{\sqrt{I}^{(d/2)-1}} \times \frac{I_{(d/2)-1}(n\sqrt{2I}\|K^{-1/2}x\|)}{(n\|K^{-1/2}x\|)^{(d/2)-1}}$$

is efficient, where $I_\nu(\cdot)$ is a modified Bessel function.

Remark 9.1.1. Despite its formidable appearance, obtaining samples for this distribution is actually quite simple. In the proof we show that it is obtained as the result of randomly mean shifting our original Gaussian noise vector over the surface of a hyperellipsoid. To obtain a sample, first generate U, a uniform random variable over the surface of the unit radius d-dimensional hypersphere centered at the origin. One way to generate U is to generate a d-dimensional Gaussian sample Z with identity covariance matrix Id. Then return $U = Z/\|Z\|$. (Note that any random variable with a spherically symmetric distribution can be used instead of Z for this purpose.) We then generate a sample of the noise distribution N and return $Y = N + \sqrt{2I}K^{1/2}U$. Y will have the density $q_n(\cdot)$.

Note that the *only* information about the set contained in the $\{q_n\}$ sequence is the value I!

Proof. We suppose that $I(x) = x^T K^{-1}x/2$. K^{-1} is symmetric as is $K^{-1/2}$. We suppose that $A \subset \mathcal{R}^d$ is closed and $\inf_{x\in A} I(x) = I(A) = I$. Let U be a random variable uniform on the surface of the d-dimensional unit radius hypersphere S_d. Denote the area of the surface as A_d. Let us define a biasing distribution as

$$q_n(x) = \frac{1}{A_d}\int_{S_d} p((x - \sqrt{2I}K^{1/2}u)\sqrt{n})\sqrt{n}^d du,$$

which corresponds to randomly shifting the mean value of the distribution of G to points on the hyperellipsoid $\{x : I(x) = I\}$. Now for $\theta \in \mathcal{R}^d$, define $M(\theta) = \mathbb{E}[\exp(\langle\theta, G\rangle)] = \exp\left(\frac{1}{2}\theta^T K\theta\right)$ which is the *moment generating function* of the G random variable. It is known that $I(x) = \sup_\theta[\langle\theta, x\rangle - \log M(\theta)] = \langle\theta_x, x\rangle - \log M(\theta_x)$, where $\theta_x = K^{-1}x$ and $M(\theta_x) = \exp(x^T K^{-1}x/2)$. It is easy to verify that

$$
\begin{aligned}
&p((x-\sqrt{2I}K^{1/2}u)\sqrt{n})\sqrt{n}^d\\
&= p(x\sqrt{n})\sqrt{n}^d \exp\big(n[\langle\theta_u^*, x\rangle - \log(M(\theta_u^*))]\big),
\end{aligned}
$$

where $\theta_u^* = \sqrt{2I}K^{-1/2}u$ and $\log(M(\theta_u^*)) = Iu^Tu$. Hence

$$
\begin{aligned}
&q_n(x)\\
&= \frac{1}{A_d}\int_{S_d} p((x-\sqrt{2I}K^{1/2}u)\sqrt{n})\sqrt{n}^d du\\
&= p(x\sqrt{n})\sqrt{n}^d \frac{1}{A_d}\int_{S_d} \exp\big(n[\langle\theta_u, x\rangle - \log(M(\theta_u))]\big)du\\
&= p(x\sqrt{n})\sqrt{n}^d \frac{\exp(-nI)}{A_d}\int_{S_d} \exp\big(n[\langle\sqrt{2I}K^{-1/2}u, x\rangle]\big)du
\end{aligned}
$$

Note that $\langle\sqrt{2I}K^{-1/2}u, x\rangle = \sqrt{2I}\langle u, K^{-1/2}x\rangle$. Also due to the spherical symmetry of U, we have that $n\langle U, K^{-1/2}x\rangle$ is equal in distribution to $n\|K^{-1/2}x\|U_1$, where U_1 is the one of the marginal random variables of U; that is, $U = (U_1, U_2, \ldots, U_d)$. The probability density of U_i, $f_U(u)$ is known to be [27, Theorem 3.1]

$$
f_U(u) = \frac{\Gamma(\frac{d}{2})}{\Gamma(\frac{d-1}{2})\sqrt{\pi}}(1-u^2)^{(d/2)-(3/2)} \quad -1 \le u \le +1.
$$

Therefore

$$
\begin{aligned}
&\frac{1}{A_d}\int_{S_d} \exp\big(n[\langle\sqrt{2I}K^{-1/2}u, x\rangle]\big)du\\
&= \frac{1}{A_d}\int_{S_d} \exp(n\sqrt{2I}\langle u, K^{-1/2}x\rangle)du\\
&= \mathbb{E}[\exp(n\sqrt{2I}\langle U, K^{-1/2}x\rangle)]\\
&= \mathbb{E}[\exp(n\sqrt{2I}U_1\|K^{-1/2}x\|)]\\
&= \int_{-1}^{1} \exp(n\sqrt{2I}u\|K^{-1/2}x\|)f_U(u)du.
\end{aligned}
$$

now change variables using $u = \cos(\theta)$

$$
\begin{aligned}
&= \frac{\Gamma(\frac{d}{2})}{\Gamma(\frac{d-1}{2})\Gamma(\frac{1}{2})}\\
&\times \int_0^{\pi} \exp(n\sqrt{2I}\|K^{-1/2}x\|\cos(\theta))\sin^{d-2}(\theta)d\theta\\
&= \frac{\Gamma(\frac{d}{2})\sqrt{\pi}\sqrt{2}^{(d/2)-1}}{\Gamma(\frac{1}{2})\sqrt{I}^{(d/2)-1}}\frac{I_{(d/2)-1}(n\sqrt{2I}\|K^{-1/2}x\|)}{(n\|K^{-1/2}x\|)^{(d/2)-1}}
\end{aligned}
$$

from [34, Eq. 8.431 3] we obtain

$$= b\frac{I_{(d/2)-1}(n\sqrt{2I}\|K^{-1/2}x\|)}{(n\|K^{-1/2}x\|)^{(d/2)-1}} \tag{9.2}$$

where $I_\nu(\cdot)$ is a modified Bessel function. (We apologize for the abuse of notation. $I(\cdot)$ is the large deviation rate function; $I_\nu(\cdot)$ is the modified Bessel function; I is the numerical value of the large deviation rate function at a minimum rate point of the set A. We plead that this is standard notation and that it would be possibly even more confusing to change it.) We now have the form of the biasing distribution that we stated in the theorem.

$$q_n(x) = p(x\sqrt{n})\sqrt{n}^d \exp(-nI)b\frac{I_{(d/2)-1}(n\sqrt{2I}\|K^{-1/2}x\|)}{(n\|K^{-1/2}x\|)^{(d/2)-1}}.$$

To verify that this choice is efficient, we have to compute

$$\begin{aligned} F_{q_n} &= \int_A \left(\frac{p(y\sqrt{n})\sqrt{n}^d}{q_n(y)}\right)^2 q_n(y)dy \\ &= \frac{\exp(2nI)}{b^2}\int_A \frac{(n\|K^{-1/2}x\|)^{d-2}}{I^2_{(d/2)-1}(n\sqrt{2I}\|K^{-1/2}x\|)} q_n(x)dx. \end{aligned}$$

Since $\inf_{x\in A} I(x) = I$, we have that

$$\begin{aligned} A &\subset \{x : \frac{1}{2}x^T K^{-1}x \geq I\} \\ &= \{x : \frac{1}{2}\langle K^{-1/2}x, K^{-1/2}x\rangle \geq I\} \\ &= \{x : \|K^{-1/2}x\|^2 \geq 2I\} \\ &= \{x : \|K^{-1/2}x\| \geq \sqrt{2I}\} \\ &= C. \end{aligned}$$

Hence (since the Bessel function $I_\nu(x)$ is monotonically increasing in x (this can be deduced from the fact that in [2, p. 375, Eq. 9.6.10] is given a power series for this function with all positive coefficients)),

$$\begin{aligned} &F_{q_n} \\ &\leq \frac{\exp(2nI)}{b^2}\int_C \frac{(n\|K^{-1/2}x\|)^{d-2}}{I^2_{(d/2)-1}(n\sqrt{2I}\|K^{-1/2}x\|)} q_n(x)dx \\ &\leq \frac{\exp(2nI)}{b^2 I^2_{(d/2)-1}(n\sqrt{2I}\|K^{-1/2}x\|)}\int_C (n\|K^{-1/2}x\|)^{d-2} q_n(x)dx. \end{aligned}$$

It is easy to verify that the integral has at most polynomial growth in n as n grows large. Also for large arguments we have that the Bessel function

$$I_\nu(z) \approx \frac{\exp(z)}{\sqrt{2\pi z}}.$$

Hence, with a little algebra, we get

$$\limsup_{n\to\infty} \frac{1}{n} \log F_{q_n} \le -2I.$$

Since the variance is greater than zero, we must have that $\liminf_{n\to\infty} \frac{1}{n} \log F_{q_n} \ge -2I$ and hence we deduce

$$\lim_{n\to\infty} \frac{1}{n} \log F_{q_n} = -2I.$$

Therefore, this choice of biasing distribution sequence is efficient in the small Gaussian setting regardless of the number (one, several, countably infinite, or uncountably infinite) of minimum rate points of the set. □

Example 9.1.1. Let $X = (X_1, X_2)$ be a two-dimensional Gaussian random variable, with independent standard Gaussian marginals. Suppose that we are interested in $\rho = P(X \in A)$, where A is the complement of a disk of radius 10 centered at the origin. Every point on the boundary of the circle of radius 10 is thus a minimum rate point. It is easy enough to analytically compute the probability in this case, giving

$$\begin{aligned}\rho &= \int_{10}^{\infty} z \exp(-z^2/2) dz \\ &= \exp(-50) \approx 1.93 \times 10^{-22}.\end{aligned}$$

We of course wish to simulate this probability as a test of our theory developed above. We use the density given in the previous theorem, with $n = 1,\ d = 2,\ K = Id,\ I = 50$, or

$$q(x) = \exp(-\|x\|^2/2) \frac{\exp(-50)}{2\pi} I_0(10\|x\|).$$

To generate samples from this distribution, we first generate samples uniform on the boundary of the disk of radius 10 as $U = (10\cos(\theta), 10\sin(\theta))$, where θ is a uniform $[0, 2\pi]$ random variable. We then generate $Y = (Y_1, Y_2)$ as $Y = X - U$. Our estimate then appears as

$$\hat{\rho} = \frac{1}{k} \sum_{l=1}^{k} 1_{\{\|Y^{(l)}\| > 10\}} \frac{\exp(50)}{I_0(10\|Y^{(l)})\|}.$$

Using the standard random number generators of Matlab, we chose k =20,000 and obtained a value $\hat{\rho} = 1.8962 \times 10^{-22}$, giving an error of less than 2%.

We should remark that there is a degree of freedom here that we haven't really discussed. We chose $n = 1, K = Id$, but we could have chosen n to be any integer and $K = nId$. We get exactly the same estimators no matter what the choice of n is. What is happening is that we have an asymptotic theory based upon large n. However, any one simulation problem has a fixed noise variance that we need to emulate. We have only assumed that our noise vector N can be expressed as $N = G/\sqrt{n}$, where G has covariance K. Clearly we can choose n to be whatever we desire as long as we scale K appropriately also.

Example 9.1.2 (Simulation of ISI and the Linear Problem). Suppose that we have a digital signal corrupted by an intersymbol interference channel and white Gaussian noise. The received sample appears as

$$R_m = B_m - \sum_{k=1}^{l} a_k B_{m-k} + N_m,$$

where $B_m \in \{-1, 1\}$ is the information bit at time m, and N_m is the i.i.d. normal, mean zero, variance σ^2, noise sample at time m. This signal is then passed through an inverting filter to give

$$Y_m = B_m + \sum_{k=0}^{L} i_k N_{m-k}.$$

We assume here that a perfect inverting FIR filter exists, which of course cannot mathematically be true. We assume that for large enough L, this is a good approximation.

Therefore to simulate an error, we wish to bias the noise samples $N = (N_m, N_{m-1}, \dots, N_{m-L})$. Suppose for now that $B_m = +1$. We wish to mean shift N to the minimum rate point (unique) of the error set

$$A = \{n : \sum_{k=0}^{L} i_k n_{m-k} < -1\}.$$

In this setting $I(n) = \frac{1}{2\sigma^2}\|n\|^2$. To find the minimum rate point we need to minimize $\sum_{k=0}^{L} n_{m-k}^2$ subject to $\sum_{k=0}^{L} i_k n_{m-k} < -1$. We do this with Lagrange multipliers by finding the critical point of $J(n) = \sum_{k=0}^{L} n_{m-k}^2 + \lambda \sum_{k=0}^{L} i_k n_{m-k}$, which leads to

$$n_{m-j} = -\frac{\lambda}{2} i_j.$$

Solving for the constraint gives us

$$n_A = -\frac{i}{\|i\|^2}.$$

Therefore we should bias our input random noises so that they have n_A as their mean value.

This problem (in a much more general setting) is considered in [80]. Here our solution is presented from a large deviation theory perspective as opposed to a direct minimization of the estimator variance. The end results match up well in the simple cases where they can be compared.

Example 9.1.3 (The Decision Feedback Equalizer). We are interested in finding the error probability caused by a decision feedback equalizer operating in an intersymbol interference channel in the presence of white Gaussian noise. As in the previous section, we receive the following bit sample

$$R_m = B_m + \sum_{k=1}^{l} a_k B_{m-k} + N_m,$$

where $B_m \in \{-1, 1\}$ is the information bit sent at time m, the $\{a_k\}$ are the ISI channel coefficients, and $\{N_m\}$ are the i.i.d. Gaussian, zero mean, variance σ^2, noise samples. We then use our past bit decisions and our current channel model (which is updated using an adaptive filtering algorithm) to try to cancel the current samples ISI, giving us the output

$$\hat{R}_m = R_m - \sum_{k=1}^{l} \hat{a}_k B_{m-k}.$$

We then take the sign of this result to get our current bit decision,

$$\hat{B}_m = sign(\hat{R}_m).$$

For simplicity in the following analysis, we assume that the adaptive system is in steady state and locked to the true channel model (which in particular implies that $a_k = \hat{a}_k \ \forall \ k$).

The problem with these types of receiver structures is that they employ a type of nonlinear feedback. If an error is made in a previous bit decision, the receiver is far more likely to make another error since the ISI instead of being perfectly cancelled is actually made worse. This causes a "burstiness" to the errors appearing in these systems. Simulating these systems is far more complicated due to this characteristic behavior (see, e.g., Al-Qaq et al. [1]).

What we propose to do is to look at d transmitted bits at a time. We denote a given vector of d transmitted bits as $b = (b_1, b_2, \ldots, b_d)$. We then consider the $2^d - 1$ possible error patterns that can occur for those d bits. Any one such error pattern is denoted as $e = (e_1, e_2, \ldots, e_d)$. We propose to compute the probability of each error pattern.

For a given transmitted b and any e, there is a certain set $A_{be} \subset \mathcal{R}^d$ such that if the noise vector $n = (n_1, n_2, \ldots, n_d) \in A_{be}$, the error pattern e will be produced. We wish to find the minimum rate points of this *set*. Hence we need to to minimize

$$I(m) = \frac{1}{\sigma^2} \sum_{i=1}^{d} m_i^2 \quad \text{with respect to} \frac{|b - \hat{b}|}{2} = e,$$

where again b is the transmitted bit sequence, $\hat{b}$ the estimated sample sequence at the receiver, and e is the error pattern of which we wish to simulate the probability.

Note that, to achieve the error pattern e, we need

$$\hat{b}_i = b_i(1 - 2e_i),$$

The estimated received bit appears as

$$\hat{b}_i = sign(b_i + \sum_{k=1}^{d} a_k b_{i-k} - \sum_{k=1}^{d} \hat{a}_k \hat{b}_{i-k} + m_i)$$

and with $a_k = \hat{a}_k$ we get

$$\hat{b}_i = sign(b_i + \sum_{k=1}^{d} a_k (b_{i-k} - \hat{b}_{i-k}) + m_i)$$

$$\hat{b}_i = sign(b_i + 2\sum_{k=1}^{f} a_k b_{i-k} e_{i-k} + m_i).$$

We denote the *channel offset* at the receiver as $o = (o_1, o_2, \dots, o_d)$, where

$$o_i = b_i + 2\sum_{k=1}^{f} a_k b_{i-k} e_{i-k}.$$

This implies

$$\hat{b}_i = sign(o_i + m_i).$$

To minimize $I(m)$ subject to the constraint, we can take $m_i = 0$ if

$$\hat{b}_i = b_i(1 - 2e_i) = sign(o_i);$$

otherwise we must take

$$m_i = -o_i,$$

which leads to

$$m_i = -o_i \frac{\left| sign(o_i) - \hat{b}_i \right|}{2}.$$

This choice of m, denote it as m_{be}, is the (unique) minimum rate point of the set A_{be}. To simulate, we then would employ white Gaussian noise with mean m and individual component variance σ^2 to simulate the ISI channel.

After computing the minimum rate point m_{be} for each data and error pattern of b and e, we should first recognize that those patterns with the smallest norm for m_{be} will have the largest contributions to the overall symbol error rate. For a given data and error pattern, our estimator appears as

$$\hat{P}(b,e) = \frac{1}{k}\sum_{i=1}^{k} 1_{\{\hat{b^{(i)}}=b^{(i)}(1-2e^{(i)})\}} \times \prod_{l=1}^{d} \exp\Big(\frac{(n_l^{(i)} - m_{be,l})^2 - (n_l^{(i)})^2}{2\sigma^2}\Big).$$

The overall bit error rate estimate is

$$\text{BER} = \sum_{\forall\ b \forall\ e} \frac{\hat{P}(b,e)\sum_{j=1}^{d} e_j}{2^d d}.$$

To test our theory above we choose a channel with ISI coefficients $a = (\frac{1}{2}, -\frac{1}{4}, \frac{1}{10})$ and white Gaussian noise with zero mean and variance $\sigma^2 = 0.04$. We are interested in the probability, that a specific bit pattern produces a specific error pattern. We choose $b = (-1\ 1\ \ -1\ 1)$ and $e = (1\ 0\ 1\ 1)$. We compute the channel offset as

$$o = \left(-1, 0, -\frac{1}{2}, -\frac{1}{5}\right).$$

Knowing the distribution of the Gaussian noise, we can compute (in closed form) the probability that each decoded bit follows the specific error pattern giving us the overall probability for this (b, e) combination as

$$\begin{aligned} P(b,e) &= P(\hat{b}_1 = 1)P(\hat{b}_2 = 1)P(\hat{b}_3 = 1)P(\hat{b}_4 = -1) \\ &= P(n_1 > 1)P(n_2 > 0)P(n_3 > 0.5)P(n_4 < 0.2) \\ &= (2.867 \times 10^{-7})(0.5)(6.21 \times 10^{-3})(0.8413) \\ &= 7.488 \times 10^{-10}. \end{aligned}$$

The minimum rate point of the set is calculated as $m_{be} = (1\ 0\ 0.5\ .2)$. Now we simulate the ISI channel using Gaussian noise with mean m_{be} and variance $\sigma^2 = 0.04$. Using $k = 25,000$, we obtain a probability estimate of $\hat{P}(b,e) = 7.632 \times 10^{-10}$, which is an error of less than 2% from the true value.

To find the overall bit error ratio we use the same scheme to compute the probability of all combinations of bit and error patterns. Supposing that all data bit patterns occur with the same probability, we calculate that the actual bit error rate is $BER = 3.6336 \times 10^{-7}$.

Simulating all (b, e) combinations by shifting to the minimum rate point and using $k = 25,000$ simulation runs for each combination, gives us an estimated bit error rate of $\hat{BER} = 3.6401 \times 10^{-7}$, giving an error of approximately 0.2% from the true value.

Example 9.1.4. To present a problem with more multidimensional character, we consider trying to estimate

$$\rho = P\Big(\sum_{j=1}^{3} \cos(N_j) + \sum_{k=1}^{3}\sum_{l=1}^{3} \cos(N_k - N_l) + N_4 > 13\Big),$$

where $N_1, N_2, \ldots, N_4$ are mean zero, variance 1/100 i.i.d. Gaussian random variables. Hence we have a four-dimensional biasing problem. We choose an identity covariance matrix for the random vector G in (9.1), or $K = Id$. To match up with our problem then we must choose $n = 100$ to give the correct variance for the N vector. We should note that the point $(g_1, g_2, g_3, g_4) = (0, 0, 0, 1)$ gives $\sum_{j=1}^{3} \cos(g_j) + \sum_{k=1}^{3} \sum_{l=1}^{m-1} \cos(g_k - g_l) + g_4 = 13$. Furthermore, $I(g) = g^T g/2 = 1/2$. Thus we know that I need not be chosen bigger than 1/2. We choose (after a few trials) the value of $I = .45$.

In Fig. 9.2, we plot out the estimate as a function of the number of runs for up to 8 million runs. The final value 7.6155×10^{-25} is (empirically) accurate to ±20% with 95% probability. Of course a direct Monte Carlo simulation of this probability would have required on the order of 10^{28} simulation runs.

Example 9.1.5 (8-PSK Revisited). In Example 5.2.14, we considered estimating the symbol error probability of an 8-PSK system. Since there were two minimum rate points, we utilized a mixture of Gaussian distributions as our efficient choice.

In this setting we can also use the universal sampling distribution for Gaussian problems. To utilize this distribution, we need only know the value of the rate function at the minimum rate point. We find that $I = \frac{2}{N_0} r^2$, where $r = \sqrt{E/2}\,\sin(\pi/8)$ $(n = 1)$. In Fig. 9.3 we see a typical simulation output for the $E/N_0 = 50$ case. We obtain the estimate of 1.3362×10^{-4} with a $k^* =$18,766 (for 5% error with probability .90).

Comparing the two methods, we see that we need about four times as many simulation runs to use the universal strategy as for when we used the also efficient 2 mean shift method.

Example 9.1.6. Let $N = (N_1, N_2, \ldots, N_5)$ be a vector with i.i.d. mean zero variance one normal components. Define

$$A = N_1^2 + N_2^2 + N_3^2$$

and

$$B = N_4^2 + N_5^2 + 1.$$

We are interested in

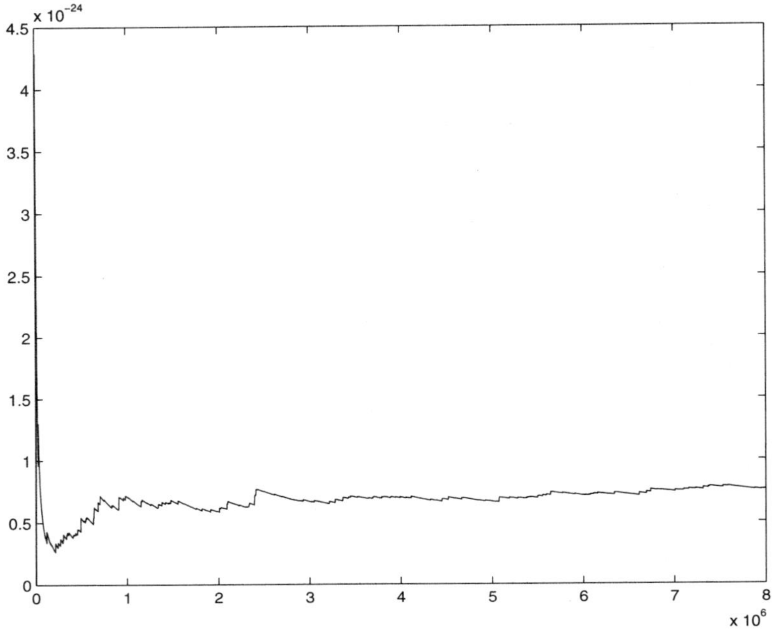

Fig. 9.2. Estimator output as a function of simulation runs.

$$\rho = P(\frac{A}{B} > t).$$

Thus our set of interest can be written as

$$\begin{aligned} E &= \{x \in \mathcal{R}^5 : \frac{x_1^2 + x_2^2 + x_3^2}{x_4^2 + x_5^2 + 1} > t\} \\ &= \{x \in \mathcal{R}^5 : x^T M x > 1\}, \end{aligned}$$

where $M = diag(\frac{1}{t}\frac{1}{t}\frac{1}{t} - 1 - 1)$. We seek to minimize $I(x)$ over the boundary of E. A moment's reflection reveals that this minimum is given by any $x \in \mathcal{R}^5$ that satisfies $x^T x = t$ with $x_4 = x_5 = 0$. Thus $I = \frac{t}{2}$. The estimate for $d = 5$ then appears as

$$\hat{\rho}(k) = \frac{1}{k}\sum_{j=1}^{k} 1_{\{Y^{(j)T} M Y^{(j)} > 1\}} \frac{\exp\left(\frac{t}{2}\right)\frac{4}{3\sqrt{\pi}}\left(\|Y^{(j)}\|\sqrt{\frac{t}{4}}\right)^{3/2}}{I_{3/2}(\|Y^{(j)}\|\sqrt{t})}.$$

We implemented this estimator and obtained the following results.

For the case of $t = 9.28$, we obtained $k^* =$109,434 (5% error with probability .95) with estimate $\hat{\rho}(k^*) = 0.0027$. The set was "hit" by 4.16% of the variables generated under the sampling distribution.

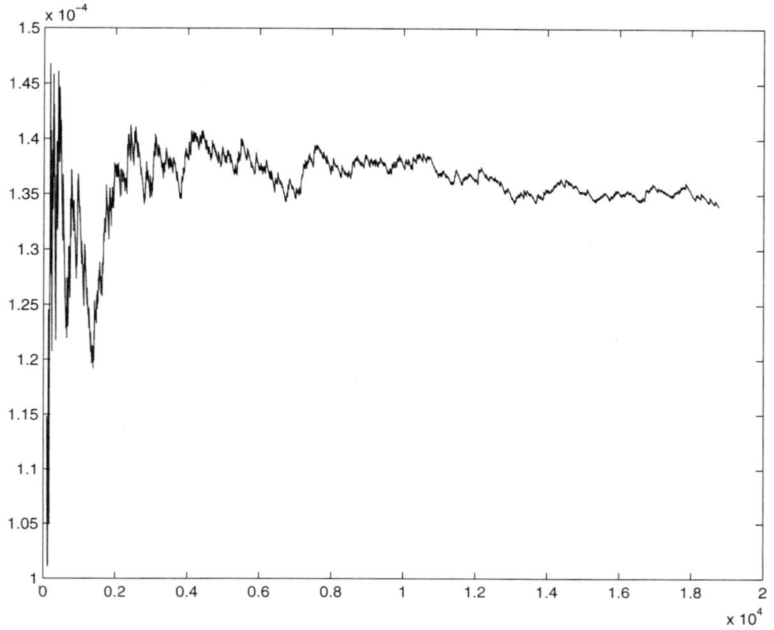

Fig. 9.3. A plot of the estimator $\hat{\rho}(k)$ as a function of the number of simulation runs k ($E/N_0 = 50$). The simulation was terminated at the computed value of $k^* = $18,766 (for a 5% error with .90 probability criterion). The estimate given of the probability is 1.3362×10^{-4}.

For the case of $t = 30$, we had $k^* = $1,714,165 (5% error with probability .95) with estimate $\hat{\rho}(k^*) = 4.735 \times 10^{-8}$. We had a hit rate on the set of 0.77%.

An observation that one can make in this problem is that the solution to the optimization problem tells us that the fourth and fifth dimensions do not need to be biased ($x_4 = x_5 = 0$). Thus we can reduce the dimensionality of the problem by not biasing those dimensions and using a universal simulation distribution on the first three dimensions. In other words, we can randomly shift the distribution only in the first three dimensions (x_1, x_2, x_3). For the other two dimensions (x_4, x_5) we just use the unbiased standard normal distributions of the original problem. Thus $Y = (Y_1\ Y_2\ Y_3\ N_4\ N_5)$. For this case our estimate appears as

$$\hat{\rho}(k) = \frac{1}{k}\sum_{j=1}^{k} 1_{\{Y^{(j)T}MY^{(j)}>1\}}$$
$$\times \frac{\exp\left(\frac{t}{2}\right)\frac{2}{\sqrt{\pi}}\left(\sqrt{(Y_1^{(j)})^2+(Y_2^{(j)})^2+(Y_3^{(j)})^2}\sqrt{\frac{t}{4}}\right)^{1/2}}{I_{1/2}(\sqrt{(Y_1^{(j)})^2+(Y_2^{(j)})^2+(Y_3^{(j)})^2}\sqrt{t})}.$$

Implementing this three-dimensional estimator (and keeping the same confidence interval criteria to set k^*), we obtained a substantial reduction in the number of simulation runs.

For example when $t = 9.28$, k^* =28,346, the estimator is $\hat{\rho}(k^*) = 0.0027$ with a hit rate of 17.63%.

For the case when $t = 30$, $k^* =$ 157,725, the estimator is $\hat{\rho}(k^*) = 4.589 \times 10^{-8}$ with a hit rate of 8.75%.

10. Universal Simulation Distributions

Fate laughs at probabilities. *E. Aram*

If I have not seen as far as others, it is because giants were standing on my shoulders. *Anon.*

In the previous chapter, we saw that in the Gaussian setting we could describe completely a sequence of simulation distributions that was efficient and depended on only one scalar parameter, the minimum rate point of the set. In this chapter we consider carrying out these techniques in the general non-Gaussian setting. Due to the special form of the multidimensional Gaussian distribution and its level sets, we could perform a closed form integration of the exponentially shifted distributions over the boundary of the level set. In general, this sort of attack which obtains a closed form solution for the sequence of simulation distributions will not be easily available. In this chapter we present an alternate methodology leading to a new class of simulation distributions, which we call *universal distributions.* We discuss their use and propose and debate some numerical techniques needed for employing them.

10.1 Universal Distributions

The key to proving universal efficiency for a continuous mixture distribution is to get around the problem of trying to find the minimum rate points and/or "important" points of the set E. Instead of shifting probability mass to these points, we instead shift probability mass to the surface of the level set associated with the minimum rate points of the set. Thus the simulation distributions derived in this way do not really depend on the set geometry. They depend only on the scalar value of the rate function at the minimum rate points; that is, $I(E)$. We discover that this "universality" is a two-edged sword, in that (not surprisingly) simulation distributions that are not closely tailored to the set geometry do not perform as well as when we are able to tailor them. We discuss this issue at some length in the next section.

We consider our usual large deviation theory setting. We suppose that the standard assumptions A1 through A4 hold for the sequence of $\mathcal{R}^d$-valued random variables $\{f_n(Y_n)\}$ and thus for a set $E \subset \mathcal{R}^d$ of interest we have

$$\lim_{n\to\infty} P(\{f_n(Y_n)\} \in E) = -I(E) = -\inf_{x\in E}\sup_{\theta}[\langle\theta, x\rangle - \phi(\theta)].$$

We suppose that $0 < I(E) < \infty$. Consider the associated level set

$$L = \{x : I(x) \leq I(E)\}.$$

If the standard assumptions are satisfied, we know that L is a closed, compact, convex set. Let $B = \partial L$ be the boundary (or the frontier) of L. Consider the sequence of simulation distributions given by

$$dQ_n(z) = \left[\int_B \exp\bigl(n[\langle\theta_v, z\rangle - \phi_n(\theta_v)]\bigr)\pi(dv)\right] dP_n(z), \tag{10.1}$$

where π is a probability measure with support on B. We call Q_n a *universal simulation distribution.* Let m_B denote the "surface area measure" on the set B.

Theorem 10.1.1. *Suppose ϕ is strictly convex and $m_B \ll \pi$. Then the sequence of simulation distributions $\{Q_n\}$ given in (10.1) is efficient.*

Proof. Since ϕ is an essentially smooth, closed, strictly convex function, we have by Theorem 26.5 of [68] that the gradient mapping $\nabla\phi : \overset{\circ}{D}_{\phi} \to \overset{\circ}{D}_I$ is one-to-one and continuous in both directions. Furthermore, the convex conjugate function I is an essentially smooth, closed, strictly convex function.

For some positive number r, define $L^r = \{x : I(x) < I(E) + r\}$. L^r is an open set since it is the inverse image of an open set through the continuous function I. Clearly $L \subset L^r \subset D_I$. Now $\overset{\circ}{D}_I$ is the largest open set contained in D_I. Thus $L^r \subset \overset{\circ}{D}_I$. Hence, $L \subset \overset{\circ}{D}_I$.

Let $y \in L \subset \overset{\circ}{D}_I$. Associated with each y is a θ_y such that $I(y) = \langle\theta_y, y\rangle - \phi(\theta_y)$ and also $\nabla\phi(\theta_y) = y$. Consider the inverse map $(\nabla\phi)^{-1} : \overset{\circ}{D}_I \to \overset{\circ}{D}_{\phi}$, defined by $(\nabla\phi)^{-1}(y) = \theta_y$. As stated above, this is a continuous map. Hence, as y varies over the compact set L, θ_y varies continuously over $\overset{\circ}{D}_{\phi}$. The image of a compact set under a continuous map is compact. Since the image is a subset of $\mathcal{R}^d$, in particular this means that the image of L under the inverse map is closed and bounded.

Now for every $y \in L$, $\phi_n(\theta_y) \to \phi(\theta_y)$. We know that as y varies over L, $\|\theta_y\|$ is bounded. Since the ϕ_n are convex functions, they converge uniformly on any bounded subset of $\overset{\circ}{D}_{\phi}$ (by Theorem 10.8 of [68]). Thus, for every $\epsilon > 0$, there exists an N_ϵ such that for all $n > N_\epsilon$,

$$|\phi_n(\theta_y) - \phi(\theta_y)| < \epsilon \quad \forall\ y \in L.$$

Define

$$\mathcal{S}_\epsilon(y) = \{v \in B : \langle\theta_v, (y - v)\rangle > -\epsilon\}.$$

If a convex set (L in our case) has a unique tangent hyperplane at each point on the surface, then the normal to the surface must be a continuous function of the point. Consequently, $\langle \theta_v, y - v \rangle$ is a continuous function of v that vanishes at $v = y$. It follows that $\mathcal{S}_\epsilon(y)$ is nonempty and open and hence has positive surface measure.

By the continuity of the normal vectors on this manifold it is easy to see that $m_B(\mathcal{S}_\epsilon(y)) > 0$ for all $y \in B$. Furthermore, the mapping defined by $y \to m_B(\mathcal{S}_\epsilon(y))$ is continuous. Thus we claim that

$$\inf_{y \in B} m_B(\mathcal{S}_\epsilon(y)) > l_\epsilon > 0.$$

Suppose not. Since B is compact, there exists a sequence $\{z_n\} \subset B$ such that $z_n \to z$, and $m_B(\mathcal{S}_\epsilon(z_n)) \to 0$. Since B is compact, $z \in B$. Also by continuity of the $y \to m_B(\mathcal{S}_\epsilon(y)$ map, we must have that $m_B(\mathcal{S}_\epsilon(z)) = 0$. This is a contradiction and hence the claim is shown.

Note also that we assume $m_B \ll \pi$. One of the characterizations of absolute continuity between measures is the following [7, Theorem 5.10]. For all $l > 0$, there exists a $\delta_l > 0$ such that $\pi(S) < \delta_l$ implies that $m_B(S) < l$. An equivalent statement is, for all $l > 0$, there exists a $\delta_l > 0$ such that $m_B(S) > l$ implies that $\pi(S) > \delta_l$. Thus, for any $l > 0$, there exists a $\delta_l > 0$ such that

$$\inf\{\pi(S) : S \subset B, m_B(S) > l\} > \delta_l > 0.$$

To prove efficiency, we proceed by directly upper bounding F_n for this family of distributions. Let $f_n^{-1}(E) = \{z : f_n(E) \in E\}$. Hence, for $n > N_\epsilon$,

$$\begin{aligned}
&F_n \\
&= \int_{f_n^{-1}(E)} \Big[\frac{dP_n}{dQ_n}(z)\Big]^2 dQ_n(z) \\
&= \int_{f_n^{-1}(E)} \Big[\int_B \exp\big(n[\langle \theta_v, z\rangle - \phi_n(\theta_v)]\big)\pi(dv)\Big]^{-2} dQ_n(z) \\
&= \int_{f_n^{-1}(E)} \Big[\int_B \exp\big(n[\langle \theta_v, v\rangle - \phi_n(\theta_v)]\big) \\
&\quad \times \exp(n\langle \theta_v, z - v\rangle)\pi(dv)\Big]^{-2} dQ_n(z) \\
&\le \int_{f_n^{-1}(E)} \Big[\int_{\mathcal{S}_\epsilon(z)} \exp\big(n[\langle \theta_v, v\rangle - \phi_n(\theta_v)]\big) \\
&\quad \times \exp(n\langle \theta_v, z - v\rangle)\pi(dv)\Big]^{-2} dQ_n(z) \\
&\le \int_{f_n^{-1}(E)} \Big[\int_{\mathcal{S}_\epsilon(z)} \exp\big(n[\langle \theta_v, v\rangle - \phi(\theta_v) - \epsilon]\big) \\
&\quad \times \exp(n\langle \theta_v, z - v\rangle)\pi(dv)\Big]^{-2} dQ_n(z) \\
&\le \exp(2n\epsilon)\exp\big(-2nI(E)\big)
\end{aligned}$$

$$\times \int_{f_n^{-1}(E)} \frac{1}{\left[\int_{\mathcal{S}_\epsilon(z)} \exp(n\langle\theta_v, z-v\rangle)\pi(dv)\right]^2} dQ_n(z)$$

$$\leq \exp(2n\epsilon)\exp\bigl(-2nI(E)\bigr)$$

$$\times \int_{f_n^{-1}(E)} \frac{1}{\left[\int_{\mathcal{S}_\epsilon(z)} \exp(-n\epsilon)\pi(dv)\right]^2} dQ_n(z)$$

$$= \exp(4n\epsilon)\exp\bigl(-2nI(E)\bigr)$$

$$\times \int_{f_n^{-1}(E)} \frac{1}{\left[\int_{\mathcal{S}_\epsilon(z)} \pi(dv)\right]^2} dQ_n(z)$$

$$\leq \exp(4n\epsilon)\exp\bigl(-2nI(E)\bigr)\delta_{l_\epsilon}^{-2} \int_{f_n^{-1}(E)} dQ_n(z)$$

$$\leq \exp(4n\epsilon)\exp\bigl(-2nI(E)\bigr)\delta_{l_\epsilon}^{-2}.$$

Therefore

$$\limsup_{n\to\infty} \frac{1}{n}\log F_n \leq -2I(E) + 4\epsilon.$$

Since $\epsilon > 0$ is arbitrary, we have

$$\lim_{n\to\infty} \frac{1}{n}\log F_n = -2I(E),$$

and the sequence of simulation distributions is efficient. □

10.2 The input formulation is not efficient

We obtain the universal distribution as a mixture of exponential shifts. The mixture parameter can be thought of as being the point on the level set of the large deviation rate function $\{x : I(x) = I(E)\}$. We obtain a representation of the universal distribution by integrating over this mixture parameter as in (10.1) of the previous section. Of course, this is a type of "output" formulation. There is another unbiased importance sampling estimate that we can consider, the so-called "input" formulation. Here we first generate a random sample $V = v$ from the surface of the level set and then generate a sample $Z = z$ with exponential parameter θ_v. The weight function for the pair (v, z) in the importance sampling estimator is given by

$$\exp\bigl(-n[\langle\theta_v, z\rangle - \phi_n(\theta_v)]\bigr)\frac{dm_B}{d\pi}(v),$$

where we assume that $m_B \ll \pi$ (as is needed in Theorem 10.1.1). This weight function operates on the pair of generated random variables (V, Z). It has the big advantage that it is not necessary to integrate out the mixture parameter in closed form (a nontrivial task in most cases). Unfortunately, this "input" estimator is *not* efficient as we see by the following simple example.

Example 10.2.1. Suppose we are interested in estimating

$$\begin{aligned}\rho &= P\Big(\frac{1}{n}\sum_{i=1}^{n} X_i \in (-\infty, -m] \cup [m, \infty)\Big)\\ &= P\big(Y_n \in (-\infty, -m] \cup [m, \infty)\big)\end{aligned}$$

where the $\{X_i\}$ are i.i.d. standard Gaussian $\mathcal{R}$-valued random variables and Y_n is a mean zero, variance $1/\sqrt{n}$ random variable with probability density denoted as $p_n(\cdot)$. Denote the real set $(-\infty, -m] \cup [m, \infty)$ as E. In this one dimensional setting, the "surface" of the level set of interest is just the two points $\{-m, m\}$. We can thus choose π to be a symmetric Bernoulli measure on the two points $\{-1, 1\}$. Our input estimator thus appears as

$$\hat{\rho}_n = \frac{1}{k}\sum_{j=1}^{k} 1_{\{Y_n^{(j)} \in (-\infty,-m]\cup[m,\infty)\}} \frac{p_n(Y_n^{(j)})}{p_n(Y_n^{(j)} + mV)\pi(V)}.$$

We can then calculate

$$\begin{aligned}F_n &\\ &= \sum_{v=\{-1,1\}} \int_E \left(\frac{p_n(x)}{p_n(x+mv)\pi(v)}\right)^2 p_n(x+mv)dx\pi(v)\\ &= 2\exp\Big(\frac{m^2 n}{2}\Big)\\ &\quad\times\left[\int_E \exp(-mxn)p_n(x)dx + \int_E \exp(mxn)p_n(x)dx\right]\\ &= 2\exp(m^2 n)\\ &\quad\times\left[\int_E p_n(x+m)dx + \int_E \exp(mxn)p_n(x-m)dx\right]\\ &\geq 2\exp(m^2 n)\end{aligned}$$

which goes to infinity as $n \to \infty$. Thus the input formulation not only is not efficient but actually performs much worse than a simple direct Monte Carlo estimate!

10.3 An Adaptive Strategy to Increase Hit Rate

When using a universal simulation distribution, the convergence rate of the estimator is often slow due to the low hit rate. The reason for the small hit rate is due to the fact that our set E may intersect the "minimum rate" level set $\{x : I(x) = I(E)\}$ at only a small number of minimum rate points.

We must also keep in mind that the set of consideration E may have a very complicated shape, and it may be difficult, if not impossible, to analytically describe or determine.

Our idea in this section is to propose a methodology to deal with these two problems. The first notion is to find suitable locations near the set E on the level set of rate I, which may or may not be close to the true rate $I(E)$. In order to find these suitable locations, which we call *hit points*, we start the process of generating samples from a universal simulation distribution with parameter I. We tabulate all the values of the samples generated on the level set $\{x : I(x) = I\}$ that produce a hit on the set. Recall that generating a simulation sample has two steps. One, we generate a value on the level set. Then we generate a sample from a distribution that has this value as its exponential shift parameter. In the Gaussian setting, we need only add a zero mean Gaussian random variable to this parameter. This is due to the fact that exponential shifts are equivalent to mean shifts for the normal distribution. We don't tabulate the value if there is no hit. Once we have found a suitable number of hit points, say N, we stop the process and start a new "adaptive" process to estimate ρ_n. Our simulation distribution Q_n is of the form (10.1) with continuous and discrete components where we choose $\pi(x) = \frac{1}{N+1}\sum_{i=1}^{N}\delta(x - \theta_{v_i}) + \frac{1}{N+1}m_B(x)$. The $\{v_i\}$ are the hit points, $\delta(\cdot)$ is the Dirac delta function, and m_B again is the surface area measure on B as defined in the previous section. It is easy to see that $m_B \ll \pi$, so this distribution is indeed efficient when $I = I(E)$.

Example 10.3.1. In this example, we consider letting $p_n(x)$ be a two-dimensional Gaussian density with covariance $K = \frac{\sqrt{5}}{24}id_2$ and $n = 5$. (We purposely choose a small value of n to illustrate a possible problem.) For a set $E = \{x : |x_1| \geq 1\}$, we get results for the "adapted" and non-"adapted" universal simulation distribution with known rate equal to 12, plotted in Fig. 10.1. The k^* plotted are the number of simulation runs needed to achieve a 10% relative error with 95% confidence.

So far we have not discussed the fundamental problem of how we might find the minimum rate $I(E)$ for a given set of interest. In general this is a nontrivial problem. We propose an adaptive rate-finding process driven by the empirical hit rate for our estimator. Ideally, our "adaptive" simulation distribution should have most, if not all, of its probability centered at the minimum rate points of the set E. For n sufficiently large, we should be able to achieve a hit rate of 50% at any of the minimum rate points, because the tangent of the hypersphere of rate I will look essentially flat near each of the minimum rate points.

To start the rate-finding algorithm, we start with an initial guess I_0 of the rate. We run the simulation using the adaptive simulation distribution from above, with the additional constraint that, in order to save simulation time, we quit looking for hit points after a sufficiently long time, say M attempts. The reason for quitting looking for hit points is that if we are too far away from the set E, we will almost never get a hit with our current rate, which

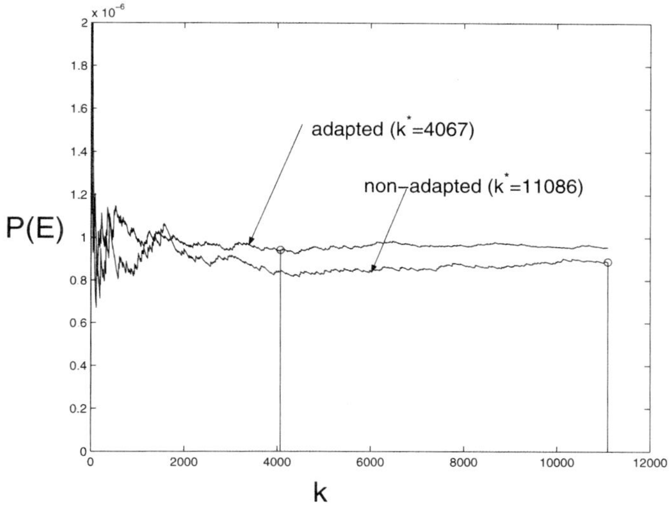

Fig. 10.1. Results of nonadapted universal simulation distribution vs. universal simulation distribution with adaptive location finding with known rate equal to 12. The estimate is plotted as a function of the number of simulation runs, k for the two methods.

would cause our simulation to take an inordinate amount of time to run. So we take the Mth "bad" location, and assume that we will not be able to achieve a 50% hit rate with it, which will cause us to move closer to the true rate $I(E)$. After we have generated a desired N number of locations for the simulation distribution, we run the simulation for a sufficiently long time (say $100-200$ iterations), to estimate the hit rate $\kappa_n(j)$ for our distribution with current rate I_j. If $\kappa_n(j)$ is not 50%, we try to move closer to the true rate $I(E)$ by putting

$$I_{j+1} = I_j + \mu(0.5 - \kappa_n(j))I_j,$$

for a sufficiently small μ, say .01, to avoid over- or undershooting $I(E)$.

For sufficiently large n, we get that the hit rate of our simulation distribution should be nearly zero at a rate less than $I(E)$ and nearly one at a rate greater than $I(E)$. This is because for a lower rate than $I(E)$ we should never expect to hit the set, and for a higher rate (with the adaptive simulation distribution) we should expect to always hit the set. So this iterative process should work very well for sufficiently large n, because once we get close to $I(E)$, we should stay close to it.

In Fig. 10.2 are the results for our above example, plotting the locations $\{v_i\}$ of our distribution against the set E and the hypersphere of rate $I(E) = 12$.

Note that, for this example, we miss the true level set of rate $I(E)$. The reason for this phenomenon is that we have chosen a small value for $n = 5$

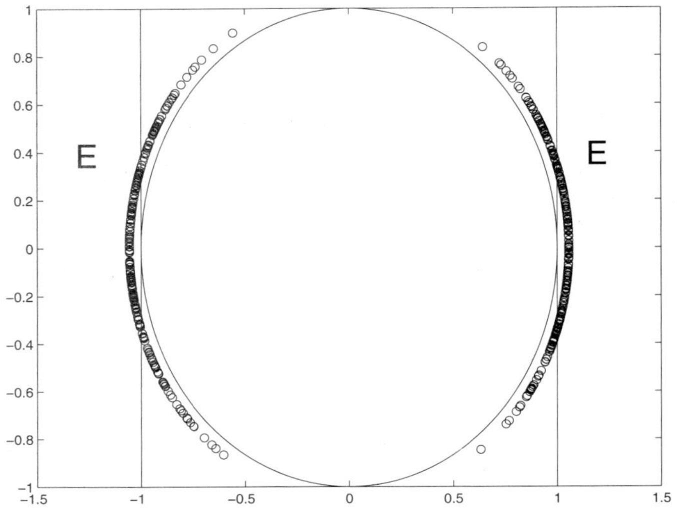

Fig. 10.2. Plot of locations $\{v_i\}$ of the adapted universal simulation distribution with unknown rate, driven by target hit rate of 50 percent. The error set is $E = \{x : |x| \geq 1\}$. The value of I found adaptively was 13.5249. The true value of I is 12, which corresponds to the circle of radius 1.

and are not really operating in the asymptotic regime. This causes a hit rate of less than 50% for each hit point on the true level set. Empirically, we find it is better to use 45% as a choice of optimum hit rate target for our adaptive process. See Fig. 10.3 for the results of using a target hit rate of 45%.

When n is larger, everything works exactly as the asymptotic theory indicates. See Fig. 10.4 for the results of increasing n to 500 in locating the minimum rate points.

Example 10.3.2. Here we give a concrete example of a non-Gaussian setting where we have a closed form expression for the universal distribution. We consider the simple Gaussian mixture $Y_n = \sum_{i=1}^n A_i X_i$, where $\{X_i\}$ are i.i.d. Gaussian vectors with identity covariance matrix and $\{A_i\}$ are i.i.d. random variables taking value a or b with equal probability and $|b| > |a|$.

Since

$$\begin{aligned}\phi_n(\theta) &= \frac{1}{n} \log \mathbb{E}[\exp(\langle \theta, \sum_{i=1}^n A_i X_i \rangle)] \\ &= \log \mathbb{E}[\exp(\langle \theta, A_i X_i \rangle)] \\ &= \log \mathbb{E}[\mathbb{E}[\exp(\langle \theta, AX \rangle)|A]] \\ &= \log \mathbb{E}[\exp(\frac{\|\theta\|^2}{2} A^2)].\end{aligned}$$

Hence $\phi_n(\theta) = \phi(\theta)$ for all n and furthermore is spherically symmetric. We can thus compute $\nabla \phi(\theta)$ via the following argument.

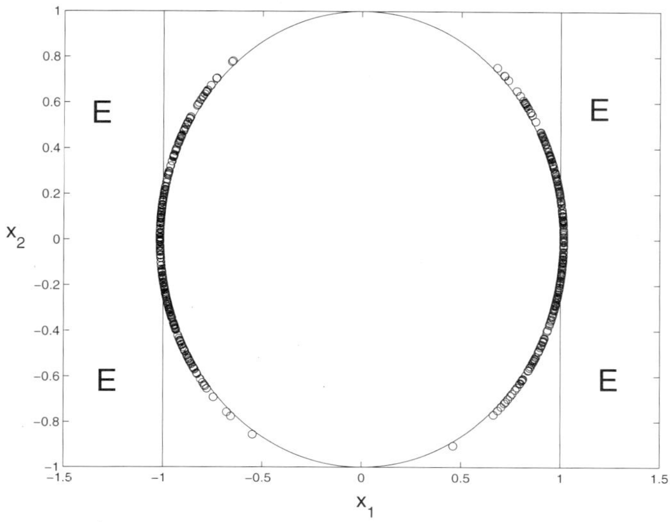

Fig. 10.3. Plot of locations $\{v_i\}$ of adapted universal simulation distribution with unknown rate, driven by empirically chosen target hit rate of 45%. The error set is $E = \{x : |x| \geq 1\}$. The value of I found adaptively was 12.3963. The true value of I is 12, which corresponds to the circle of radius 1.

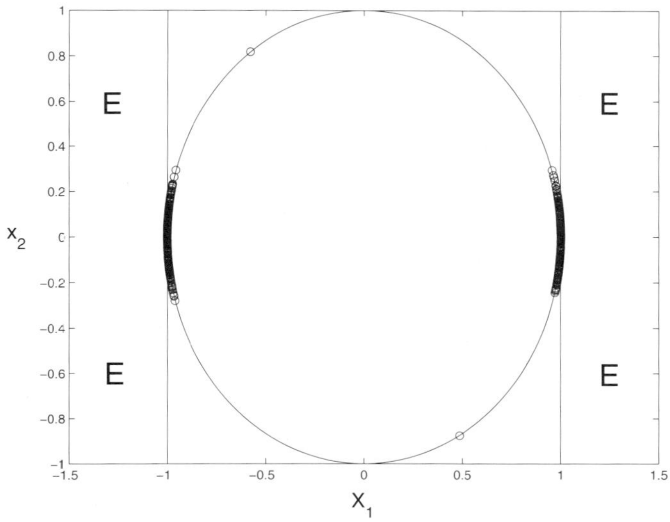

Fig. 10.4. Plot of hit points of adapted universal simulation distribution with unknown rate, using lower noise variance (increasing n to 500) to find better hit points near the minimum rate points, driven by target hit rate of 50%. The adapted value of I was found to be 12.0384. The true value of I is 12.

First, let $\psi(r) = \phi(r\frac{\theta}{\|\theta\|}) = \log \mathbb{E}[\exp(\frac{r^2 A^2}{2})]$. Then

$$\nabla\phi(\theta) = \frac{d\psi(r)}{dr}|_{r=\|\theta\|}(\frac{\theta}{\|\theta\|}) = \frac{\mathbb{E}[A^2 \exp(\frac{\|\theta\|^2}{2}A^2)]}{\mathbb{E}[\exp(\frac{\|\theta\|^2}{2}A^2)]}\theta.$$

Now consider $\nabla\phi(\boldsymbol{re})$ where $\boldsymbol{e}$ is a unit vector.

$$\begin{aligned}\nabla\phi(\boldsymbol{re}) &= \frac{a^2\exp(\frac{r^2a^2}{2}) + b^2\exp(\frac{r^2b^2}{2})}{\exp(\frac{r^2a^2}{2}) + \exp(\frac{r^2b^2}{2})}\boldsymbol{re} \\ &= \left[a^2 + \frac{b^2 - a^2}{\exp(\frac{r^2}{2}(a^2-b^2)) + 1}\right]\boldsymbol{re}.\end{aligned}$$

This is an increasing function for all b, a and hence has an inverse. Since $\boldsymbol{e}$ is arbitrary, $\nabla\phi$ is invertible with inverse $\nabla\phi^{-1}$. Furthermore, since $\nabla\phi(\theta_\nu) = \nu$, $\theta_\nu = \nabla\phi^{-1}(\nu)$.

Due to the spherical symmetry, any minimum distance point of E from the origin (such points exist if E is closed and $I(E) < \infty$) will also be a minimum rate point. Let z^* correspond to any such point. Again due to the spherical symmetry, $\|\nabla\phi^{-1}(z^*)\|$ has the same value no matter which minimum rate point we choose.

Therefore we have

$$\begin{aligned}&dQ_n(z) \\ &= \int_B \exp(n[\langle\nabla\phi^{-1}(\nu), z\rangle])\pi(d\nu)\mathbb{E}[\exp(\frac{\|\nabla\phi^{-1}(z^*)\|^2A^2}{2})]^{-n}dP_n(z).\end{aligned}$$

Note also that $\int_B \exp(n[\langle\nabla\phi^{-1}(\nu), z\rangle])\pi(d\nu) = \mathbb{E}[\exp(n\|\nabla\phi^{-1}(z^*)z\|U_1)]$, where U_1 is a random variable uniform on the surface of the unit sphere $\mathcal{R}^d$. From (9.2) of the previous chapter we have

$$\mathbb{E}[\exp(n\|\nabla\phi^{-1}(z^*)z\|U_1)] = c\frac{I_{(d/2)-1}(n|\nabla\phi^{-1}(z^*)||z|)}{(n|z|)^{(d/2)-1}},$$

where

$$c = \frac{\Gamma(\frac{d}{2})\sqrt{\pi}2^{(d/2)-1}}{\Gamma(\frac{1}{2})|\nabla\phi^{-1}(z^*)|^{(d/2)-1}}.$$

Therefore the final form for the universal distribution is

$$dQ_n(z) = c\frac{I_{(d/2)-1}(n|\nabla\phi^{-1}(z^*)||z|)}{(n|z|)^{(d/2)-1}} \times \exp(-n\frac{|\nabla\phi^{-1}(z^*)|^2}{2})dP_n(z).$$

10.4 Notes and Comments

At the time of writing of this book, the jury is still out on the usefulness of the universal distributions. They are plagued by two big problems: the low hit rates that they typically achieve which of course slows convergence and the difficult analytical problem of how to compute the weight function in closed form.

The low hit rate seems to be reparable via the use of the adaptive scheme presented in the text. The analytical weight function problem is still a research problem that needs to be addressed. Many other questions regarding the applicability of this method to a range of common simulation problems including for example; high dimensionality, path-dependent rare events, small noise inputs with state-dependent dynamics remain completely open. Hopefully the readers of this text can provide some of the answers regarding the potential usefulness (or uselessness!) of this method.

11. Rare Event Simulation for Level Crossing and Queueing Models

Those who can, do. Those who can't, simulate. *Anon.*

11.1 Simulation of Level Crossing Probabilities

For every integer n, let $Z_{p,n}$ be a random variable taking values in some complete separable metric space $\mathcal{S}_n$. Let P_n be the probability measure induced by $Z_{p,n}$ on $\mathcal{S}_n$. Instead of directly simulating $Z_{p,n}$, we choose to simulate with another $\mathcal{S}_n$-valued random variable $Z_{q,n}$ which in turn induces a probability measure on $\mathcal{S}_n$, Q_n. Let f_n be an $\mathcal{R}$-valued measurable function on the space $\mathcal{S}_n$; that is $f_n : \mathcal{S}_n \to \mathcal{R}$.

Define

$$\phi_n(\theta) = \frac{1}{n} \log \mathbb{E}[\exp(\theta f_n(Z_{p,n}))].$$

We assume $\phi(\theta) = \lim_n \phi_n(\theta)$ exists for all $\theta \in \mathcal{R}$, where we allow ∞ both as a limit value and as an element of the sequence $\{\phi_n(\theta)\}$. We also assume ϕ is differentiable on D^o_ϕ, and that there exists a value $\theta^* \in D^o_\phi$ such that $\theta^* > 0$, $\phi(\theta^*) = 0$, and $\phi'(\theta^*) > 0$. As usual we define

$$I(x) = \sup_\theta [\theta x - \phi(\theta)].$$

Now let $T_{p,M} = \inf\{n : f_n(Z_{p,n}) > M\}$; that is, $T_{p,M}$ is the time of the first level crossing of size M. We are interested in the probabilities

$$\rho_M = P(T_{p,M} < \infty).$$

In general, the direct estimation of a rare event by a direct Monte Carlo simulation is difficult since for good relative accuracy a large number of simulation runs must be taken. In many settings, a direct simulation is even impossible because each realization of the event $T_M = \infty$ (which often occurs with positive probability) would require the simulation of the whole infinite sequence $Z_{p,1}, Z_{p,2}, \ldots$. Consequently we search for good importance sampling techniques.

Let $T_{q,M} = \inf\{n : f_n(Z_{q,n}) > M\}$. We wish to choose the $\{Q_i\}$ so that the event $\{T_{q,M} < \infty\}$ occurs with probability one. Let B be a Borel set in $\mathcal{S}_n$. For each n, we wish to define a pair of measures on $\mathcal{S}_n$ by

$$\tilde{P}_n(B) = P(Z_{p,n} \in B; T_{p,M} = n)$$
$$\tilde{Q}_n(B) = P(Z_{q,n} \in B; T_{q,M} = n).$$

Note that these are not probability measures. We further assume that for each n, the measure $\tilde{P}_n$ is absolutely continuous with respect to $\tilde{Q}_n$ and hence the (Radon–Nikodym) derivative

$$w_n(z) = \frac{d\tilde{P}_n}{d\tilde{Q}_n}(z)$$

exists.

Our proposed estimator is

$$\hat{\rho_M} = \frac{1}{k}\sum_{j=1}^{k} w_{T^{(j)}_{q,M}}(Z^{(j)}_{q,T^{(j)}_{q,M}}).$$

This estimator is unbiased since

$$\mathbb{E}_Q[w_{T_{q,M}}(Z_{q,T_{q,M}})] = \sum_{m=1}^{\infty} \int w_m(z) dQ_m(z|T_{q,M} = m)P(T_{q,M} = m),$$

where $Q_m(z|T_{q,M} = m)P(T_{q,M} = m) = \tilde{Q}_m(z)$,

$$\begin{aligned}
&= \sum_{m=1}^{\infty} \int w_m(z)P(T_{q,M} = m)dQ_m(z|T_{q,M} = m) \\
&= \sum_{m=1}^{\infty} \int w_m(z)d\tilde{Q}_m(z) \\
&= \sum_{m=1}^{\infty} \int d\tilde{P}_m(z) \\
&= \sum_{m=1}^{\infty} \int dP_m(z|T_{p,M} = m)dzP(T_{p,M} = m) \\
&= \sum_{m=1}^{\infty} P(T_{p,M} = m) \\
&= P(T_{p,M} < \infty) \\
&= \rho_M.
\end{aligned}$$

As always, we have $k\,Var(\hat{\rho}_M) = F_{M,Q} - \rho_M^2$.

Let's see if we can upper bound the variance rate of this estimator and perhaps find an efficient estimation strategy.

It would be wonderful to be able to continue and solve this problem in this very general setting but unfortunately the author doesn't know how. To continue, let's make another simplifying assumption. We suppose that the occurrence or not of the event $\{T_{p,M} = n\}$ is completely determined by knowledge of $Z_{p,n}$. In other words, there exists some function $g_n : \mathcal{S}_n \to \{0, 1\}$ such that the event $\{T_{p,M} = n\}$ occurs if and only if $g_n(Z_{p,n}) = 1$. For example, this would be true if $Z_{p,n} = (X_1, X_2, \ldots, X_n)$.

In this setting, $d\tilde{P}_n(z) = dP_n(z)1_{\{z:g_n(z)=1\}}$. We use the same "stopping time function" for the simulation random variables; that is, the event $\{T_{q,M} = n\}$ occurs if and only if $g_n(Z_{q,n}) = 1$. We then have

$$w_n(z) = \frac{dP_n}{dQ_n}(z)1_{\{z:g_n(z)=1\}}.$$

We show that we can get arbitrarily close to an efficient simulation strategy. For any $\epsilon \in (0, \theta^*)$ define

$$dQ_{n,\epsilon}(z) = dP_n(z) \exp((\theta^* - \epsilon) f_n(z)) \exp\bigl(-n\phi_n(\theta^* - \epsilon)\bigr).$$

We assume that under this measure $P(T_{q,M} < \infty) = 1$. This last assumption is a "reasonability" assumption; that is, we assume that we have chosen a biasing distribution that is indeed forcing (or biasing) the sample trajectory in the correct direction.

Then (since $P(T_{q,M} < \infty) = 1$),

$$\begin{aligned}
&F_{M,Q} \\
&= \sum_{m=1}^{\infty} \int_{\{z:g_m(z)=1\}} w_m(z)^2 dQ_{m,\epsilon}(z | T_{q,M} = m) P(T_{q,M} = m) \\
&= \sum_{m=1}^{\infty} \exp\bigl(2m\phi_n(\theta^* - \epsilon)\bigr) \int_{\{z:g_m(z)=1\}} \exp\bigl(-2(\theta^* - \epsilon) f_n(z)\bigr) dQ_{m,\epsilon}(z) \\
&\leq \exp\bigl(-2(\theta^* - \epsilon)M\bigr) \sum_{m=1}^{\infty} \exp\bigl(2m\phi_n(\theta^* - \epsilon)\bigr) \int_{\{z:g_m(z)=1\}} dQ_{m,\epsilon} \\
&\leq \exp\bigl(-2(\theta^* - \epsilon)M\bigr) \sum_{m=1}^{\infty} \exp\bigl(2m\phi_n(\theta^* - \epsilon)\bigr) \\
&\leq K_\epsilon \exp\bigl(-2(\theta^* - \epsilon)M\bigr),
\end{aligned}$$

where $K_\epsilon < \infty$ is a constant depending only on ϵ. (This constant is less than infinity since $\phi_n(\theta^* - \epsilon) < 0$.) Thus

$$\limsup \frac{1}{M} \log F_{M,Q} \leq -2(\theta^* - \epsilon).$$

Since we may choose ϵ as close to zero as we wish, we can get arbitrarily close to an efficient simulation strategy.

Example 11.1.1. A common application of level crossing probabilities is in the evaluation of sequential likelihood ratio hypothesis testing. Consider the following hypothesis testing problem.

Suppose $\{P_i\}$ are i.i.d. Poisson with parameter $\lambda = 2$, and $\{N_i\}$ are i.i.d. standard normal. The two hypotheses H_0 and H_1 for our observations $\{Y_i\}$ that we need to choose between are

$$\begin{aligned} H_0 : Y_i &= N_i - P_i \\ H_1 : Y_i &= N_i + P_i. \end{aligned}$$

We employ a sequential test strategy. Let $S_k = \sum_{i=1}^{k} Y_i$. If S_k exceeds $+M$ before it goes below $-M$, we announce H_1; otherwise we announce H_0. We are interested in the error probabilities, which due to symmetry are equal,

$$\rho = P(\text{Error}|H_0 \text{ is true}) = P(\text{Error}|H_1 \text{ is true}).$$

We should note that this problem doesn't fit exactly into our original simple level crossing simulation problem. In that problem, there was only one level with which to concern ourselves. Suppose that H_0 is true. The level crossing problem is to consider $\rho_M = P(T_{p,M} < \infty)$. It is fairly obvious that $\rho_M \geq \rho$. Also for large M, these two probabilities should be fairly close to each other. Hence our philosophy is to use the biasing distributions associated with ρ_M for the simulation of the probability ρ.

The moment generating function for the difference of the Gaussian and the Poisson random variables is

$$M(\theta) = \exp\Big(\frac{\theta^2}{2} + 2\big(\exp(-\theta) - 1\big)\Big),$$

and numerically solving the equation $\log M(\theta^*) = 0$ implies $\theta^* \approx 1.833$.

Explicitly our estimator is

$$\hat{\rho}_M = \frac{1}{n} \sum_{j=1}^{n} \prod_{i=0}^{T_M^{(j)}} \frac{p(Y_i^{(j)})}{q(Y_i^{(j)})} = \frac{1}{n} \sum_{j=1}^{n} \exp\left(\theta^\star \sum_{i=0}^{T_M^{(j)}} Y_i^{(j)}\right). \tag{11.1}$$

In Fig. 11.1 we plot the estimator as a function of the number of simulation runs for 40,000 runs. The convergence is almost instantaneous.

11.2 Single-Server Queue

We first derive some fundamental equalities about the $G/G/1$ queue. We start with the waiting time process. Let W_n denote the waiting time of the nth customer, $\{A_n\}$ denote the interarrival time sequence, and $\{B_n\}$ denote the service time sequence. Then it is easy to see that the waiting-time sequence follows *Lindley's recursion*

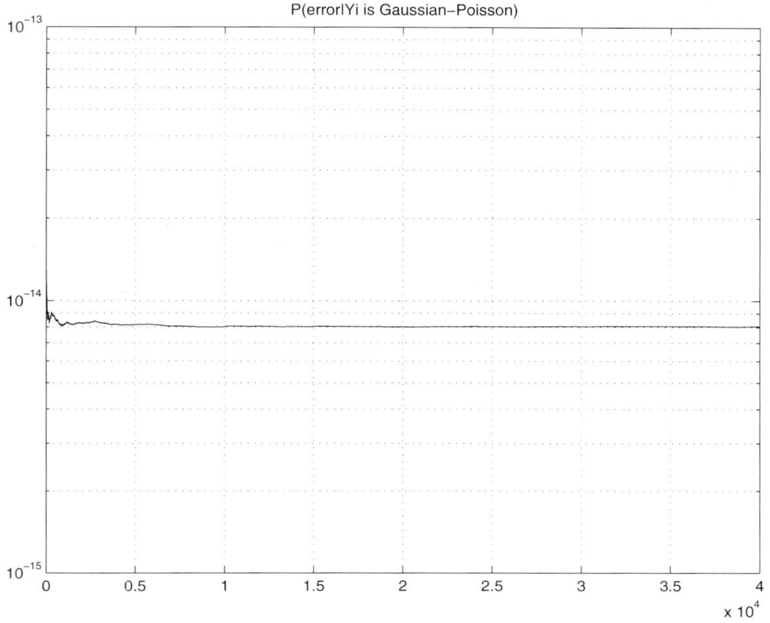

Fig. 11.1. The estimator as a function of the number of simulation runs.

$$W_{n+1} = \max(0, W_n + B_n - A_{n+1}).$$

For simplicity we take $W_0 = 0$. It is frequently assumed that the sequences of interarrival times and service times are i.i.d. (giving rise to a $GI/GI/1$ queue). In this case if $\rho = \mathbb{E}[B_n]/\mathbb{E}[A_n] < 1$, then the queue is stable and W_n converges in distribution to W the steady-state or equilibrium waiting time. We would then be interested in $P(W > x)$ for large values of x.

Getting back to our general $G/G/1$ setting, define $X_{n+1} = B_n - A_{n+1}$. Define $Y_0 = 0$ and $Y_n = \sum_{i=1}^{n} X_i$. Then

Theorem 11.2.1.

$$W_n = \max(Y_n, Y_n - Y_1, Y_n - Y_2, \ldots, Y_n - Y_{n-1}, 0).$$

Proof. By the Lindley recursion equation the positive increments of $\{W_n\}$ are at least those of the $\{Y_n\}$. Thus

$$W_n - W_{n-k} \geq Y_n - Y_{n-k} \quad k = 0, 1, \ldots, n.$$

Since $W_{n-k} \geq 0$, $W_n \geq Y_n - Y_{n-k}$ for $k = 0, 1, \ldots, n$, which gives the "greater than" part of the equality. For the converse, we argue that either $W_n = Y_n$ or $W_n = Y_n - Y_{n-k}$ for some k. The first case occurs if $Y_k \geq 0$ for all $k \leq n$. Otherwise, $W_l = 0$ for some $l \leq n$ and letting k be the last such l, the Lindley recursion equation yields $W_n = Y_n - Y_{n-k}$, and hence the theorem statement. $\square$

For the rest of this section, we concentrate on the $GI/GI/1$ queue. Define $E_n = \max_{0 \le k \le n} Y_k$ and $E = \max_{0 \le k < \infty} Y_k$. In the $GI/GI/1$ setting, $Y_n - Y_{n-k} = \sum_{i=n-k+1}^{n} X_i$ has the same distribution as Y_k, and thus W_n has the same distribution as E_n. In the stable queue situation we expect that as $n \to \infty$, W_n converges to W which would have the same distribution as E, the maximum of the random walk (which has negative drift). We note that $P(W > M) = P(E > M) = P(T_M < \infty)$, where T_M is the first time that the random walk exceeds M.

We know what the large deviation asymptotics of T_M are. From the section on level crossing probabilities, we have

$$\begin{aligned}
\phi_n(\theta) &= \frac{1}{n} \log \mathbb{E}[\exp(\theta Y_n)] \\
&= \log \mathbb{E}[\exp(\theta X_{n+1})] \\
&= \log \mathbb{E}[\exp(\theta(B_n - A_{n+1}))] \\
&= \log M_B(\theta) + \log M_A(-\theta),
\end{aligned}$$

where $\theta \in \mathcal{R}$ and $M_B(\cdot)$, $M_A(\cdot)$ are the moment generating functions for the service time distributions and interarrival distributions, respectively. If the queue is stable, there exists a value of θ, call it $\theta^* > 0$, such that $\phi(\theta^*) = 0$ and $\phi'(\theta^*) > 0$.

As usual we define

$$I(x) = \sup_\theta [\theta x - \phi(\theta)],$$

and

$$\rho_M = P(W > M) = P(T_M < \infty),$$

and we know that,

$$\lim_{M \to \infty} \frac{1}{M} \log P(T_m < \infty) = -\theta^*.$$

Instead of simulating the i.i.d. $\{X_i\}$ (with associated probability measure P) directly, we simulate instead an i.i.d. sequence $\{\tilde{X}_i\}$ with associated probability measure Q. The distribution Q should be chosen so that dP/dQ can be computed and $\mathbb{E}_Q[\tilde{X}_1] > 0$. This last condition implies that $P_Q(T_M < \infty) = 1$, and so all of our simulations are of finite length.

Our proposed estimator is thus

$$\hat{\rho_M} = \frac{1}{k} \sum_{j=1}^{k} \prod_{i=1}^{T_M^{(j)}} \frac{dP}{dQ}(\tilde{X}_i^{(j)}).$$

As always, we have $k\,Var(\hat{\rho}_M) = F_{M,Q} - \rho_M^2$. Let's see if we can evaluate the variance rate of this estimator and perhaps find an efficient estimation strategy. For the sake of simplicity, we define the probability density (or mass function) associated with X_1 as p and the simulation probability density (or mass function) for $\tilde{X}_1$ as q. Define

$$c(\theta) = \log \int \frac{p(x)^2}{q(x)} \exp(\theta x)dx.$$

We suppose that there exists a $\tilde{\theta} > 0$ such that $c(\tilde{\theta}) = 0$, and from the previous section we know

$$\lim_{M\to\infty} \frac{1}{M} \log F_{M,Q} = -\tilde{\theta}.$$

Suppose we choose $q(x) = p(x)\exp(\theta^*)$ (recall that the moment generating function of X_1 is $M(\theta)$ and $M(\theta^*) = 1$). Then

$$\begin{aligned}
\exp\bigl(c(\theta)\bigr) &= \int \frac{p(x)^2}{q(x)} \exp(\theta x)dx \\
&= \int p(x)\exp(-\theta^* x)\exp(\theta x)dx \\
&= M(\theta - \theta^*).
\end{aligned}$$

The equation $c(\tilde{\theta}) = 0$ implies that $\tilde{\theta} = 2\theta^*$, which implies that this choice of simulation distribution is efficient.

Example 11.2.1 (M/M/1 Queue). For the $GI/GI/1$ queue, the equation $M(\theta^*) = 1$ is equivalent to $M_A(-\theta^*)M_B(\theta^*) = 1$. Suppose that for an $M/M/1$ queue we have an arrival rate of λ and a service rate of μ (with $\lambda < \mu$ to ensure stability). Thus

$$M_B(\theta) = \frac{\mu}{\mu - \theta} \qquad \text{for } \theta < \mu$$

and

$$M_A(-\theta) = \frac{\lambda}{\lambda + \theta} \qquad \text{for } \theta > -\lambda.$$

The distribution of the $\{X_i\}$ is easily obtained by the convolution

$$\begin{aligned}
p_x(z) &= p_b(z) * p_a(-z) \\
&= \mu\exp(-\mu z)U(z) * \lambda\exp(\lambda z)U(-z) \\
&= \frac{\mu\lambda}{\lambda + \mu} \exp(-\mu z)\exp\bigl((\lambda + \mu)\min(0, z)\bigr).
\end{aligned}$$

Solving for θ^* yields $\theta^* = \mu - \lambda$. Hence the probability density distribution of the simulation random variables is

$$\begin{aligned} p_{\tilde{x}}(z) &= \frac{p_x(z)\exp(\theta^* z)}{M(\theta^*)} \\ &= \frac{\mu\lambda}{\lambda+\mu}\exp(-\mu z)\exp\big((\lambda+\mu)\min(0,z)\big)\exp\big((\mu-\lambda)z\big) \\ &= \frac{\mu\lambda}{\lambda+\mu}\exp(-\lambda z)\exp\big((\lambda+\mu)\min(0,z)\big). \end{aligned}$$

Thus $p_{\tilde{x}}$ can be obtained from p_x by switching μ and λ. Therefore, in essence, our simulation strategy is to simulate an unstable queue with arrival rate μ and service rate λ. We could do this or just directly simulate the associated random walk.

As an illustration, we simulate to obtain $\rho_M = P(W > M)$ when $M = 100$, $\lambda = .8$, and $\mu = 1$. In Fig. 11.2, we plot the current value of the estimator as a function of the number of simulation runs. K^* is the point at which we empirically claim that we have 2% error with 95% confidence. In Fig. 11.3 we give a representative plot of the associated random walk exceeding the threshold. Remember this will occur with probability one eventually. Hence, in the simulation, we keep on generating random variables until we do cross the threshold.

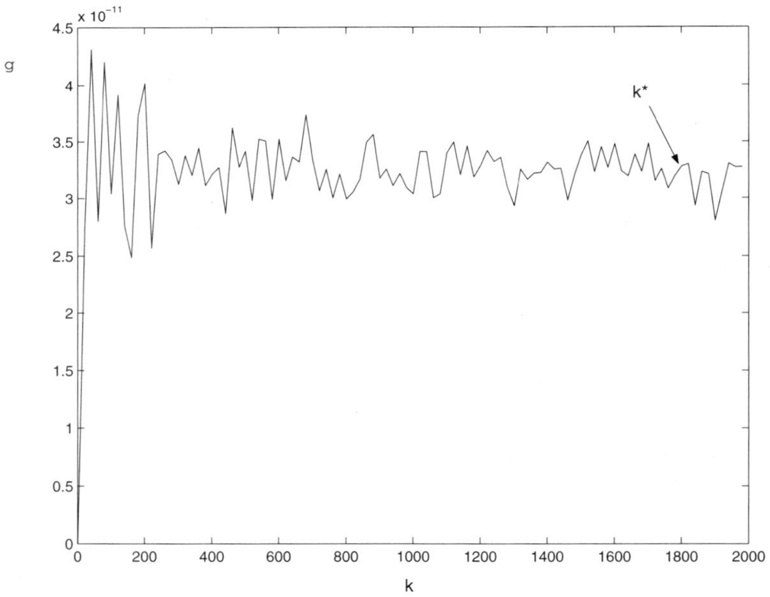

Fig. 11.2. The estimator as a function of the number of simulation runs.

The above analysis leans heavily on the fact that in the $GI/GI/1$ queue, the waiting time has the same distribution as the maximum of a reversed

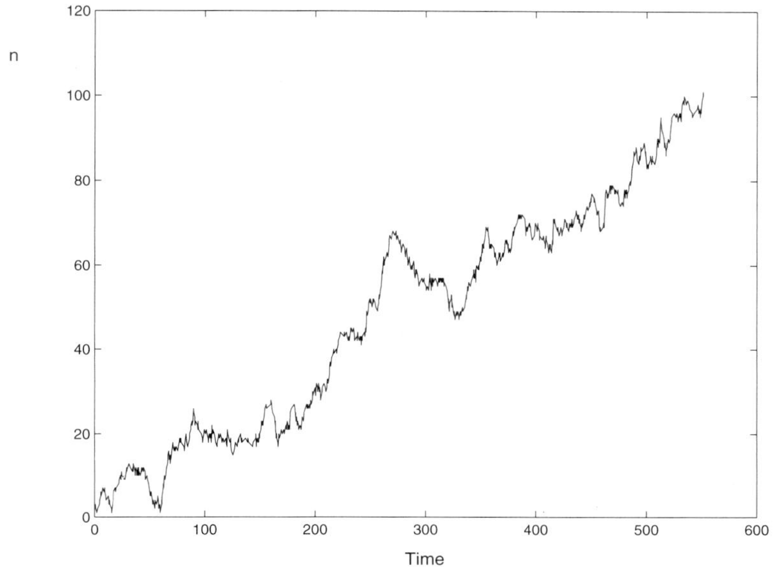

Fig. 11.3. A representative simulation of the associated random walk exceeding the threshold, 100.

random walk with negative drift. Unfortunately, this relationship doesn't occur in more general settings (e.g. multiserver queues, queues with finite buffers, or in networks of queues). We now consider a technique that works directly with the queueing process.

Suppose that we denote the system state at time s by X_s. We assume that there is a particular state, call it 0, such that the process returns to state 0 infinitely often and that, upon hitting state 0, the stochastic evolution of the system is independent of the past and has the same distribution as if the process were started in state 0. Such a system is called *regenerative* [17]. An ergodic Markov chain is an example of a regenerative system, where we can take state 0 to be any state. Arrivals at an empty $GI/GI/1$ queue constitute regenerating points for a regenerative system.

Suppose we are interested in estimating the expected time (starting in state 0) until a rare event occurs, for example estimating $\mathbb{E}_0[\tau_F]$ where τ_F is the first time that the system enters into a "failure" set of states F. Let τ_0 be the first return time to state 0.

First, note that

$$\tau_F = \min(\tau_F, \tau_0) + (\tau_F - \tau_0)1_{\{\tau_0 < \tau_F\}}.$$

Now, applying the regenerative property at time τ_0 shows that on the set $\{\tau_0 < \tau_F\}$, $(\tau_F - \tau_0)$ is conditionally independent of $1_{\{\tau_0 < \tau_F\}}$. Therefore taking the expectation of the above expression and rearranging terms gives us

$$\mathbb{E}[\tau_F] = \frac{\mathbb{E}[\min(\tau_F, \tau_0)]}{P(\tau_F < \tau_0)}.$$

Let us denote $\gamma = P(\tau_F < \tau_0)$. Hence our problem now becomes one of estimating γ, which is a rare event problem. Thus quantities such as the mean time to failure and the mean time until buffer overflow can be estimated by using importance sampling to estimate γ and standard simulation to estimate $\mathbb{E}[\min(\tau_F, \tau_0)]$. (Most simulations of $\min(\tau_F, \tau_0)$ are just τ_0.)

Let's see how this theory would apply to a queueing problem. Suppose we wish to estimate the mean time until a $GI/GI/1$ queue reaches a queue length of n; that is, we wish to estimate $\mathbb{E}[\tau_n]$. We assume (as an initial condition) that customer number zero has just arrived at an empty queue. We thus need to simulate $\gamma_n = P(\tau_n < \tau_0)$. For $k \geq n$ define $Z_k(n) = \sum_{j=1}^{k} A_j - \sum_{j=1}^{k-n} B_{j-1}$. For $\tau_n < \tau_0$, we must have that $Z_k(n) \leq 0$ for some $k \geq n$.

Let's think about this a bit more deeply. Define $K_0 = \min\{k : Z_k(0) > 0\}$. The cycle begins with the arrival of job 0 at time $t = 0$. Recalling that A_k is the interarrival time between the kth and $(k-1)$th jobs and B_k is the service time for job k, we see that job K_0 is the first job to arrive to find an empty queue and an idle server. In other words, it is the first job of the second cycle. Let $\tilde{K}_n = \min\{k \geq n : Z_k(n) < 0\}$ and $K_n = \min\{K_0, \tilde{K}_n\}$. Hence the event $\{K_n = \tilde{K}_n\}$ is the event that the queue backlog exceeds n jobs during the cycle, and on this event the arrival of job $\tilde{K}_n$ causes the queue backlog to exceed n for the first time. Thus $\gamma_n = P(K_n = \tilde{K}_n) = P(\tilde{K}_n < K_0)$.

Again let us assume that there exists a $\theta^* > 0$ such that $M_A(-\theta^*)M_B(\theta^*) = 1$. We can upper bound γ_n in the following fashion,

$$\begin{aligned}
\gamma_n &= \mathbb{E}_P[1_{\{\tilde{K}_n < K_0\}}] \\
&= \mathbb{E}_Q[\frac{dP}{dQ} 1_{\{\tilde{K}_n < K_0\}}] \\
&= \mathbb{E}_Q[\frac{dP}{dQ} \sum_{k=n}^{\infty} 1_{\{\tilde{K}_n = k, K_0 > k\}}] \\
&= \sum_{k=n}^{\infty} \mathbb{E}_Q[\frac{dP}{dQ} 1_{\{\tilde{K}_n = k, K_0 > k\}}].
\end{aligned}$$

To determine the occurrence of the event $\{\tilde{K}_n = k, K_0 > k\}$ we need only have knowledge of the random variables $B_0, B_1, \ldots, B_{k-n-1}, A_1, \ldots, A_k$. Thus we need only write down the likelihood ratio for those random variables. Therefore we can now choose dQ in terms of dP on that set to be

$$dQ = \frac{\prod_{j=1}^{k-n} \exp(\theta^* b_{j-1})}{M_B(\theta^*)^{k-n}} \frac{\prod_{j=1}^{k} \exp(-\theta^* a_j)}{M_A(-\theta^*)^k} dP$$

$$\frac{dP}{dQ} = \prod_{j=1}^{k-n} \exp(-\theta^* b_{j-1}) M_B(\theta^*)^{k-n} \prod_{j=1}^{k} \exp(\theta^* a_j) M_A(-\theta^*)^k. \tag{11.2}$$

Hence,

$$\begin{aligned}
\gamma_n &= \sum_{k=n}^{\infty} \mathbb{E}_Q[\exp\left(\theta^*[\sum_{j=1}^{k} A_j - \sum_{j=1}^{k-n} B_{j-1}]\right) \\
&\quad \times M_B(\theta^*)^{k-n} M_A(-\theta^*)^k 1_{\{\tilde{K}_n = k, K_0 > k\}}] \\
&\le \sum_{k=n}^{\infty} \mathbb{E}_Q[M_B(\theta^*)^{k-n} M_A(-\theta^*)^k 1_{\{\tilde{K}_n = k, K_0 > k\}}] \\
&= M_B(\theta^*)^{-n} \sum_{k=n}^{\infty} \mathbb{E}_Q[1_{\{\tilde{K}_n = k, K_0 > k\}}] \\
&= M_B(\theta^*)^{-n} \mathbb{E}_Q[\sum_{k=n}^{\infty} 1_{\{\tilde{K}_n = k, K_0 > k\}}] \\
&= M_B(\theta^*)^{-n} \mathbb{E}_Q[1_{\{\tilde{K}_n < K_0\}}] \\
&\le M_B(\theta^*)^{-n}.
\end{aligned}$$

Getting a good lower bound is a little more difficult. For simplicity, we assume that the service times are bounded, in other words, that $P(B_1 < b) = 1$ for some constant $b \in [0, \infty)$. (This is really just a technical condition and it can be removed with a bit of analysis.)

Then on the event $\{\tilde{K}_n = k, K_0 > k\}$ for $k \ge n+1$, we must have

$$\sum_{j=1}^{k} A_j - \sum_{j=1}^{k-n} B_{j-1} = [\sum_{j=1}^{k-1} A_j - \sum_{j=1}^{k-n-1} B_{j-1}] + A_k - B_{k-n-1} \ge -b.$$

(The inequality is trivial for $k = n$.) Thus

$$\begin{aligned}
\gamma_n &= \sum_{k=n}^{\infty} \mathbb{E}_Q[\exp\left(\theta^*[\sum_{j=1}^{k} A_j - \sum_{j=1}^{k-n} B_{j-1}]\right) \\
&\quad \times M_B(\theta^*)^{k-n} M_A(-\theta^*)^k 1_{\{\tilde{K}_n = k, K_0 > k\}}] \\
&\ge \exp(-\theta^* b) M_B(\theta^*)^{-n} \sum_{k=n}^{\infty} \mathbb{E}_Q[1_{\{\tilde{K}_n = k, K_0 > k\}}] \\
&= \exp(-\theta^* b) M_B(\theta^*)^{-n} P_Q(\tau_n < \tau_0).
\end{aligned}$$

But since under the Q distribution the queue is unstable, we can guarantee that $P_Q(\tau_n < \tau_0)$ is bounded away from zero as $n \to \infty$. This establishes the final result:

$$\lim_{n\to\infty} \frac{1}{n} \log \gamma_n = -\log M_B(\theta^*) = \log M_A(-\theta^*).$$

Suppose that we generate twisted random variables from Equation (11.2). We know that $P(K_n < \infty) = 1$. The estimate is of the form

$$\hat{\gamma}_n = \frac{1}{k}$$

$$\sum_{j=1}^{k} \frac{dP}{dQ}(B_0^{(j)}, B_1^{(j)}, \ldots, B^{(j)}_{\tilde{K}_n^{(j)}-n+1}, A_1^{(j)}, A_2^{(j)}, \ldots, A^{(j)}_{\tilde{K}_n^{(j)}}) 1_{\{K_n^{(j)} < K_0^{(j)}\}}.$$

The computation of an upper bound for the expectation of the square of the summand F_n is straightforward (following directly from the upper bound derivation for γ_n),

$$\begin{aligned} F_n &= \mathbb{E}_P[\frac{dP}{dQ} 1_{\{\tilde{K}_n < K_0\}}] \\ &\leq M_B(\theta^*)^{-n} \mathbb{E}_P[1_{\{\tilde{K}_n < K_0\}}] \\ &= M_B(\theta^*)^{-n} \gamma_n. \end{aligned}$$

Hence

$$\begin{aligned} \lim_{n\to\infty} \frac{1}{n} \log F_n &\leq -\log M_B(\theta^*) + \lim_{n\to\infty} \frac{1}{n} \log \gamma_n \\ &= -2 \log M_B(\theta^*), \end{aligned}$$

which implies that this estimator is efficient.

11.3 Notes and Comments

This chapter is but a brief inadequate introduction to the huge and important application area of queueing networks. There is a vast literature on this subject utilizing many large deviation theory type techniques that are beyond the scope of this text; see e.g. the excellent survey paper [38].

One of the first to use large deviation theory techniques to come up with good simulation distributions for networks of queues is [65] The basics given in the chapter involving the simulation of a single queue is due in the present form to J.S. Sadowsky [69].

12. Blind Simulation

God not only plays dice. He also sometimes throws them where they can't be seen. *Stephen Hawking*

Even a blind pig finds an acorn once in a while. *Anon.*

12.1 Introduction

Consider the following problem. One is allowed to observe the output of a very complicated system. The system could be a computer model whose internal workings would contain many random number generators and various interacting subsystems. Alternatively one could have a physical device consisting of complicated possible nonlinear electronics from which one can sample data. It may even just be a string of numbers taken as data from some experiment. The simulation problem is to estimate the probability of some output sequence event. Because of the system's complexity, it is not known nor would it be feasible analytically to derive the underlying probability law of the experiment. This is what we call a *blind simulation problem*: we have to estimate the probability of an event without complete knowledge of the underlying probability law. A very large class of practical simulation problems is blind.

One great advantage of direct Monte Carlo simulation is that it *is* blind. One can count the relative frequency of an event of interest and obtain an estimate of the probability of that event with little knowledge of the underlying probability law (other than an ergodic assumption). In contrast, all of the importance sampling simulation strategies that we have so far considered require exact knowledge of the underlying probability law (in order to compute the weight functions). In this chapter, we try to extend the ability of importance sampling techniques (with their associated attribute of large variance reduction) to handle blind problems. We present some blind (and partially blind) strategies that will allow one to proceed with the same confidence in an unknown probability law setting as one does with a classical direct Monte Carlo.

The simplest setting imaginable would be to suppose that the output of the system could be characterized as an i.i.d. sequence of random variables

$\{X_1, X_2, \dots\}$. We further suppose it is desired to estimate a probability of the form

$$\rho_n = P(\sum_{i=1}^{n} X_i > 0).$$

All we have access to is the random sequence. We do not know its distribution. We do, however, suppose that we can measure very accurately any desired moments of the X_i sequence. We may suppose then that we know $\mathbb{E}[s(X_1)]$ for any desired $s(\cdot)$. In particular we may know (by setting $s(x) = x$) that $\mathbb{E}[X_1] = m < 0$ and hence that ρ_n is converging exponentially fast to zero.

How would one go about estimating ρ_n? If ρ_n is very small, a direct Monte Carlo is going to take a large number of samples. Is there a better strategy? Mean shifting or variance scaling are two important attack strategies. One could simply add (or multiply) constants to the observed sequence and have the required shifts. The problem here then is computation of the weight functions. One could collect data and estimate the density of X. This line of attack leads also to problems in that one needs good density estimates in the extreme tail regions (precisely where they are the most difficult to obtain).

Suppose one could generate from the observed sequence, multiplicatively shifted random variables instead. By this we mean, we generate random variables that have densities of the form $p(x)h(x)/\int p(y)h(y)dy$, where p is the original density and h is some nonnegative function chosen by the designer. The problem of computing the weights is much alleviated since the weight function will be proportional to the known h function.

How does one go about generating multiplicatively shifted random variables? We propose an acceptance-rejection method as follows. Suppose $h : \Re \to [0,1]$. One then generates i.i.d. random variables uniform on the interval $[0,1]$, $\{U_1, U_2, \dots\}$. Keep rejecting the $\{X_i\}$ until $U_i \le h(X_i)$. If the inequality holds, set $Y = X_i$. Let us calculate the probability law of the accepted random variables.

$$\begin{aligned} P\big(X \text{ accepted and } X \in [z, z+dz]\big) &= h(z)p(z)dz \\ P\big(X \text{ accepted}\big) &= \int h(z)p(z)dz \\ P\big(Y \in [z, z+dz]\big) &= P\big(X \in [z, z+dz] | X \text{ accepted}\big) \\ &= \frac{h(z)p(z)dz}{\int h(z)p(z)dz}. \end{aligned}$$

Thus the $\{Y_j\}$ accepted random variable sequence has a multiplicative shift.

Of course lots of random variables are rejected (and hence wasted). Is this going to be more efficient than direct Monte Carlo, mean, or variance scaling? How much worse is it than our known efficient simulation methods? We provide some answers to these questions in the succeeding sections.

12.2 Development

In this section we analyze the performance of some blind simulation strategies for various dependency models.

12.2.1 I.I.D. Sum Case

Suppose we are interested in estimating

$$\rho = P\left(\sum_{i=1}^{n} X_i > 0\right),$$

where the $\{X_i\}$ are i.i.d. $\mathcal{R}$-valued random variables with (unknown) density $p(\cdot)$.

Remark 12.2.1. The assumption that the random variables are $\mathcal{R}$-valued isn't essential; the same arguments will go through with minor modifications for the $\mathcal{R}^d$-valued case. The principal additional assumption needed for the multidimensional case is that the set of interest has a known dominating point. In the scalar setting treated here, zero will always play the role of known dominating point of the set. This additional necessary assumption may or may not be reasonable in a given multidimensional setting depending on the problem. We do not discuss the multidimensional case any further in the chapter.

Let $h : \mathcal{R} \to [0,1]$ be given. We suppose that we have available Ln i.i.d. samples of the $\{X_i\}_{i=1}^{Ln}$. We further suppose that the mean of these random variables is negative and the value of $a = \int h(x)p(x)dx$ is known. The prior knowledge of the a assumption is relaxed later.

We accept or reject each element of the $\{X_i\}$ sequence to generate a $\{Y_i\}_{i=1}^{N}$ sequence, where N is the number of acceptances. In other words, sequentially for each X_l, we accept it with probability $h(X_l)$ or toss it away (reject it) with probability $1 - h(X_l)$. If it is accepted we place it as the next element of the $\{Y_i\}$ sequence. The a priori probability of accepting a sample is easily seen to be a. Therefore N is a binomial random variable with parameter (Ln, a). Define

$$K = \lfloor \frac{N}{n} \rfloor,$$

where $\lfloor x \rfloor$ is the greatest integer less than or equal to x. K is the number of simulation runs we can obtain from the accepted sequence. It is easy to verify that the distribution of the $\{Y_i\}$ is independent of the value of N (or K). Furthermore each of the $\{Y_i\}$ are independent and have density $p(x)h(x)/a$.

We have to try to make a fair comparison between a direct Monte Carlo estimator using the Ln $\{X_i\}$ samples and an importance sampling estimator based on the available Kn $\{Y_i\}$ samples (where K is random). K is almost always much less than N. We expect the variance of the respective estimators to be decreasing as the inverse of the number of simulation runs N for direct Monte Carlo, K for importance sampling. Thus the importance sampling estimator starts off with a big disadvantage since it will usually employ far fewer simulation runs. We discover that this disadvantage will be more than compensated for by the gains to be expected by using importance sampling.

Partially Blind Estimator–I.I.D. Case. For simplicity in notation, let $Y_i^{(j)} = Y_{i+(j-1)n}$, $i = 1, \ldots, n$, $j = 1, \ldots, K$. We propose the following estimator for p,

$$\hat{p}_L = \frac{1}{K} \sum_{j=1}^{K} 1_{\{\sum_{i=1}^{n} Y_i^{(j)} > 0\}} \frac{a^n}{\prod_{i=1}^{n} h(Y_i^{(j)})}.$$

Our notation is to use a subscript L to indicate the size of the original population of samples before doing the acceptance-rejection procedure. Note that for a random sum of K i.i.d. random variables $\{Z_k\}$, where K is independent of the $\{Z_k\}$,

$$\begin{aligned} &Var(G(K) \sum_{k=1}^{K} Z_k) \\ &= (\mathbb{E}[Z_1])^2 \, Var(KG(K)) + Var(Z_1)\mathbb{E}[KG(K)^2]. \end{aligned}$$

Therefore letting $G(K) = 1/K$ in (12.1), we have

$$Var(\hat{p}_L) = \mathbb{E}[\frac{1}{K}] \, Var(1_{\{\sum_{i=1}^{n} Y_i^{(j)} > 0\}} \frac{a^n}{\prod_{i=1}^{n} h(Y_i^{(j)})}).$$

We now note that (for large L and $a > 0$)

$$\begin{aligned} \mathbb{E}[\frac{1}{K}] &\approx \mathbb{E}[\frac{n}{N}] \\ &\approx \mathbb{E}[\frac{n}{N+1}] \\ &= n \sum_{j=0}^{Ln} \binom{Ln}{j} a^j (1-a)^{Ln-j} \frac{1}{j+1} \\ &= \frac{n}{a(Ln+1)} \sum_{j=0}^{Ln} \frac{(Ln+1)!}{(Ln-j)!(j+1)!} a^{j+1} (1-a)^{Ln-j} \\ &= \frac{n}{(Ln+1)a} [1 - (1-a)^{Ln+1}] \\ &\approx \frac{1}{La} \end{aligned}$$

or

$$Var(\hat{\rho}_L) = \frac{1}{La}\, Var(1_{\{\sum_{i=1}^n Y_i^{(j)} > 0\}} \frac{a^n}{\prod_{i=1}^n h(Y_i^{(j)})}).$$

Therefore the effect of the acceptance-rejection procedure is merely to introduce a $1/a$ factor into the variance of the estimator.

To compare with previous chapters, we have that $L\, Var(\hat{\rho}_L) = F_h(\hat{\rho}_L) - \rho^2$, which implies that

$$F_h(\hat{\rho}_L) = \frac{1}{a}\mathbb{E}_P[1_{\{\sum_{i=1}^n Y_i^{(j)} > 0\}} \frac{a^n}{\prod_{i=1}^n h(Y_i^{(j)})}].$$

Hence as $n \to \infty$, we have that

$$\lim_{n\to\infty} \frac{1}{n} \log F_h(\hat{\rho}_L) = -\log(\inf_\theta \int_{-\infty}^{\infty} \frac{\exp(\theta y) p(x) a}{h(x)} dx),$$

where we note that the $1/a$ factor gets washed out in the logarithm limit operations of computing the rate.

Completely Blind Estimator–I.I.D. Case. Now let us consider changing the form of our estimator to something that does not require prior knowledge of the parameter a. We propose the following estimator,

$$\tilde{\rho}_L = (\frac{K}{L})^n \frac{1}{K} \sum_{j=1}^{K} \frac{1_{\{\sum_{i=1}^n Y_i^{(j)} > 0\}}}{\prod_{i=1}^n h(Y_i^{(j)})}.$$

Note that

$$\begin{aligned}
\mathbb{E}[\tilde{\rho}_L] &= \mathbb{E}_K[\mathbb{E}[\frac{K^{n-1}}{L^n} \sum_{j=1}^{K} \frac{1_{\{\sum_{i=1}^n Y_i^{(j)} > 0\}}}{\prod_{i=1}^n h(Y_i^{(j)})} | K]] \\
&= \mathbb{E}_K[\frac{K^{n-1}}{L^n} \frac{K\rho}{a^n}] \\
&= \rho \frac{1}{a^n} \mathbb{E}[\frac{K^n}{L^n}].
\end{aligned}$$

To evaluate the final expectation in the above expression, we have

$$\begin{aligned}
\mathbb{E}[\frac{K^n}{L^n}] &= \mathbb{E}[\frac{(Kn)^n}{(Ln)^n}] \\
&\approx \mathbb{E}[\frac{N^n}{(Ln)^n}] \\
&= \mathbb{E}[(\frac{1}{Ln} \sum_{j=1}^{Ln} 1_{\{X_j \text{ is accepted}\}})^n] \\
&\xrightarrow[L\to\infty]{} a^n.
\end{aligned}$$

The last limit follows since the average is bounded (less than one actually) and it converges to a almost surely. Therefore the expectation converges by the dominated convergence theorem. Hence

$$\lim_{L\to\infty} \mathbb{E}[\tilde{\rho}_L] = \rho$$

and $\tilde{\rho}$ is asymptotically unbiased.

The variance calculation is a little more complicated.

Lemma 12.2.1. *Let $\{Z_i\}$ be i.i.d Bernoulli zero-one random variables with $P(Z_1 = 1) = z$. Then*

$$\lim_{L\to\infty} L\,Var[(\frac{1}{nL}\sum_{j=1}^{nL} Z_j)^n] = nz^{2n-2}z(1-z)$$

Proof.

$$nL\,Var[(\frac{1}{nL}\sum_{j=1}^{nL} Z_j)^n]$$

$$= \frac{nL}{(nL)^{2n}}\left[\mathbb{E}[(\sum_{j=1}^{nL} Z_j)^{2n}] - \mathbb{E}[(\sum_{j=1}^{nL} Z_j)^n]^2\right]$$

$$= \frac{nL}{(nL)^{2n}}\left[\sum_{j=0}^{2n}\frac{S(2n,j)(nL)!z^j}{(nL-j)!} - (\sum_{j'=0}^{n}\frac{S(n,j')(nL)!z^{j'}}{(nL-j')!})^2\right]$$

See [47, p. 107] where $S(\cdot,\cdot)$ is defined to be the Stirling numbers of the second kind.

$$= \frac{1}{(nL)^{2n-1}}[z^{2n}((nL)(nL-1)\cdots(nL-2n+1)$$
$$-(nL)(nL)(nL-1)(nL-1)\cdots(nL-n+1)(nL-n+1))$$
$$+z^{2n-1}C + o(L^{2n-2})],$$

where $C = S(2n, 2n-1) - 2S(n,n)S(n,n-1) = n^2$ and using the identity $S(k, k-1) = \binom{k}{2}$

$$\to_{L\to\infty} n^2 z^{2n-2} z(1-z).$$

This completes the proof of the lemma. □

Hence, using (12.1) (with $G(K) = \frac{1}{a^n K}(\frac{K}{L})^n$),

$$
\begin{aligned}
L\,Var(\tilde{\rho}_L) &= (\mathbb{E}[1_{\{\sum_{i=1}^n Y_i^{(j)}>0\}}\frac{a^n}{\prod_{i=1}^n h(Y_i^{(j)})}])^2 L\,Var((\frac{K}{L})^n\frac{1}{a^n}) \\
&+ Var(1_{\{\sum_{i=1}^n Y_i^{(j)}>0\}}\frac{a^n}{\prod_{i=1}^n h(Y_i^{(j)})})\mathbb{E}[\frac{1}{a^{2n}}\frac{K^{2n-1}}{L^{2n-1}}] \\
&\underset{L\to\infty}{\longrightarrow} \frac{1}{a}Var(1_{\{\sum_{i=1}^n Y_i^{(j)}>0\}}\frac{a^n}{\prod_{i=1}^n h(Y_i^{(j)})}) + \rho^2 n(1-a)/a.
\end{aligned}
$$

To compare with the results of previous chapters we need to define our variance rate in terms of asymptotically large L. Hence $\lim_{L\to\infty} L\,Var(\tilde{\rho}_L) = F_h(\tilde{\rho}_L) - \rho^2$.

Therefore, since the second term is negligible (except in the case where h can be chosen to make the estimator efficient), we have

$$
\lim_{n\to\infty}\frac{1}{n}\log F_h(\tilde{\rho}_L) = -\log(\inf_{\theta}\int_{-\infty}^{\infty}\frac{\exp(\theta y)p(x)a}{h(x)}dx).
$$

The Optimal Rejection Function–I.I.D. Case. We now consider the problem of choosing the best h. Can it always be chosen so that we have an efficient estimator? Consider first the special case where $p(\cdot)$, the density function of the data, has compact support. Consider an h of the form

$$
h(x) = k\,\exp(\theta_0)x,
$$

where k is chosen so that h will satisfy $0 \le h(x) \le 1$ for all x in the support of p. It is easy to verify that the accepted random variables are the efficient twisted simulation random variables. Hence, in this case, even though our estimator is blind, it is efficient! This means that with a blind simulation strategy, we can obtain the same variance rate results as with the optimal quick simulation importance sampling strategy. In other words, by throwing away data samples according to some rule, we can achieve orders of magnitude improvement in simulator speed.

If p is not of compact support, then no optimal choice of h exists. We may consider h of the form

$$
h_B(x) = \frac{1}{B}\min[\exp(\theta_0)x, 1].
$$

As $B \to \infty$, we can (in the usual case) get arbitrarily close to the optimal performance of a quick simulation strategy.

Of course, there is a fly in the ointment in the above argument. How does one choose $\theta = \theta_0$ for the above choices of h in a blind problem? One possibility is to make the search for a good θ be done adaptively from the data. One property of the optimal θ, θ_0, is that the mean value of the optimal twisted distribution will be zero. One could monitor the accepted samples, block by block calculate sample averages, and recursively compute successive values of θ via a stochastic approximations method. This can be viewed as front end processing and should not affect the overall calculations of the variance rate in any case. We employ this methodology in the following example.

Example 12.2.1. We suppose that a researcher is confronted with the task of computing $P\big(\sum_{i=1}^{n} X_i > n\big)$ for some collection of i.i.d. random variables $\{X_i\}$ of unknown distribution. For the purposes of this example we take $\{X_i\}$ to be standard normals which will allow closed form expressions for the true probability. Of course, in our estimation procedure, we nowhere make use of this knowledge. We consider acceptance-rejection functions of the form $h(x) = 1$ for $x > \theta$ and $h(x) = \exp(\theta x - \theta^2)$ for $x \leq \theta$. We generate data from the $\{X_i\}$ and vary θ until the accepted data have a mean of 1. We do this by generating accepted blocks of data of length 10. We then vary θ according to the rule $\theta_{k+1} = \theta_k + \mu(1 - (\sum_{i=1}^{10} Y_i)/n)$. With $\mu = .1$, this algorithm converges very quickly (well under 100 iterations) and we empirically find that $\theta = 1.4$ is a good choice. We then fix this value of θ and begin our estimation procedure with estimator $\tilde{\rho}$. We set K =10,000. The empirical reject probability K/L is empirically found to be .269. We denote the direct Monte Carlo using the same number of generated random variables (in fact using the same random variables) as ρ_d. The true value we denote as $\rho = .5erfc(\sqrt{n/2})$. The results:

$$
\begin{array}{llll}
n = 10 & \tilde{\rho} = 8.07 \times 10^{-4} & \rho_d = 9.16 \times 10^{-4} & \rho = 7.83 \times 10^{-4} \\
n = 20 & \tilde{\rho} = 3.88 \times 10^{-6} & \rho_d = 2.68 \times 10^{-5} & \rho = 3.87 \times 10^{-6} \\
n = 30 & \tilde{\rho} = 1.92 \times 10^{-8} & \rho_d = 0 & \rho = 2.16 \times 10^{-8} \\
n = 40 & \tilde{\rho} = 9.80 \times 10^{-11} & \rho_d = 0 & \rho = 1.27 \times 10^{-10}.
\end{array}
$$

Of course the values of zero for the last two cases indicate that there were no sample sums that averaged over 1, not surprisingly given the small probabilities. Thus we see that the blind simulation method can be a very powerful tool for estimating rare events in such situations.

12.2.2 Direct-Twist Markov Chain Method

The key fact that we have learned in the previous section is that rejection methods enter into the asymptotic variance calculation only through their effect on the accepted random variables. The efficiency of the acceptance-rejection procedure (i.e., the proportion of random variables that get rejected) gets washed out in the calculation of the speed factors.

Suppose we are now interested in calculating

$$\rho = P\Big(\sum_{i=1}^{n} g(X_i) > 0\Big),$$

where $\{X_i\}$ is a (nicely behaved) countable state space Markov chain. Without loss of generality, we assume that the state space of the chain is $\mathcal{Z}$ with transition probabilities $p_{ij} = P\big(X_{k+1} = j | X_k = i\big)$ and $g(\cdot)$ is some function mapping from the state space $\mathcal{Z}$ into $\Re$. We suppose that for the purposes of simulation we can stay at a state i and generate possible transitions (according to p_{ij}) as many times as we wish. This is certainly the case in many

Monte Carlo problems where a computer model of the system is available. We suppose that when the previous accepted value is i, we accept j with probability $h(j)c(i)$. We suppose that h and c are nonnegative functions and their product satisfies $0 \leq h(j)c(i) \leq 1$. The probability of accepting the next sample when we are at state i is simply

$$a(i) \doteq c(i) \sum_j p_{ij} h(j).$$

The accepted sequence of samples is a Markov chain in its own right with transition structure

$$q_{ij} \doteq \frac{p_{ij} h(j) c(i)}{a(i)} = \frac{p_{ij} h(j) c(i)}{c(i) \sum_j p_{ij} h(j)} = \frac{p_{ij} h(j)}{\sum_j p_{ij} h(j)}.$$

Interestingly enough the c dependence drops away. This means that we have freedom to choose c to satisfy other criteria. To accept as many random variables as possible, we would choose $c(i) = 1/\sup_j [h(j)]$.

There is a variety of interesting problems from the point of view of practical simulation that arises here. To compute the importance sampling weights we must know $a(i)$ (or equivalently $\sum_j p_{ij} h(j)$). This is a more formidable problem than in the i.i.d. case since we must know this sum for all i. We can (as in the i.i.d. case) estimate it from the data but we still have to store all these estimates.

What should be our choice for h? Unfortunately, the quick simulation optimal choice is not a simple exponential times the original transition density. It instead leads us to $h(j) = \exp\big(\theta_0 g(j)\big) r(j)$, where $r(\cdot)$ is the principal eigenvector of the large deviation Markov operator T_{θ_0} defined in Equation (3.11). Without a priori knowledge of p_{ij}, we have no way of knowing what form r is.

The calculation of variance rates for various suboptimal choices of h could now be embarked upon using our knowledge of large deviation theory and the techniques of the previous chapters.

13. The (Over-Under) Biasing Problem in Importance Sampling

It isn't that they can't see the solution. It is that they can't see the problem. *G. K. Chesterton*
I was gratified to be able to answer promptly, and I did. I said I didn't know. *Mark Twain*

In the previous chapters we have said a lot about how to choose good importance sampling biasing distributions. We have found that in many interesting situations easily computable efficient choices exist. All in all we have painted quite a rosy picture of the problem of rare event simulation. In this chapter, we indicate some of the (dangerous) problems that can face the importance sampling practitioner and how one might go about guarding against these dangers.

First of all, we should note that an efficient biasing scheme will not always be available. If we put enough restrictions on the class of biasing distributions that we are going to allow, then clearly we can make this class so restrictive that there won't be any good sampling distributions from which to choose. For example, if we have a functional of a Markov Chain type problem and we insist that only i.i.d. biasing distributions can be used, then an efficient alternative will not in general be available.

It is very often the case that we have some parametric class of sampling distributions from which to choose. The problem of choosing a sampling distribution then becomes a simpler problem of choosing a (possibly multidimensional) parameter. Changing the value of the parameter will typically change where the sampling distribution puts its mass in the observation space. We typically want to put more sampling distribution probability mass on the rare event set of interest. It is hoped that by controlling the parameter, we can to some extent choose where and how much of this mass to put on the set.

Choosing the biasing distributions must be done with care. Whenever we bias the distribution, we pay a certain cost. It is hoped we will gain back more than this cost by sampling in the correct region. To get an idea of the cost of biasing, consider the following scenario.

Suppose we wish to estimate $\rho = P\big((X_1, X_2) \in A_1 \times \mathcal{R}^d\big)$, where X_1 and X_2 are independent $\mathcal{R}^d$-valued random variables and $A_1 \subset \mathcal{R}^d$. Obviously $\rho = P\big(X_1 \in A_1\big)$ but not realizing this, we decide that we are going to bias (X_1, X_2) with probability density $p_1(x_1)p_2(x_2)$ with the random quantity (Y_1, Y_2) with associated probability density $q_1(y_1)q_2(y_2)$. We can think of X_2 as a "nuisance" random variable that really has nothing to do with the probability of the event that we are trying to measure. Our importance sampling estimator is thus

$$\hat{\rho} = \frac{1}{k}\sum_{j=1}^{k} 1_{\{(Y_1,Y_2)\in A_1\times\mathcal{R}^d\}} \frac{p_1(Y_1)p_2(Y_2)}{q_1(Y_1)q_2(Y_2)}.$$

As always $k\,Var(\hat{\rho}) = F_q - \rho^2$, where

$$\begin{aligned} F_q &= \int\int_{A_1} \frac{p_1^2(y_1)p_2^2(y_2)}{q_1(y_1)q_2(y_2)} dy_1 dy_2 \\ &= \left(\int_{A_1} \frac{p_1^2(y_1)}{q_1(y_1)} dy_1\right)\left(\int \frac{p_2^2(y_2)}{q_2(y_2)} dy_2\right). \end{aligned}$$

For any nonnegative scalar random variable Z, by Jensen's inequality $\mathbb{E}[1/Z] \geq 1/\mathbb{E}[Z]$. Thus

$$\begin{aligned} \gamma &= \int \frac{p_2^2(y_2)}{q_2(y_2)} dy_2 \\ &= \mathbb{E}[p_2(X_2)/q_2(X_2)] \\ &\geq 1/\mathbb{E}[q_2(X_2)/p_2(X_2)] \\ &= 1/\int q_2(y_2)dy_2 \\ &= 1/1 \\ &= 1. \end{aligned}$$

Therefore biasing this "nuisance" random variable X_2 inevitably causes us to suffer a higher estimator variance. We can think of γ as a cost multiplier associated with the unnecessary biasing of a random variable.

Sometimes these nuisance variables can be very well hidden. For example, suppose we are interested in $\rho = P(X \in A)$, where X is a random variable with probability density p. We generate a sequence of random variables $Y_1, Y_2, \ldots, Y_2n$ with probability density q to use for an importance sampling simulation. We can construct a family of estimators as follows (for all n that divide k),

$$\hat{\rho}_n = \frac{1}{k/n}\sum_{i=1}^{k/n}\left(\frac{1}{n}\sum_{j'=1}^{n} 1_{\{Y_{(i-1)n+j'}\in A\}} \prod_{j=1}^{n} \frac{p(Y_{(i-1)n+j})}{q(Y_{(i-1)n+j})}\right).$$

For each n, this is an unbiased estimator. The variance of the estimator is given by the variance of the summand times n/k. The variance of the summand, Var(summand), is

$$
\begin{aligned}
&Var(\text{summand}) \\
&= \int \left(\frac{1}{n} \sum_{j'=1}^{n} 1_{\{y_j \in A\}} \right)^2 \prod_{j=1}^{n} \frac{p(y_j)^2}{q(y_j)} dy_1 y_2 \cdots y_n \\
&= \frac{1}{n^2} \int \sum_{r=1}^{n} \sum_{i=1}^{n} 1_{\{y_r \in A\}} 1_{\{y_i \in A\}} \prod_{j=1}^{n} \frac{p(y_j)^2}{q(y_j)} dy_1 y_2 \cdots y_n \\
&= \frac{1}{n^2} \sum_{i=1}^{n} \int 1_{\{y_i \in A\}} \prod_{j=1}^{n} \frac{p(y_j)^2}{q(y_j)} dy_1 y_2 \cdots y_n \\
&+ \frac{1}{n^2} \int \sum_{r \neq i, r=1}^{n} \sum_{i=1}^{n} 1_{\{y_r \in A\}} 1_{\{y_i \in A\}} \prod_{j=1}^{n} \frac{p(y_j)^2}{q(y_j)} dy_1 y_2 \cdots y_n .
\end{aligned}
$$

Now define the nuisance factor $\gamma = \int p(x)^2/q(x)dx$ and $\gamma_A = \int_A p(x)^2/q(x)dx$. Then

$$
\begin{aligned}
Var(\text{summand}) &= \frac{1}{n^2} \sum_{i=1}^{n} \gamma^{n-1} \gamma_A + \frac{1}{n^2} \sum_{r \neq i, r=1}^{n} \sum_{i=1}^{n} \gamma^{n-2} \gamma_A^2 \\
&= \frac{1}{n} \gamma^{n-1} \gamma_A + \frac{n(n-1)}{n^2} \gamma_A^2 \gamma^{n-2}.
\end{aligned}
$$

Therefore

$$
\begin{aligned}
k\, Var(\hat{\rho}_n) &= n\, Var(\text{summand}) \\
&= \gamma^{n-1} \gamma_A + (n-1) \gamma_A^2 \gamma^{n-2}.
\end{aligned}
$$

We know that the nuisance parameter $\gamma \geq 1$. This implies that the best choice for n is $n = 1$.

In the discussion so far, we have assumed that the nuisance variables are independent of the event of interest. It can also be the case that an event may depend very weakly upon a certain random variable. Biasing this random variable will lead to an increase in the estimator variance given (almost) by a multiplication by γ. There are cases, for example, involving the simulation of long trajectories of Markov chains, where this problem is so acute that one can show that the optimal biasing distribution choice is to almost not bias at all and to use a simulation distribution very close to that of the direct Monte Carlo [3].

14. Tools and Techniques for Importance Sampling

You need only two tools. WD-40 and duct tape. If it doesn't move and it should, use WD-40. If it moves and shouldn't, use the tape. *Anon.*

If you put garbage in a computer nothing comes out but garbage. But this garbage, having passed through a very expensive machine, is somehow ennobled and none dare criticize it. *Anon.*

14.1 Adaptive Importance Sampling

All of the simulation distribution families that we have seen and worked with in this book are some sort of parametric family parameterized by some vector parameter θ. The mean shift method can be thought of as a family of distributions parameterized by the mean shift. The same is true for the variance scaling method. Of course the exponential shifts have the exponential shift parameter. A mixture of m exponential shifts can be thought of as a parametric class where we have the parameters $\theta_1, \theta_2, \ldots, \theta_m, p_1, p_2, \ldots, p_m$, where the $\{\theta_i\}$ are the (vector) exponential shifts for each element of the mixture and the $\{p_i\}$ are the scalar mixture weights (they must sum to one which implies there are really only $m - 1$ free mixture weight parameters). The universal simulation distributions (once we select the "covering measure" π) are characterized by a single scalar parameter. In fact this is their major selling point.

The large deviation methodology that we have espoused for choosing the parameter values is to compute the variance rate function for the given simulation problem as a function of the parameter. We then choose the parameter values that maximize the variance rate function. In many situations with the exponential shifts, mixtures of exponential shifts, or with the universal distributions, we have found that such a procedure in fact will give rise to an efficient simulation strategy. The main drawback to this procedure is that it requires that we have a lot of knowledge of the set structure and the relevant variance rate large deviation asymptotics of the simulation problem. In many situations, we have some very complicated rare event simulation problem. We may decide (sometimes just by intuition) that we want to use some family of parametric simulation distributions. How do we go about choosing good parameter values without trying to analyze completely the underlying large deviation theory?

In the next few subsections we discuss the main methodologies currently known and in use for attacking this problem.

14.1.1 Empirical Variance Minimization

Let $\theta \in \mathcal{R}^d$ be a parameter that we wish to optimize. The idea here is that we start off with some initial value of the parameter. We perform some simulation runs with this parameter value until we can get a good estimate of the empirical variance. We then (adaptively) change the parameter value (computing an empirical variance for each value) to try to minimize the estimator variance.

Suppose our estimate for k runs is

$$\hat{\rho}(\theta) = \frac{1}{k}\sum_{j=1}^{k} g(Y^{(j)})\frac{dP}{dQ}(Y^{(j)}, \theta)$$

with empirical mean square value given by

$$\hat{F}(\theta) = \frac{1}{k}\sum_{j=1}^{k} g(Y^{(j)})^2(\frac{dP}{dQ}(Y^{(j)}, \theta))^2.$$

Note that this is not the empirical variance but rather the mean square term. The empirical variance would be given by this term minus the estimate squared ($\hat{\rho}(\theta)^2$). Since we know that all the estimates are unbiased we choose to just minimize the mean square term. If $(\frac{dP(y)}{dQ(Y^{(j)},\theta)})^2$ is twice differentiable in the parameter θ we can utilize an algorithm based on the multidimensional version of Newton's method to minimize the empirical variance. Suppose $\theta \in \mathcal{R}^d$. Then define

$$\hat{F}'(\theta) = \frac{1}{k}\sum_{j=1}^{k} g(Y^{(j)})^2\nabla(\frac{dP}{dQ}(Y^{(j)}, \theta))^2$$

and

$$\hat{F}''(\theta) = \frac{1}{k}\sum_{j=1}^{k} g(Y^{(j)})^2\nabla^2(\frac{dP}{dQ}(Y^{(j)}, \theta))^2.$$

The algorithm can then be written

$$\theta_{n+1} = \theta_n - (\hat{F}''(\theta_n))^{-1}\hat{F}'(\theta_n).$$

The above algorithm requires that we be able to take two derivatives of the weight function with respect to the parameter. In many situations these derivatives are not available and one must try to optimize without calculating them explicitly. The classical method for gradient-free stochastic optimization is the Kieffer-Wolfowitz finite difference stochastic approximation algorithm [48]. This algorithm in our setting would be

$$\theta_{n+1} = \theta_n - \delta \triangle(\hat{F})(\theta_n),$$

where

$$\triangle(\hat{F})(\theta) = (\triangle(\hat{F})_1(\theta), \triangle(\hat{F})_2(\theta), \ldots, \triangle(\hat{F})_d(\theta)),$$

and

$$\triangle(\hat{F})_i(\theta) = \frac{\hat{F}(\theta_n + ce_i) - \hat{F}(\theta_n - ce_i)}{2c},$$

where δ and c are small "stepsize" parameters and e_i is a vector in $\mathcal{R}^d$ with a one in the ith position and zeros elsewhere. Note that for a "one-sided" alternative we could also use

$$\triangle(\hat{F})_i(\theta) = \frac{\hat{F}(\theta_n + ce_i) - \hat{F}(\theta_n)}{c}.$$

Obviously, $\triangle(\hat{F})(\theta)$, either the one-sided or two-sided version, is an approximation to the gradient of $\hat{F}$ at the point θ. To use this algorithm requires that we compute the empirical variance at $2d$ (for the two-sided) or d (for the one-sided) new values of the parameter to go to the next step. (That is, we would have to do this many simulations.) In large-dimensional problems this could be an almost overwhelming computational effort. An interesting variant of the Kieffer–Wolfowitz algorithm has been proposed in [77], called the Simultaneous Perturbation stochastic approximation. This algorithm is given as

$$\theta_{n+1} = \theta_n - \delta \tilde{\triangle}(\hat{F})(\theta_n),$$

where

$$\tilde{\triangle}(\hat{F})(\theta) = (\tilde{\triangle}(\hat{F})_1(\theta), \tilde{\triangle}(\hat{F})_2(\theta), \ldots, \tilde{\triangle}(\hat{F})_d(\theta)),$$

and

$$\tilde{\triangle}(\hat{F})_i(\theta) = \frac{\hat{F}(\theta_n + cb_n) - \hat{F}(\theta_n - cb_n)}{2b_{ni}},$$

where, as before, c and δ are small step-size parameters and $\{b_n\}$ is an i.i.d. sequence of $\mathcal{R}^d$-valued random vectors. Typically $b_n = (b_{n1}, b_{n2}, \ldots, b_{nd})$ is chosen to have its d components to be *i.i.d.* Bernoulli $\{-1, 1\}$ random variables. See [77] for other possible choices.

The main problem with empirical variance minimization is that if we start the parameter search with an extremely underbiased parameter, we get very few or no hits on the set of interest. This means that the computed empirical variance may very well be tiny which in turn causes the algorithm to behave as if it were near its optimal value.

For these reasons, it is the author's opinion that empirical variance minimization should only be attempted as a last resort. It has many practical and theoretical problems that preclude its use except possibly as a last ditch effort to optimize the simulation parameters.

Example 14.1.1. Consider the simple estimation problem

$$\rho = P(\frac{1}{10}\sum_{i=1}^{10}(X_i) > 10),$$

where $\{X_i\}$ are scalar i.i.d. standard normal random variables. Let θ be a scalar mean shift parameter and denote $q(x,\theta)$ as the variance one, mean θ normal probability density.

The usual estimate for k simulation runs is

$$\rho(\theta) = \frac{1}{k}\sum_{j=1}^{k} 1_{\{\sum_{i=1}^{10}(Y_i^{(j)})>100\}} \prod_{i=1}^{10} \frac{p(Y_i^{(j)})}{q(Y_i^{(j)},\theta)}$$

with associated empirical variance given by

$$F(\theta) = \frac{1}{k}\sum_{j=1}^{k} 1_{\{\sum_{i=1}^{10}(Y_i^{(j)})>100\}} \left(\prod_{i=1}^{10} \frac{p(Y_i^{(j)})}{q(Y_i^{(j)},\theta)}\right)^2.$$

The scalar Newton's algorithm is

$$\theta_{n+1} = \theta_n - (F''(\theta_n))^{-1}F'(\theta_n),$$

where the empirical first and second derivatives may be computed directly as

$$\begin{aligned} F'(\theta) &= \frac{1}{k}\sum_{j=1}^{k} 1_{\{\sum_{i=1}^{10}(Y_i^{(j)})>100\}} (20\theta - 2\sum_{i=1}^{10} Y_i^{(j)}) \\ &\quad \times \exp\left(-\sum_{i=1}^{10}(Y_i^{(j)})^2 + \sum_{i=1}^{10}(Y_i^{(j)}-\theta)^2\right) \\ F''(\theta) &= \frac{1}{k}\sum_{j=1}^{k} 1_{\{\sum_{i=1}^{10}(Y_i^{(j)})>100\}} (20 + (\sum_{i=1}^{10} 2(Y_i^{(j)}-\theta))^2) \\ &\quad \times \exp\left(-\sum_{i=1}^{10}(Y_i^{(j)})^2 + \sum_{i=1}^{10}(Y_i^{(j)}-\theta)^2\right). \end{aligned}$$

In Fig. 14.1, we see that the hit rate becomes perilously close to zero when the parameter value is much less than 9. Empirically, the algorithm has a lot of trouble converging for initial values of θ smaller than 9. Each value of θ was computed with 10,000 samples. When $\theta = 8.75$ we only obtained a single hit during the simulation (the expected number of hits for this sample size is computed to be approximately .39). We recommend great care in using empirical variance minimization and that before starting such a Newton's method iteration, users first investigate the hit rate on the set and assure themselves that it is sufficiently high before attempting this procedure.

For overbiasing, the pull-in range of the algorithm is much better. Here we start with many hits on the set of interest and thus we are not plagued with the underbiasing problem of artificially low empirical variance values. In the table below, we give some sample results. Again, each new value of θ in the iteration was computed with 10,000 points. (We also computed the table with 100 and 1000 and observed little variation from the results presented.)

Initial value $\theta = 12$	40 iterations to reach 10.005
Initial value $\theta = 15$	115 iterations to reach 10.005
Initial value $\theta = 20$	241 iterations to reach 10.005

Thus the lesson to be gained here is that if you are going to use empirical variance minimization, make sure that you initialize with an overbiased parameter. One can assure oneself of this just by monitoring the hit rate.

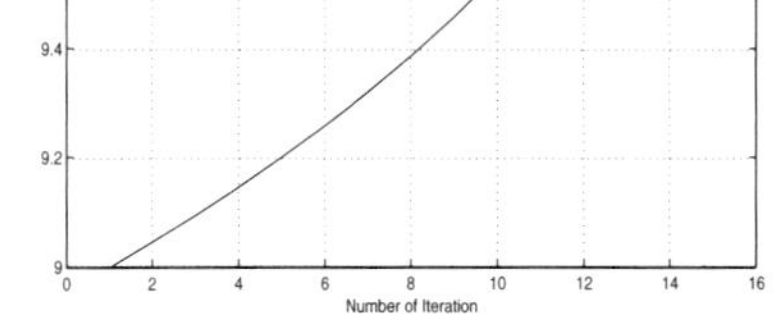

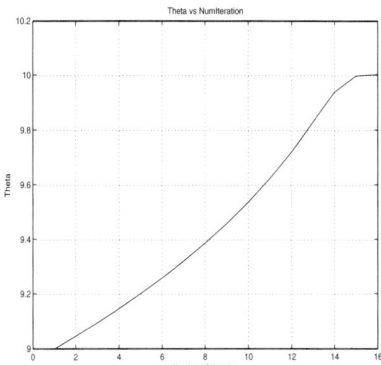

Fig. 14.1. Adaptive IS.

14.1.2 Exponential Shifts and the Dominating Point Shift Property

If we choose to use the family of exponential shifts with parameter θ, we can make use of the dominating point shift property to derive an adaptive algorithm. Recall that the dominating point shift property basically says that we should choose the parameter θ so that the probability mass of the simulation distribution is centered on the dominating point (assumed to be unique) of the set under consideration. In the i.i.d. sum case, this property is called the the mean value property, which says that we choose the exponential shift whose mean value is equal to the dominating point of the set.

Suppose we are interested in simulating to obtain estimates of the probability,

$$\rho_n = P\big(\frac{f_n(Z_{p,n})}{n} \in E\big),$$

where E is a Borel set such that $\overset{\circ}{E} \neq \emptyset$, $\bar{E} = \overline{\overset{\circ}{E}}$, $0 < I(E) < \infty$, and E has a unique dominating point ν. If P_n is the probability measure associated with the random variable $Z_{p,n}$, then the family of exponential shifts

$$dQ_n = \frac{\exp\big(\langle \theta_\nu, f_n(z)\rangle\big) dP_n(z)}{\exp\big(n\phi_n(\theta_\nu)\big)}$$

is efficient. How do we find (or approximate) θ_ν?

One possible approximation method would be to compute the empirical mean of the simulation distributions and adjust the parameter θ until it equals the dominating point (assumed known) ν. An algorithm performing this task might be

$$\theta_{n+1} = \theta_n + \delta(\nu - \frac{1}{k}\sum_{j=1}^{k} f_n(Z_{q,n}^{(j)})),$$

where δ is the step-size parameter.

Example 14.1.2. We continue with our example of the previous subsection. Suppose we are interested in simulating

$$\rho = P\big(\frac{1}{10}\sum_{i=1}^{10}(X_i) > 10\big),$$

where $\{X_i\}$ are scalar i.i.d. standard normal random variables. Again let θ be a scalar mean shift parameter and denote $q(x, \theta)$ as the variance one, mean θ normal probability density. The dominating point of the set of interest is obviously 10. Using the mean shift property, our algorithm thus becomes

$$\theta_{n+1} = \theta_n - \delta\left(\frac{1}{k}\left(\sum_{j=1} kY_\theta^{(j)} - 10\right)\right),$$

where $\{Y_\theta^{(j)}\}$ are i.i.d. samples taken from the density $q(\cdot, \theta)$. The initial value does not have to be chosen carefully here and the "pull-in" range of the algorithm is essentially infinity. In Fig. 14.2 we see a typical iteration where we start off at $\theta = 0$.

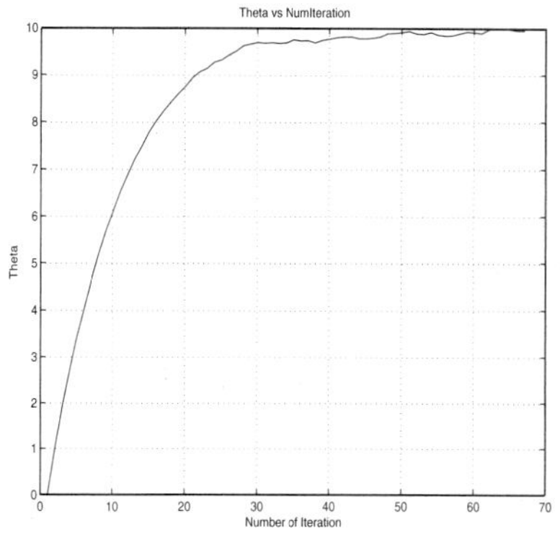

Fig. 14.2. The mean value property method.

In situations where the dominating point is not known or doesn't exist additional complications arise. Obviously, if there are several "important points" that need to be covered and they are known, we can use the above procedure to fix the parameter set for a finite mixture of exponential shifts. If the minimum rate point structure of the set is completely unknown, then some other attack strategy (perhaps using universal simulation distributions) will have to be tried.

14.2 Hit Rate Considerations

The probability that we hit the set during a simulation run is called the *hit rate*. We denote the hit rate as κ. It is the author's opinion that this statistic is one of the most underused, misunderstood, dangerous (!), and valuable diagnostic tools available to the simulation practitioner. In many situations, it can be an indicator that your methodology is sick. In other situations, it might be just one of several indicators that the simulation is proceeding more or less on track. Whenever the author does a simulation, as a matter of course, hit rate statistics are collected. Let's consider a little bit what might be going on with this statistic.

Many researchers have tried to develop an intuition for good simulation distributions by considering the optimal (but impractical) simulation distribution of Equation (4.2). In the setting of estimating a set probability, this distribution is

$$q_{opt}(x) = \frac{p(x)1_E(x)}{\rho}, \tag{14.1}$$

where $\rho = P(E)$ is the quantity of interest. This simulation distribution depends directly on ρ, the quantity that we wish to estimate and thus is not a practical choice. The important point to note here is that the hit rate κ of this simulation distribution is one. This fact has led many practitioners into the trap of thinking that good simulation distributions should have high hit rates (or hit rates as close to one as possible). This philosophy, of course, leads to overbiasing and very large estimator variance.

We now consider what the hit rate might be when we use certain families of efficient simulation distributions.

14.2.1 Hit Rates for a Single Exponential Shift

We are interested in simulating $\rho = P\big(f_n(Z_{p,n}) \in nE\big)$. Suppose the set of interest E has a minimum rate point ν (we don't assume that it is a dominating point) and we decide to use the simulation distribution,

$$dQ_n(z) = \exp\big(-n\phi_n(\theta_\nu)\big)\exp\big(\langle\theta_\nu, f_n(z)\rangle\big)dP_n(z).$$

Of course, we must have that

$$\kappa_n = \int_E dQ_n(z).$$

It is quite normal (bad pun intended) to assume that Q_n (the probability measure associated with the random variable $\{Z_{q,n}\}$) should satisfy some sort of Central Limit theorem. In particular, we assume that

$$\frac{f(Z_{q,n}) - n\nu}{\sqrt{n}} \to G,$$

where the convergence is in distribution and G is a zero mean Gaussian vector with positive definite covariance matrix V. Denote the probability density of the $\mathcal{R}^d$-valued Gaussian vector G as Φ. Thus, for large n,

$$\begin{aligned}
\kappa_n &= P(f(Z_{q,n}) \in E) \\
&\approx P(G\sqrt{n} + n\nu \in E) \\
&= \int_{(E/\sqrt{n})-\sqrt{n}\nu} \Phi(x)dx.
\end{aligned}$$

This gives us quite a lot of information. What this would mean in the one-dimensional case is that the hit rate should be around .5. In higher-dimensional settings, everything depends on the local shape of the set E near the point ν_E and how elliptical the Gaussian distribution ϕ is (i.e., how far away from being a diagonal matrix V would be). In Fig. 14.3, we present a variety of possible scenarios. The circles and ellipses are supposed to represent an equi-probability contour of the limiting Gaussian distribution ϕ with covariance V. In the upper left drawing, V is diagonal, and E is a half-space. In this setting, we would expect a κ near .5. In the upper right drawing, we see with the same limiting distribution, how the local shape of E might give us very different hit rates. In the lower two drawings, we show how the eccentricity and the orientation of the major and minor axes can also heavily influence the hit rate.

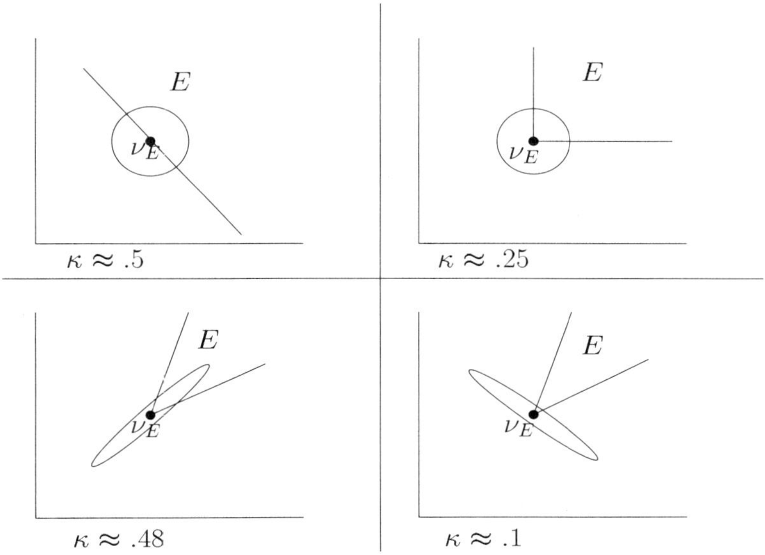

Fig. 14.3. Various hit rate scenarios.

How one might use the hit rate to diagnose a simulation depends quite a lot on the information disposable to the practitioner. If she has a good notion of the set geometry and the eccentricity of the local limiting normal distributions, a lot of diagnostic information is available from the hit rate. In other situations, the practitioner may possess very little of this type of information and the hit rate information is not very useful. The author's experience is that if he is getting a hit rate of more than .6 or so, or less than .1, he begins to feel a horrible sense of foreboding.

14.2.2 Hit Rates for the Universal Distributions

The universal distributions have been put forth as a panacea for situations where the structure of the set of interest E is unknown. We know that these distributions are indeed efficient. However, unfortunately, in situations where there is only a single or just a few important points, they give rise to slowly converging estimators. The reason is quite simple: they spend a lot of time "looking" in nonproductive regions of the space. One of the manifestations of this unproductive search time translates into low hit rates. In this section we consider exactly what sort of hit rates we might expect in a given situation.

Specifically, we consider the family of simulation distributions given by Theorem 9.1.1, $\{q_n\}$. Consider a very nice set $E = \{x \in \mathcal{R}^d : x_1 > \rho\}$ which of course is just a hyperplane at distance ρ from the origin. We wish to determine the hit rate on this set. The results here will be of much greater validity than one might think given such a simple setting. First note that we are using the universal alternatives for the Gaussian setting. However, if there is a Central Limit Theorem in operation (which will almost always be the case), we will obtain the same asymptotic hit rate results for this Gaussian problem as we would for the more general universal simulation distributions given by Equation (10.1). Also the fact that our set of interest is a hyperplane has very little to do with our results. What is important is that the set is "flat" or has zero curvature at the dominating point.

Thus we wish to compute the hit probability,

$$\int_E q_n(x)dx.$$

Now we know that q_n has the same density as $Y = \frac{Z}{\sqrt{n}} + \rho U$ and

$$f_U(u) = \frac{\Gamma(\frac{d}{2})}{\Gamma(\frac{d-1}{2})\sqrt{\pi}}(1-u^2)^{((d-1)/2)-1}.$$

We consider the easiest case: take $K = id$, where K is the covariance matrix of Z and $r = I(E) = \frac{\rho^2}{2}$. Note also that

$$\begin{aligned}
&\int_E q_n(x)dx \\
&= \int_E P(\frac{1}{\sqrt{n}}Z > \rho - x_1\rho) f_U(x_1)dx_1 \\
&= \int_E \Phi^{-1}(\sqrt{n}\rho(1-x_1)) f_U(x_1)dx_1 \\
&= \int_{-1}^{1} \Phi^{-1}(\sqrt{n}\rho(1-x)) \frac{\Gamma(\frac{d}{2})}{\Gamma(\frac{d-1}{2})\sqrt{\pi}}(1-x^2)^{((d-1)/2)-1}dx \\
&= \frac{\Gamma(\frac{d}{2})}{\Gamma(\frac{d-1}{2})\sqrt{\pi}} \int_0^2 \Phi^{-1}(\sqrt{n}\rho x)x^{((d-1)/2)-1}(2-x)^{((d-1)/2)-1}dx \\
&= \frac{\Gamma(\frac{d}{2})}{2\Gamma(\frac{d-1}{2})\sqrt{\pi}} \int_0^2 erfc(\frac{\sqrt{n}\rho u}{\sqrt{2}})u^{((d-1)/2)-1}(2-u)^{((d-1)/2)-1}du,
\end{aligned}$$

the last line following since $\Phi^{-1}(x) = \frac{1}{2}erfc(\frac{x}{\sqrt{2}})$.

The integral above is of the form $\int_0^\infty h(\lambda u)f(u)du$, where

$$\lambda = \frac{\sqrt{n}\rho}{\sqrt{2}}$$
$$h(t) = erfc(t)$$

and

$$f(t) = \begin{cases} t^{((d-1)/2)-1}(2-t)^{((d-1)/2)-1} & \text{if } 0 \le t \le 2 \\ 0 & \text{otherwise.} \end{cases}$$

Note also that as $n \to \infty$, $\lambda \to \infty$. Also we have $f(t) \approx 2^{((d-3)/2)}t^{((d-3)/2)}$ as $t \to 0^+$. Furthermore, $h(t) \approx \frac{1}{\sqrt{\pi}}\frac{1}{t}\exp(-t^2)$ as $t \to \infty$.

From Theorem 4.4 case 2 of [10] with $d = 1, \nu = 2, r_0 = 1, c_{00} = \frac{1}{\sqrt{\pi}}, q = 0, a_0 = \frac{d-3}{2}$, and $p_{00} = 2^{((d-3)/2)}$, the asymptotic expansion of the integral (subject to some technical conditions which we verify) is

$$(\frac{\sqrt{n}\rho}{\sqrt{2}})^{-1-a_0}2^{((d-3)/2)}M[h;\frac{d-1}{2}], \tag{14.2}$$

where $M[h;s]$ is the Mellin transform of h at s. From [26],

$$M[h;s] = \frac{1}{\sqrt{\pi}}\frac{1}{s}\Gamma(\frac{1}{2}s+\frac{1}{2}) \qquad Re(s) > 0.$$

We can rewrite f as $f(t) = 2^{((d-3)/2)}t^{((d-3)/2)}g(\frac{t}{2})$ where

$$g(t) = \begin{cases} (1-t)^{((d-3)/2)} & \text{if } 0 \le t \le 1 \\ 0 & \text{otherwise.} \end{cases}$$

The Mellin transform of g is given by

$$M[g;s] = B(\frac{d-1}{2}, s) \qquad Re(s) > 0,$$

where $B(x,y)$ is the Beta function. A simple scaling operation gives

$$M[g(\frac{t}{2});s] = 2^s B(\frac{d-1}{2}, s) \qquad Re(s) > 0,$$

and thus

$$M[f;s] = 2^{((d-3)/2)}2^{(s+\frac{d-3}{2})}B(\frac{d-1}{2}s + \frac{d-3}{2}) \qquad Re(s) > -\frac{d-3}{2}.$$

Hence $M[f;1-s]$ and $M[h;s]$ are holomorphic on the strip $0 < Re(s) < \frac{d-1}{2}$ and nonzero for all $d \ge 2$, which is a requisite for applying Theorem 4.4 of [10].

Substituting our expression for $M[h;s]$ into Equation (14.2) we obtain,

$$\int_0^2 erfc(\frac{\sqrt{n}\rho u}{\sqrt{2}})u^{((d-1)/2)-1}(2-u)^{((d-1)/2)-1}du$$
$$\approx (\frac{\sqrt{n}\rho}{\sqrt{2}})^{-(d-1)/2} 2^{((d-3)/2)} \frac{1}{\sqrt{\pi}} \frac{2}{d-1} \Gamma(\frac{d+1}{4}),$$

and thus for large n,

$$\int_E q_n(x)dx \approx \frac{2^{(3d-7)/4}}{\rho^{(d-1)/2}\pi(d-1)} \frac{\Gamma(\frac{d}{2})\Gamma(\frac{d+1}{4})}{\Gamma(\frac{d-1}{2})} n^{-(d-1)/4}. \tag{14.3}$$

In Fig. 14.4, we give the empirical hit rate and the theoretical value (computed from (14.3)) for the case of $d = 4$, $\rho = 1$ as we vary n. To compute each point of the empirical hit rate we computed 1,000,000 independent samples from the universal distribution and counted the number of times we hit the set E.

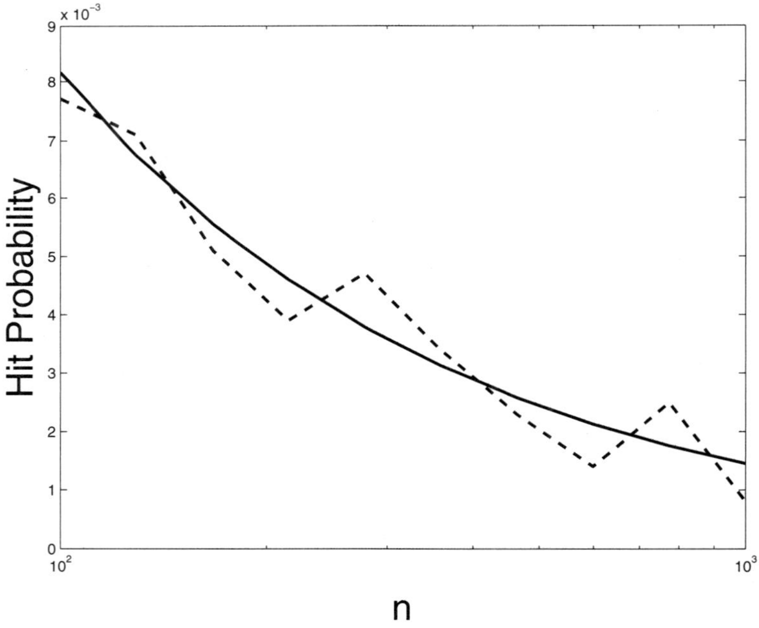

Fig. 14.4. Plot theoretical (solid line) and empirical (dotted line) hitrates for dimension $d = 4$, $\rho = 1$.

14.3 Efficient Biasing of Functions of Independent Random Sequences

In many situations, we have some function of two (or more) independent random sequences. Suppose $\{Z_{1,n}\}$ is an $\mathcal{R}^d$-valued random variable sequence and $\{Z_{2,n}\}$ is an $\mathcal{R}^{d'}$-valued sequence independent of each other whose logarithm moment generating function sequences

$$\phi_{i,n}(\cdot) = (1/n)\log \mathbb{E}[\exp(\langle \cdot, Z_{i,n}\rangle)] \quad i = 1, 2$$

each satisfy the standard assumptions A1 through A4. We can then consider the pair $\{(Z_{1,n}, Z_{2,n})\}$ as an $\mathcal{R}^{d+d'}$-valued sequence. Its logarithm moment generating function sequence $\{\phi_n(\theta, \theta') = \phi_{1,n}(\theta) + \phi_{2,n}(\theta')\}$ for $\theta \in \mathcal{R}^d$ and $\theta' \in \mathcal{R}^{d'}$. $\{\phi_n\}$ then also satisfies the standard assumptions. Hence by Theorem 3.2.1 $\{(Z_{1,n}, Z_{2,n})\}$ satisfies a large deviation principle with rate function

$$\begin{aligned} I(z, z') &= \sup_{\theta,\theta'}[\langle \theta, z\rangle + \langle \theta', z'\rangle - \phi_1(\theta) - \phi_2(\theta')] \\ &= I_1(z) + I_2(z'), \end{aligned}$$

where $I_i(\cdot)$, $i = \{1, 2\}$ is the rate function for the sequence $\{Z_{i,n}\}$. Now suppose we are interested in $\rho_n = P\big(F(Z_{1,n}/n, Z_{2,n}/n) \in A\big)$, where F is some continuous mapping from $\mathcal{R}^{d+d'}$ into $\mathcal{R}^{\hat{d}}$, and $\bar{A} = \bar{A^o}$. Let $C = \{(z, z') : F(z, z') \in A\}$. Then it is true that $\bar{C} = \bar{C^o}$, and $\rho_n = P\big((Z_{1,n}/n, Z_{2,n}/n) \in C\big)$. From the Contraction Principle, Theorem 3.5.1, we may deduce the asymptotics of the probability sequence ρ_n in terms of the rate function I. To simplify the argument, let us suppose that the set C has a dominating point (z_0, z_0'). Then

$$\begin{aligned} \lim_{n\to\infty} \frac{1}{n}\log \rho_n &= -\inf_{(z,z')\in C} I(z, z') \\ &= -I(z_0, z_0') \\ &= -\inf_{(z,z')\in C} I_1(z) + I_2(z') \\ &= -(I_1(z_0) + I_2(z_0')). \end{aligned}$$

We know that an efficient simulation strategy exists in this setting by choosing as the biasing distributions, the exponential shift of the original distributions. This means that we simulate with the distributions

$$\begin{aligned} dQ_n(z, z') &= \frac{\exp(\langle \theta_o, z\rangle + \langle \theta_o', z\rangle)dP_{1,n}(z)P_{2,n}(z')}{\exp\big(n\phi_{1,n}(\theta_o)\big)\exp\big(n\phi_{2,n}(\theta_o')\big)} \\ &= \frac{\exp(\langle \theta_o, z\rangle)dP_{1,n}(z)}{\exp\big(n\phi_{1,n}(\theta_o)\big)} \frac{\exp(\langle \theta_o', z'\rangle)dP_{2,n}(z')}{\exp\big(n\phi_{2,n}(\theta_o')\big)} \\ &= dQ_{1,n}(z)Q_{2,n}(z'), \end{aligned}$$

where θ_o is the value of θ that satisfies $I_1(z_o) = \sup_\theta[\langle\theta, z_o\rangle - \phi_1(\theta)] = \langle\theta_o, z_o\rangle - \phi_1(\theta_o)$ (similarly θ'_o satisfies $I_2(z'_o) = \langle\theta'_o, z'_o\rangle - \phi_2(\theta'_o)$).

What this means is that the random variable sequences $\{Z_{1,n}\}$ and $\{Z_{2,n}\}$ can be biased *separately* and still be able to attain an efficient simulation. Our assumption of a dominating point is not essential to the argument. If there is a finite set of "important" points for the set C, we can use a convex combination of biasing distributions for each random variable sequence separately and still have an efficient simulation.

In some situations, we may be able to bias one of the random variable sequences but not the other. Hence suppose that we must choose $Q_{1,n} = P_{1,n}$ for all n. In other words, we are not allowed to bias the subscript one random variable sequence. Under the same assumptions as above, we can compute the generating function sequence

$$\begin{aligned}
&c_n(\theta, \theta') \\
&= \frac{1}{n}\log\int \exp(\langle\theta, z\rangle + \langle\theta', z'\rangle) \\
&\quad \times \left(\frac{dP_{1,n}(z)dP_{2,n}(z')}{dQ_{1,n}(z)dQ_{2,n}(z')}\right)^2 dQ_{1,n}(z)dQ_{2,n}(z') \\
&= \frac{1}{n}\log\int \exp(\langle\theta, z\rangle + \langle\theta', z'\rangle) \\
&\quad \times \left(\frac{dP_{1,n}(z)dP_{2,n}(z')}{dP_{1,n}(z)dQ_{2,n}(z')}\right)^2 dP_{1,n}(z)dQ_{2,n}(z') \\
&= \frac{1}{n}\log\int \exp(\langle\theta, z\rangle + \langle\theta', z'\rangle)\left(\frac{dP_{2,n}(z')}{dQ_{2,n}(z')}\right)^2 dP_{1,n}(z)dQ_{2,n}(z') \\
&= \phi_{1,n}(\theta) + \frac{1}{n}\log\int \exp(\langle\theta', z'\rangle)\left(\frac{dP_{2,n}(z')}{dQ_{2,n}(z')}\right)^2 dP_{1,n}(z)dQ_{2,n}(z') \\
&= \phi_{1,n}(\theta) + c_{2,n}(\theta').
\end{aligned}$$

Thus

$$\lim_{n\to\infty} c_n(\theta, \theta') = \phi_1(\theta) + c_2(\theta').$$

The variance rate function would then just be $R(z, z') = I_1(z) + R_2(z')$. Minimizing the variance rate function over the set of interest gives

$$\begin{aligned}
\inf_{(z,z')\in C} R(z, z') &= \inf_{(z,z')\in C} I_1(z) + R_2(z') \\
&= I_1(z_0) + \inf_{(z,z')\in C} R_2(z').
\end{aligned}$$

The smallest possible value for the above infimum is $2I_2(z'_0)$ which can be achieved if we choose $Q_{2,n}$ *exactly* as we did above in the situation where we could have an efficient simulation strategy.

What this means is that if we are unable to bias both sequences of random variables, we should not try to overbias one of the sequences in order to make up for the lack of ability to bias the other one.

Thus suppose we are interested in estimating

$$\rho_n = P\big(Z_{1,n} + Z_{2,n} > na\big),$$

where the two $\mathcal{R}$-valued sequences $\{Z_{1,n}\}, \{Z_{21,n}\}$ are independent. Let us suppose also that we don't know the distributions of $Z_{1,n}$ well enough to be able to easily construct efficient biasing sequences for it. Thus we decide to only bias the second $\{Z_{2,n}\}$ sequence. Our intuition would normally tell us to bias this second sequence so that the mean value of the sum of the two sequences is at na. From our above discussion, however, we see that this is incorrect and in fact leads us to overbias the second sequence. How *does* one handle this situation then? At the time of this writing this is still an open problem.

14.3.1 Sums of Independent Sequences

Suppose we are interested in $\rho_n = P\big(Z_{1,n} + Z_{2,n} > na\big)$ for two independent $\mathcal{R}$-valued sequences $\{Z_{1,n}\}$ and $\{Z_{2,n}\}$ with respective rate functions $I_1(\cdot)$ and $I_2(\cdot)$. The set C of interest is $C = \{(z, z') : z + z' > a\}$. If there is a dominating point (z_o, z_o') for this set, then the rate with which ρ_n goes to zero is given by $I(z_o, z_o') = I_1(z_o) + I_2(z_o')$ and where $z_o + z_o' = a$. Note also that

$$\begin{aligned}\rho_n &= P\big(Z_{1,n} + Z_{2,n} > na\big) \\ &\geq P\big(Z_{1,n} > x_1 n\big)P\big(Z_{2,n} > x_2 n\big) \\ &\quad \text{for all } x_1, x_2 \ \ni \ \{(x_1, x_2) : x_1 + x_2 = a\}.\end{aligned}$$

This then implies that $I(z_o, z_o') \leq I_1(x_1) + I_2(x_2)$ for all (x_1, x_2) such that $x_1 + x_2 = a$. Hence, from the above characterizations of $I(z_o, z_o')$, we must have

$$I(z_o, z_o') = \inf_{(x_1,x_2):x_1+x_2=a} I_1(x_1) + I_2(x_2).$$

Under the standard assumptions, the rate functions are differentiable, leading us to the requirement that

$$I_1'(x_1) = I_2'(x_2). \tag{14.4}$$

Recall that

$$\begin{aligned}I_i(x) &= \sup_\theta[\theta x - \phi_i(\theta)] \qquad i = \{0, 1\} \\ &= [\theta_{i,x} x - \phi_i(\theta_{i,x})] \quad\; i = \{0, 1\}.\end{aligned}$$

Due to a duality property $I_i'(x) = \theta_{i,x}$. Therefore an alternate characterization of (14.4) is the following equation,

$$\theta_{1,x_1} = \theta_{2,x_2}. \tag{14.5}$$

Suppose that we can generate exponentially shifted versions of the random variable sequences $\{Z_{1,n}\}$, $\{Z_{2,n}\}$. This suggests an adaptive procedure for finding (x_1, x_2). Choose a initial value of θ as the exponential shift parameter of both the $\{Z_{1,n}\}$ and $\{Z_{2,n}\}$ random variables. Generate these sequences (call them $\{Z_{1,n}^{(\theta)}\}$ and $\{Z_{2,n}^{(\theta)}\}$), and compute $Z_{i,n}^{(\theta)}$ $i = \{1, 2\}$ for large n. Adaptively, adjust the exponential shift parameter θ until $\mathbb{E}[Z_{1,n}^{(\theta)} + Z_{2,n}^{(\theta)}] \approx an$.

This technique can be generalized to a sum of N independent sequences of $\mathcal{R}^d$-valued random variables. Suppose $\rho_n = P\big(\sum_{r=1}^{N} Z_{r,n} \in nA\big)$ and where $A \subset \mathcal{R}^d$ and has a dominating point η. The associated rate functions of the N sequences are denoted $I_1(\cdot), I_2(\cdot), \ldots, I_N(\cdot)$. Let us denote the $\mathcal{R}^d$-valued exponential shift parameters as $\theta_1, \theta_2, \ldots, \theta_N$. The exponentially shifted random variables are such that $\lim_{n\to\infty} Z_{r,n}^{(\theta)}/n = x_{\theta,r}$ (where the convergence is almost sure). Then following the same argument as above, we find that the optimization problem becomes

$$\inf_{\{(x_1,x_2,\ldots,x_n):\sum_{i=1}^{N} x_i = \eta\}} \sum_{j=1}^{N} I_j(x_j).$$

To solve this we set up the Lagrange multiplier formulation; minimize the objective function

$$\sum_{j=1}^{N} I_j(x_j) + \lambda \sum_{i=1}^{N} x_i.$$

Taking the gradient with respect to x_j implies $\nabla I_j(x_j) = \lambda$ for all j. Thus the gradient at the optimizing points $\nabla I_j(x_j)$ should be a constant. We again use the identity that $\nabla I_j(x_j) = \theta_{x_j}$ and so the exponential shift parameters for all N sequences should be the same. Thus to find the optimal shift parameters, we merely adjust θ until we have $\sum_{r=1}^{N} x_{\theta,r} = \eta$.

Example 14.3.1. We define the following three sequences of $\mathcal{R}$-valued random variables: $\{X_{1,n}\}$ are i.i.d. exponential random variables with mean value 1; $\{X_{2,n}\}$ are i.i.d. uniform $[0, 1]$; and $\{X_{3,n}\}$ are i.i.d. standard Gaussian. Suppose we are interested in

$$\rho = P\Big(\sum_{i=1}^{n} X_{1,i} + X_{2,i} + X_{3,i} > na\Big).$$

To employ the above theory we need to investigate the exponential shifts of each of these sequences. The exponential random variables when exponentially shifted (with parameter θ) have the probability density

$$p_1^{(\theta)} = (1-\theta)\exp\bigl(-(1-\theta)x\bigr) \qquad x>0,$$

which has mean value

$$\mathbb{E}[X_1^{(\theta)}] = 1/(1-\theta).$$

The uniform when exponentially shifted has density

$$p_2^{(\theta)} = \frac{\theta\,\exp(\theta x)}{\exp(\theta)-1} \qquad 0<x<1,$$

which has mean value

$$\mathbb{E}[X_2^{(\theta)}] = \frac{\theta\,\exp(\theta)-\exp(\theta)+1}{\theta(\exp(\theta)-1)}.$$

Finally the standard Gaussian when exponentially shifted is a Gaussian with variance one and mean $\mathbb{E}[X_3^{(\theta)}] = \theta$. According to our methodology, we need to adjust the value of theta so that $\mathbb{E}[X_1^{(\theta)}] + \mathbb{E}[X_2^{(\theta)}] + \mathbb{E}[X_3^{(\theta)}] = a$, which gives the equation

$$\frac{1}{1-\theta} + \frac{\theta\,\exp(\theta)-\exp(\theta)+1}{\theta(\exp(\theta)-1)} + \theta = a.$$

When $\theta = 0$ (i.e. we do not shift at all), the expression above is 1.5. Hence for any value of $a \geq 1.5$, we can find a positive solution θ to the equation.

14.4 The Method of Conditioning

One of the principal themes of this text has been to emphasize the importance of finding the minimum rate points of rare event sets. Our simulation attack procedures change radically from the case of one simple dominating point to the case of many or even infinitely many minimum rate points or important points.

A very typical situation in communications systems is that each symbol sequence typically will have its own associated noise "error" set. The simplest example is to consider binary signaling down an additive Gaussian channel. The received signal is thus $Y = B + N$ where Y is the received scalar random variable, B is a symmetric Bernoulli $\{-1, 1\}$ random variable, and N is a zero mean Gaussian noise. We suppose that N has a small variance $1/n$. Suppose we are interested in the probability of error of such a system. In this case, we can write

$$\begin{aligned} P(\text{Error}) &= P\bigl(sgn(Y) \neq B\bigr) \\ &= P\bigl(sgn(B+N) \neq B\bigr). \end{aligned}$$

If we were going to simulate this using importance sampling, for large N, it is easy to compute and intuitive to see that the important set for the noise samples is $(-\infty,-1]\cup[1,\infty)$. In other words, only if N lies in this set can there be an error observed. Thus there are two minimum rate points $\{-1,1\}$ and our efficient simulation strategy would be to use a mixture of two normals with the minimum rate points as their respective means.

One can see here that a lot of simulation runs are wasted effort. For example, whenever $B=+1$, for an error to occur N needs to be less than -1. Thus, whenever the noise sample is taken from the $+1$ mean of the mixture, we will have very little chance of producing an error. Therefore 50% of the time we are generating the "wrong" noise samples.

The obvious solution here is to use conditioning,

$$
\begin{aligned}
P(\text{Error}) &= P\big(sgn(B+N)\neq B\big)\\
&= P\big(sgn(B+N)\neq B|B=1\big)P\big(B=1\big)\\
&\quad +P\big(sgn(B+N)\neq B|B=-1\big)P\big(B=-1\big)\\
&= P\big(sgn(1+N)\neq 1\big)P\big(B=1\big)\\
&\quad +P\big(sgn(-1+N)\neq -1\big)P\big(B=-1\big).
\end{aligned}
$$

By the assumed symmetry, we can obviously simplify this even further. However, we wish to note that by conditioning on the transmitted symbol, we have now have created two simulation problems instead of the one we had before. This disadvantage is more than offset by the fact that each of these simulation problems is simpler than the original. For example, to estimate via simulation

$$
P\big(sgn(1+N)\neq 1\big)=P\big(N\leq -1\big),
$$

we have an important set of $(-\infty,-1]$ which has a single dominating point of -1. We can simulate this probability with a single exponential shift simulation distribution (in this Gaussian setting, it is just a mean shifted Gaussian). We do not "waste" any random variables, generating them from the "wrong" segments of a mixture distribution.

This simple example illustrates a very common simulation strategy. By conditioning on a symbol sequence or any collection of dependent random variables, we can very often break a large simulation problem into several smaller manageable pieces and handle them separately.

14.5 Simulating Ergodic Systems with Memory

The general problem of how to go about simulating systems with memory is so vast that we can only hope to touch upon a few points and recommend to the reader a few generic techniques. We assume that we wish to estimate the probability of some rare event that occurs at the output of an ergodic system with memory. The idea is that the system has settled into some sort of steady-state and we wish to get an idea of the steady-state probability of visiting some rare event. This setting arises constantly in the study of many complicated communications receivers that have feedback. Feedback in a communications receiver creates dependency on past decisions and inputs. This dependency or memory significantly complicates the analysis and/or simulation of such systems.

Consider the system in Fig. 14.5. The idea is that the sequence (in i)

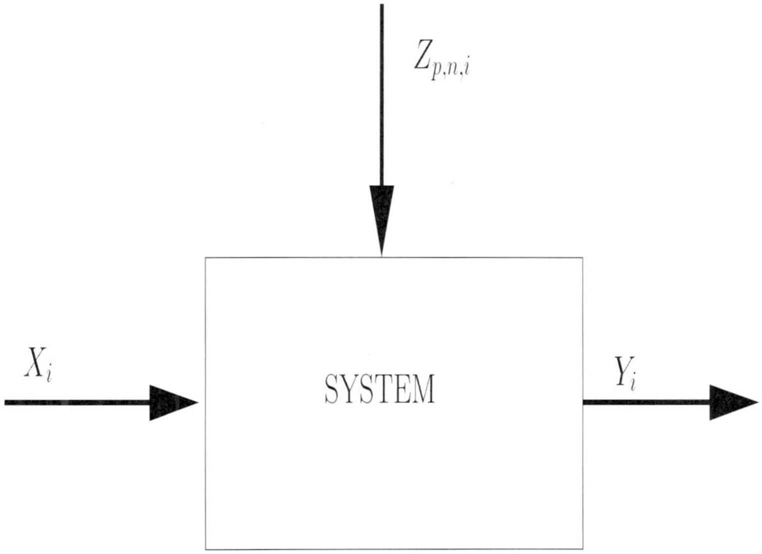

Fig. 14.5. A generic system with memory.

$\{Z_{p,n,i}\}$ is an i.i.d. sequence of $\mathcal{S}$ valued random variables obeying some sort of large deviation principle (in n). What we are really proposing to do is to study a sequence (in n) of systems with memory. In a given application, we embed our problem in such a sequence and use the asymptotic theory developed here to give guidance on how to choose the simulation distributions. In the usual communication system setting, we might think of $\{Z_{p,n,i}\}$ as some sort of "small noise" sequence. We have another "data sequence" of $\mathcal{T}$-valued random variables $\{X_i\}$ being input into the system that together with the small noise inputs creates the random variables $\{Y_n\}$. We are interested in

$$\rho = P\big(f(Y_l) \in E\big),$$

where due to the assumed stationarity, this probability does not vary with l. We think of f as being a mapping from the state space of the $\{Y_i\}$ into $\mathcal{R}^d$. E, then, is some set (event) in $\mathcal{R}^d$.

If the system is *causal* (i.e., the present output depends only upon the present and past values of the inputs), we can express ρ as

$$\rho = P\big(f(h(\dots, X_{l-m}, Z_{p,n,l-m}, X_{l-m+1}, Z_{p,n,l-m+1}, \dots X_l, Z_{p,n,l})) \in E\big),$$

where h describes the system function.

As simulation designers we now must decide how to bias the $\{Z_{p,n,i}\}$ sequence. The first fact that we must face is that even if the system has infinite memory, during a simulation, we can only bias a finite number of the random variables. We hope (this would be the usual case in a communications system) the memory dies off fairly rapidly and we can think of the system as having only m units of "effective memory." We emphasize that we will still try to simulate the infinite memory system, but we will only *bias* in each simulation run a finite number of "noise" variables. To clarify this notion a bit further: the idea is that we let the system run until it is in steady-state. Then at time $l-m$, we begin biasing the input noise samples. We continue biasing until time l at which time we check to see if the event $\{f(Y_l) \in E\}$ has occurred. Now, at this point, we would have to wait again until our system reaches steady state again. Remember, we have just artificially perturbed it away from its normal behavior with the biased noise samples. We of course did this on purpose so that we could increase occurrence frequency of the rare event. However, now this biasing sequence is coming back to haunt us. We may have to wait a long time before the system has settled back down into its steady state.

To get around this problem, we employ the so-called *twin system method* [1]. The idea is simple. During the simulation, we simulate the original system and an identical system (the twin). To start with, we apply the same inputs to both systems until we reach steady state. Then we begin biasing the inputs to the original system in order to increase the rare event hit rate. We continue to apply *unbiased* inputs to the twin system. After we have finished applying the biasing sequence to the original system and obtained the results, we then clear the original system of all its variables and replace it with a copy of the evolved twin system. This allows us to essentially take samples from the stationary distribution of the original system. We should note that successive simulation runs are no longer i.i.d. in this setting (which as we show will effect our simulation diagnostic calculations of the confidence intervals). Happily, however, the importance sampling estimates will still be unbiased.

To make this process a bit more clear, let's write down the relevant estimators. We first consider a simple Monte Carlo estimate. To help with the notation, define the semi-infinite sequence

$$\begin{aligned}\mathcal{O}_j = \dots, &X_{l-m-1+(j-1)}, Z_{p,n,l-m-1-(j-1)},\\ &X_{l-m+(j-1)}, Z_{p,n,l-m+(j-1)}, X_{l-m+1+(j-1)}, Z_{p,n,l-m+1+(j-1)}, \dots,\\ &X_{l+(j-1)}, Z_{p,n,l+(j-1)},\end{aligned}$$

and the direct Monte Carlo estimate

$$\hat{\rho}_{MC}(k) = \frac{1}{k}\sum_{j=1}^{k} 1_{\{f(h(\mathcal{O}_j))\in E\}}.$$

Note that now for each "simulation run" j, we are not generating a new set of i.i.d. random variables. In reality, there is only one simulation run and we are just taking a time average over the output values to obtain the expectation. We are simply replacing the "ensemble averages" with "time averages." This can be done here due to the assumed ergodicity of the system. Also, directly due to the ergodicity, we have

$$\lim_{k\to\infty} \hat{\rho}_{MC}(k) = \rho.$$

Due to assumed stationarity (the system is in steady state) and to the ergodicity, we also have

$$\mathbb{E}[\hat{\rho}_{MC}(k)] = \rho.$$

There are some subtleties that we must consider when modifying our Monte Carlo estimator to create the importance sampling estimator. The idea is that we are going to start biasing at time $l-m$ and try to hit the event at time l. The question becomes: How do we "load" the state from the twin system into the original system after this biasing sequence? Do we load the state values of the twin system at time l into the original system and then begin another biasing sequence? The answer to this last question in general is no. For a random sequence to be ergodic means that we can time average (one sample at a time) over the sequence and obtain in the limit the statistical expectation with respect to the stationary distribution. It does *not* mean that we can average by taking every other or every third or every m samples and obtain the desired expectation. An ergodic sequence in general is not m-ergodic. Thus the answer to how do we load the state from the twin system into the original system is that we load the state of the twin system at time $l-m+1$ and then begin another biasing sequence. Thus, after we begin biasing, the original system will increment m time units while the twin system will only increment one. Defining

$$\begin{aligned}\mathcal{B}_j = {} & \ldots, X_{l-m-1+(j-1)}, Z_{p,n,l-m-1+(j-1)}, \\ & X_{l-m+(j-1)}, Z^{(j)}_{q,n,0}, X_{l-m+1+(j-1)}, Z^{(j)}_{q,n,1}, \ldots X_{l+(j-1)}, Z^{(j)}_{q,n,m}\end{aligned}$$

mathematically, our importance sampling estimator may be written as

$$\hat{\rho}_{IS}(k) = \frac{1}{k}\sum_{j=1}^{k} 1_{\{f(h(\mathcal{B}_j))\in E\}} \prod_{r=0}^{m} \frac{dP_n}{dQ_n}(Z^{(j)}_{q,n,r}).$$

We note here that we have assumed that the random variables $\{Z^{(j)}_{q,n,r}\}$ and $\{Z_{p,n,r}\}$ for $r = 0, 1, \ldots, l-m$ are independent. We could easily modify the weight function above to allow for dependency among these random variables. We also assume that $\{Z_{q,n,r}^{(j)}\}$ is an independent sequence for $j = 1, 2, \ldots, k$, which is the usual assumption for how we generate the simulation random variables. Recall that the letter j is keeping track of the simulation run with which we are dealing.

Again due to the assumed ergodicity and stationarity,

$$\lim_{k \to \infty} \hat{\rho}_{IS}(k) = \rho,$$

and

$$\mathbb{E}[\hat{\rho}_{IS}(k)] = \rho.$$

We still have not discussed how we are going to find the important points of the error set in order to specify the simulation distributions. Very often we will now wish to use the method of conditioning and condition on a certain symbol sequence. For example, if we are biasing over $m + 1$ time units, we can condition on a set sequence of data symbols transmitted over those time units. If there are l data symbols possible to transmit at each time unit, we then would have l^{m+1} conditional subproblems to simulate. We hope that each of these subproblems can be simulated more efficiently than the global problem.

14.5.1 Simulation Diagnostics

Constructing confidence intervals is a little bit more complicated in this setting due to the fact that the output of each simulation run is *not* independent of the previous output. Our estimate is consistent (i.e., as $k \to \infty$ the estimate converges to the true value) not because of an i.i.d. law of large numbers but rather due to an assumed ergodicity relationship. (We are being a bit vague here on exactly what type of ergodicity we are assuming. It would be sufficient for our purposes to think of the system as being ergodic in the mean.)

Under a vast array of assumptions, $\hat{\rho}(k)$ suitably scaled and shifted will converge in distribution to a Gaussian law. We already know that the estimate is unbiased. Hence, to construct the confidence intervals, we need only to estimate the variance. Previously, we estimated the variance of the individual summands composing $\hat{\rho}(k)$ using $\hat{F}_n(k)$. Due to the fact that each summand, the result of each simulation run, was independent, we knew that the variance of the estimate was given by the variance of one of the summands divided by k, the number of simulation runs. In our current setting, we can no longer proceed in this fashion. Our simulation outputs are not independent.

The usual method to get around this problem is via "block" processing of our estimates (in the statistics literature this is called "batching"). Let $k = Kb$, where b, a positive integer, is the "blocksize" and K, also a positive integer, is the number of blocks. We can then form our estimates in the following way

$$\begin{aligned}\hat{\rho}_{IS}(k) &= \frac{1}{k}\sum_{j=1}^{k} 1_{\{f(h(\mathcal{B}_j))\in E\}} \prod_{r=0}^{m} \frac{dP_n}{dQ_n}(Z_{q,n,r}^{(j)}) \\ &= \frac{1}{K}\sum_{j=1}^{K} \frac{1}{b}\sum_{j'=0}^{b-1} 1_{\{f(h(\mathcal{B}_{(j-1)b+j'+1}))\in E\}} \prod_{r=0}^{m} \frac{dP_n}{dQ_n}(Z_{q,n,r}^{((j-1)b+j'+1)}) \\ &= \frac{1}{K}\sum_{j=1}^{K} \hat{\rho}_b(j).\end{aligned}$$

Now the idea is that for most systems of engineering interest, the dependence between any two outputs of the systems will decay away as that time distance between those outputs grows. Thus for sufficiently large values of the blocksize b, we can think of the successive values of $\{\hat{\rho}_b(j)\}$ as being statistically independent. We can then construct confidence intervals for $\hat{\rho}(k)$ on the basis that it is composed of the sample average of K independent replicas.

Thus define

$$\hat{\hat{F}}_n(K) = \frac{1}{K}\sum_{j=1}^{K} \hat{\rho}_b(j)^2$$

and mimicking our derivation of Equation (4.6), we find approximately that the number of simulation blocks of blocksize b needed to obtain x percent error with probability y is

$$K^*(K) = \left(\frac{t_y 100}{x}\right)^2 \left(\frac{\hat{\hat{F}}_n(K)}{\hat{\rho}_n(k)^2} - 1\right)$$

with our simulation stopping criteria being the following.

Block Sequential Stopping Criterion: Stop after K blocks of simulation runs if

$$K \geq K^*(K).$$

Depending on the situation, analytically determining a good value for b can be quite difficult. One common rule of thumb is to choose b sufficiently large so that the lag-one correlation of the sequence of batch means $\{\hat{\rho}_b(j)\}$ is less than .05. Happily in many practical settings, the amount of the memory that we need to simulate for the system is more or less obvious.

A. Convex Functions and Analysis

Convex analysis is the study of the relationships and interrelationships between convex sets and convex functions. It differs from classical analysis in that differentiability assumptions are replaced by convexity assumptions. In this appendix we collect in one spot the principal definitions and theorems that are called for in the text.

A subset C of $\mathcal{R}^d$ is said to be *convex* if $(1-\lambda)x + \lambda y \in C$ whenever $x \in C$, $y \in C$, and $0 < \lambda < 1$.

Let f be a function from a convex set C to $(-\infty, \infty]$. Then f is *convex* on C if

$$f((1-\lambda)x + \lambda y) \leq (1-\lambda)f(x) + \lambda f(y).$$

The function f is *strictly convex* from a convex set C to $(-\infty, \infty]$ if the following inequality

$$f((1-\lambda)x + \lambda y) < (1-\lambda)f(x) + \lambda f(y)$$

holds for all x, y in C.

The *effective domain* f, is defined as

$$D_f = \{x : f(x) < \infty\}.$$

A convex function f is said to be *proper* if $f(x) < \infty$ for at least one x and $f(x) > -\infty$ for every x. A convex function that is not proper is *improper*.

An extended $\mathcal{R}$-valued function f given on a set $S \subset \mathcal{R}^d$ is said be *lower semi-continuous* at a point x of S if

$$f(x) \leq \lim_{i \to \infty} f(x_i)$$

for every sequence $\{x_i\}$ in S such that x_i converges to x and the limit exists in $[-\infty, \infty]$.

Theorem A.0.1. *Let f be an arbitrary function from $\mathcal{R}^d$ to $[-\infty, \infty]$. Then f is lower semi-continuous throughout $\mathcal{R}^d$ if and only if $\{x : f(x) \leq \alpha\}$ is a closed set for every $\alpha \in \mathcal{R}$.*

Given any function f on $\mathcal{R}^d$, there exists a greatest lower semi-continuous function (not necesarily finite) majorized by f. This function is called the *lower semi-continuous hull* of f.

The *closure* of a convex function is defined to be the lower semi-continuous hull of f if f nowhere has the value $-\infty$ and is defined to be the constant function $-\infty$ if f is an improper convex function such that $f(x) = -\infty$ for some x. Either way, the closure of a convex function is another convex function denoted $\text{cl}f$. A convex function is said to be *closed* if $\text{cl}f = f$. For a proper convex function, closedness is thus the same as lower semi-continuity.

The main result about convergence of convex functions is given below. The surprising result is that the convergence of a sequence of convex functions implies that the convergence is uniform on compact subsets.

Theorem A.0.2. *Let C be an open set in $\mathcal{R}^d$ and suppose $\{f_i\}$ is a sequence of finite convex functions converging to a finite function f on C. Then the limit function f is convex and furthermore the convergence is uniform on every closed and bounded subset of C.*

Let f be any closed convex function on $\mathcal{R}^d$. Define the function f^* as

$$f^*(x^*) = \sup_x[\langle x, x^* \rangle - f(x)].$$

f^* is called the *conjugate* of f. This function is another closed convex function and in fact $(f^*)^* = f$.

A vector x^* is said be a *subgradient* of a convex function f at a point x if

$$f(z) \geq f(x) + \langle x^*, z - x \rangle, \quad \forall\ z.$$

The set of all subgradients of f at x is called the *subdifferential* of f at x and is denoted by $\partial f(x)$. The multivalued mapping $\partial f : x \to \partial f(x)$ is called the *subdifferential of f*. In general $\partial f(x)$ may be empty or it may just consist of one vector. If $\partial f(x)$ is not empty, f is said to be *subdifferentiable* at x.

Theorem A.0.3. *For any proper closed convex function f and any vector x, the following conditions on a vector x^* are equivalent to each other:*

a) $x^* \in \partial f(x)$;
b) $\langle z, x^* \rangle - f(x)$ *achieves its supremum in z at $z = x$;*
c) $f(x) + f^*(x^*) \leq \langle x, x^* \rangle$;
d) $f(x) + f^*(x^*) = \langle x, x^* \rangle$;
e) $x \in \partial f^*(x^*)$;
f) $\langle x, z^* \rangle - f^*(z^*)$ *achieves its supremum in z^* at $z^* = x^*$.*

Theorem A.0.4. *Let f be a convex function, and let x be a point where f is finite. If f is differentiable at x, then $\nabla f(x)$ is the unique subgradient of f at x. Conversely, if f has a unique subgradient at x, then f is differentiable at x.*

Theorem A.0.5. *Let f be a proper convex function on $\mathcal{R}^d$, and let S be the set of points where f is differentiable. Then S is a dense subset of $\overset{\circ}{D}_f$, and its complement in the $\overset{\circ}{D}_f$ is a set of measure zero. Furthermore, the gradient mapping $\nabla f : x \to \nabla f(x)$ is continuous on S.*

Theorem A.0.6. *Let C be an open convex set, and let f be a convex function that is finite and differentiable on C. Let $f_1, f_2, \ldots,$ be a sequence of convex functions finite and differentiable on C such that $\lim_n f_n(x) = f(x)$. Then*

$$\lim_{n \to \infty} \nabla f_n(x) = \nabla f(x), \quad \forall\, x \in C.$$

Furthermore, the mappings ∇f_n converge to ∇f uniformly on every compact subset of C.

A proper convex function f is *essentially smooth* if it satisfies the following conditions.

a) $\overset{\circ}{D}_f$ is not empty;

b) f is differentiable throughout $\overset{\circ}{D}_f$;

c) $\lim_{i \to \infty} |\nabla f(x_i)| = \infty$ whenever $\{x_i\}$ is a sequence in $\overset{\circ}{D}_f$ converging to a boundary point of $\overset{\circ}{D}_f$.

Theorem A.0.7. *Let f be a closed proper convex function. Then ∂f is a single-valued mapping if and only if f is essentially smooth. In this case, ∂f reduces to the gradient mapping ∇f; that is, $\partial f(x)$ consists of the vector $\nabla f(x)$ alone when $x \in \overset{\circ}{D}_f$, while $\partial f(x) = \emptyset$ when $x \notin \overset{\circ}{D}_f$.*

Theorem A.0.8. *Let f be a closed proper convex function. The ∂f is a one-to-one mapping if and only if f is strictly convex on $\overset{\circ}{D}_f$ and essentially smooth.*

Suppose f is a differentiable real-valued function on an open subset C of $\mathcal{R}^d$. Let D be the image of C under the gradient mapping ∇f. The *Legendre conjugate* of the pair (C, f) is defined to be the pair (D, g), where g is the function on D given by the formula:

$$g(x^*) = \langle (\nabla f)^{-1}(x^*), x^* \rangle - f((\nabla f)^{-1}(x^*)).$$

For g to be well defined (i.e., single-valued) it is sufficient (but not necessary) that ∇f be one-to-one on C. The function g here is important because it is by this formula that we usually try to compute the conjugate of a convex function.

Theorem A.0.9. *Let f be any closed proper convex function differentiable on $\overset{\circ}{D}_f$ which we assume to be nonempty. The Legendre conjugate (D, g) of $(\overset{\circ}{D}_f, f)$ is then well defined. Morever, D is a subset of D_{f^*} (namely, the range of ∇f), and g is the restriction of f^* to D.*

Theorem A.0.10. *Suppose f is closed, proper, essentially smooth, and strictly convex on $\overset{\circ}{D}_f$ which is assumed to be nonempty. Then $(\overset{\circ}{D}_{f^*}, f^*)$ is the Legendre conjugate of $(\overset{\circ}{D}_f, f)$ and $(\overset{\circ}{D}_f, f)$ is the Legendre conjugate of $(\overset{\circ}{D}_{f^*}, f^*)$. The gradient mapping ∇f is then one-to-one from the open convex set $\overset{\circ}{D}_f$ onto the open convex set $\overset{\circ}{D}_{f^*}$, continuous in both directions, and $\nabla f^* = (\nabla f)^{-1}$.*

B. A Covering Lemma

Given a nonzero point $\theta \in \mathcal{R}^d$ and a real number α, define the closed half-space $H_+(\theta, \alpha) = \{z \in \mathcal{R}^d : \langle \theta, z \rangle - c(\theta) \geq \alpha\}$ and the opposite closed half-space as $H_-(\theta, \alpha) = \{z \in \mathcal{R}^d : \langle \theta, z \rangle - c(\theta) \leq \alpha\}$.

Theorem B.0.11. *Assume the standard assumptions A1 and A2. Let K be a nonempty closed set in $\mathcal{R}^d$. Then if $0 < I(K) < \infty$, then for any positive ϵ, there exist finitely many nonzero points $\theta_1, \ldots, \theta_r$ such that $K \subseteq \cup_{i=1}^r H_+(\theta_i, I(K) - \epsilon)$.*

Proof. Define $A = \{z : I(z) \leq I(K) - \epsilon\}$, where $0 < \epsilon < I(K)$. A is nonempty, closed, and bounded and is disjoint from K. (We know that A is compact due to Lemma 3.2.3.) Let B be a closed ball containing A such that the boundary of B (∂B) and A are disjoint. Define $C = (K \cap B) \cup \partial B$.

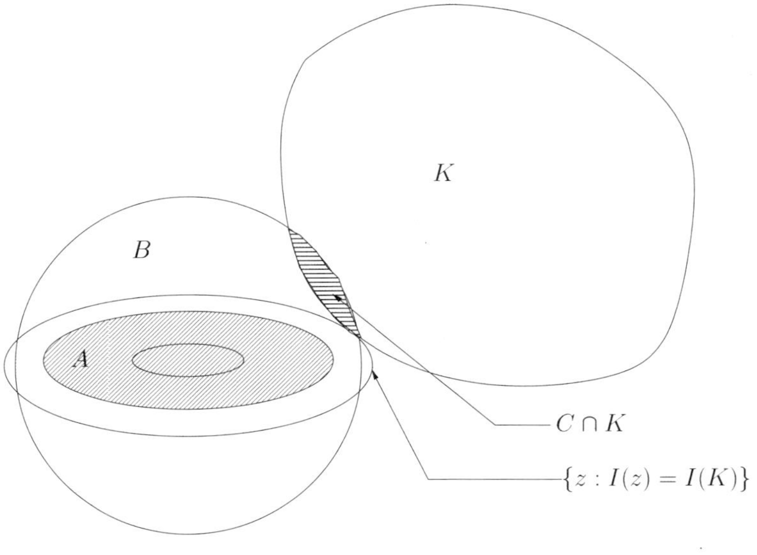

Fig. B.1. A sketch of the set relationships.

Since 0 is in the interior of the domain of c, $c(\theta)$ is continuous at 0. Thus $I(z) = \sup_\theta[\langle\theta, z\rangle - c(\theta)] = \sup_{\theta\neq 0}[\langle\theta, z\rangle - c(\theta)]$. So $A = \{z : I(z) \leq I(K) - \epsilon\}$ $= \{z : \sup_{\theta\neq 0}[\langle\theta, z\rangle - c(\theta)] \leq I(K) - \epsilon\} = \cap_{\theta\neq 0} H_-(\theta, I(K) - \epsilon)$. Therefore $A^c = \cup_{\theta\neq 0} H_+(\theta, I(K) - \epsilon)^o$. Now C is a closed set contained in A^c and we have demonstrated an open cover for it. Hence by the Heine–Borel theorem there must exist a finite subcover. Therefore, there exists $\theta_1, \ldots, \theta_r$ such that $C \subset \cup_{i=1}^r H_+(\theta_i, I(K) - \epsilon)^o \subset U = \cup_{i=1}^r H_+(\theta_i, I(K) - \epsilon)$.

Now K is a subset of $C \cup B^c$. Consider any point in B^c. Suppose that it doesn't lie in U. Hence it lies in $I = \cap_{i=1}^r H_-(\theta_i, I(K) - \epsilon)^o$. Note that I is convex since it is the intersection of convex sets. Note also that $A \subset I$. Now pick any point $y \in A$. Then the set $(y, x] = \{z : z = \lambda y + (1-\lambda)x, 0 \leq \lambda \leq 1\}$ must lie in I. Recall that B is a closed ball containing A and whose boundary is disjoint from A. Since $x \in B^c$ and $y \in A$, this implies that the set $(y, x]$ intersects ∂B at some point b. Hence $b \in I$. But $\partial B \subset C \subset U$, which is a contradiction since I and U are complements of each other. The proof is complete. □

C. Pseudo-Random Number Generator Programs

Programs in C

```
#include <stdio.h>
#define znew   (z=36969*(z&65535)+(z>>16))
#define wnew   (w=18000*(w&65535)+(w>>16))
#define MWC    ((znew<<16)+wnew )
#define SHR3  (jsr^=(jsr<<17), jsr^=(jsr>>13), jsr^=(jsr<<5))
#define CONG   (jcong=69069*jcong+1234567)
#define FIB   ((b=a+b),(a=b-a))
#define KISS  ((MWC^CONG)+SHR3)
#define LFIB4 (c++,t[c]=t[c]+t[UC(c+58)]+t[UC(c+119)]+t[UC(c+178)])
#define SWB   (c++,bro=(x<y),t[c]=(x=t[UC(c+34)])-(y=t[UC(c+19)]+bro))
#define UNI   (KISS*2.328306e-10)
#define VNI   ((long) KISS)*4.656613e-10
#define UC    (unsigned char)  /*a cast operation*/
typedef unsigned long UL;

/*  Global static variables: */
 static UL z=362436069, w=521288629, jsr=123456789, jcong=380116160;
 static UL a=224466889, b=7584631, t[256];
/* Use random seeds to reset z,w,jsr,jcong,a,b, and the table t[256]*/

 static UL x=0,y=0,bro; static unsigned char c=0;

/* Example procedure to set the table, using KISS: */
 void settable(UL i1,UL i2,UL i3,UL i4,UL i5, UL i6)
 { int i; z=i1;w=i2,jsr=i3; jcong=i4; a=i5; b=i6;
 for(i=0;i<256;i=i+1)  t[i]=KISS;
 }

/* This is a test main program.  It should compile
   and print 7  0s. */
int main(void){
int i; UL k;
settable(12345,65435,34221,12345,9983651,95746118);

for(i=1;i<1000001;i++){k=LFIB4;} printf(``%u\n'', k-1064612766U);
for(i=1;i<1000001;i++){k=SWB  ;} printf(``%u\n'', k- 627749721U);
for(i=1;i<1000001;i++){k=KISS ;} printf(``%u\n'', k-1372460312U);
for(i=1;i<1000001;i++){k=CONG ;} printf(``%u\n'', k-1529210297U);
for(i=1;i<1000001;i++){k=SHR3 ;} printf(``%u\n'', k-2642725982U);
for(i=1;i<1000001;i++){k=MWC  ;} printf(``%u\n'', k- 904977562U);
```

```
for(i=1;i<1000001;i++){k=FIB  ;} printf(``%u\n'', k-3519793928U);
              }
/*-----------------------------------------------------------
   Write your own calling program and try one or more of
   the above, singly or in combination, when you run a
   simulation. You may want to change the simple 1-letter
   names, to avoid conflict with your own choices.         */
```

Programs in Fortran

A Fortran Version of KISS

```
      function kiss()
c The  KISS (Keep It Simple Stupid) random number generator. Combines:
c (1) The congruential generator x(n)=69069*x(n-1)+1327217885, period
2^32.
c (2) A 3-shift shift-register generator, period 2^32-1,
c (3) Two 16-bit multiply-with-carry generators, period
597273182964842497>2^59
c  Overall period>2^123;  Default seeds x,y,z,w.
c  Set your own seeds with statement i=kisset(ix,iy,iz,iw).
      integer x,y,z,w
      data x,y,z,w/123456789,362436069,521288629,916191069/
      m(k,n)=ieor(k,ishft(k,n))
      x=69069*x+1327217885
      y=m(m(m(y,13),-17),5)
      z=18000*iand(z,65535)+ishft(z,-16)
      w=30903*iand(w,65535)+ishft(w,-16)
      kiss=x+y+ishft(z,16)+w
      return
      entry kisset(ix,iy,iz,iw)
      x=ix
      y=iy
      z=iz
      w=iw
      kisset=1
      return
      end
```

Programs in MATLAB:

```
%Congruential generator:  This program generates
%500 variates starting from initial value given
%by the rand function of MatLab

CONG = zeros(500,1);
CONG(1,1) = round(rand(1)*2^32);
for i=2:500
    CONG(i,1) = mod(69069*CONG((i-1),1)+1234567,2^32);
end

%Multiply-with-carry generator:  This program generates
%500 variates starting from two initial values given by
%the rand function of MatLab
```

```
MWC = zeros(500,1);
Z = round(rand(1)*2^32);
W = round(rand(1)*2^32);
for i=1:500
    Z = 36969*bitand(Z,65535)+bitshift(Z,-16);
    W = 18000*bitand(W,65535)+bitshift(W,-16);
    MWC(i,1) = mod(bitshift(Z,16,32)+W,2^32);
end


%KISS genertor:  We use the 500 values generated with the above
%congruential and multiply-with-carry generators along with 500
%new values generated by a shift register generator to give us
%500 values from KISS.

KISS = zeros(500,1);
SHR3 = zeros(500,1);
JSR = round(rand(1)*2^32);
for i=1:500
    JSR = bitxor(bitshift(JSR,17,32),JSR);
    JSR = bitxor(bitshift(JSR,-13),JSR);
    JSR = bitxor(bitshift(JSR,5,32),JSR);
    SHR3(i,1) = JSR;
end
KISS = mod(bitxor(MWC,CONG)+SHR3,2^32);
```

References

1. W. A. Al-Qaq, M. Devetsikiotis, and J. K. Townsend, "Importance sampling methodologies for simulation of communication systems with time-varying channels and adaptive filters," *IEEE Journal on Selected Areas of Communications*, Vol. 11, No. 3, pp. 317–326, April 1993.
2. M. Abramowitz and I. Stegun, *Handbook of Mathematical Functions*, ninth printing, Dover, New York, 1970.
3. S. Andradottir and D.P. Heyman, "On the choice of alternative measures in importance sampling with Markov-chains," *Operations Research*, Vol 43, No. 3, pp. 509–519, 1995.
4. R.K. Bahr and J.A. Bucklew, "Quick simulation of detector error probabilities in the presence of memory and nonlinearity," *IEEE Transactions on Communications*, Vol. 41, No. 11, November 1993.
5. R.R. Bahadur, "A note on the fundamental identity of sequential analysis," *Annals of Mathematical Statistics*, Vol. 29, pp. 534–543, 1958.
6. R. Bahadur and R.R. Rao, "On deviations of the sample mean," *Annals of Mathematical Statistics,* Vol. 31, pp. 1015–1027, 1960.
7. J.J. Benedetto, *Real Variable and Integration*, B.G. Teubner, Stuttgart, 1976.
8. P. Billingsley, *Convergence of Probability Measures*, Wiley, New York, 1968.
9. P. Billingsley, *Probability and Measure*, Wiley, New York, 1979.
10. N. Bleistien and R.A. Handelsman, *Asymptotic Expansion of Integrals*, Dover, New York, 1986.
11. P. Brémaud, *Markov Chains – Gibbs Fields, Monte Carlo Simulation, and Queues*, Springer-Verlag, New York, 1999.
12. J.A. Bucklew, *Large Deviation Techniques in Decision, Simulation, and Estimation*, Wiley-Interscience, Applied Probability and Statistics Series, New York, 1990.
13. J.A. Bucklew, P. Ney, and J.S. Sadowsky, "Monte Carlo simulation and large deviations theory for uniformly recurrent Markov chains," *Journal of Applied Probability*, Vol. 27, pp. 44–59, March 1990.
14. J.A. Bucklew and J.A. Gubner, "Input versus output analysis in the importance sampling Monte Carlo simulation of systems," *IEEE Transactions on Signal Processing*, Vol. 51, No. 1, pp. 152–159, January 2003.
15. J.A. Bucklew and R. Radeke, "On the Monte Carlo simulation of digital communication systems in Gaussian noise," *IEEE Transactions on Communications*, Vol. 51, No. 2, pp. 267–274, Febuary 2003.
16. M. Cottrell, J. Fort, and G. Malgouyres, "Large deviations and rare events in the study of stochastic algorithms," *IEEE Transactions on Automatic Control*, Vol. 28, pp. 907–920, 1983.
17. M.A. Crane and D.L. Iglehart, "Simulating stable stochastic systems III: Regenerative processes and discrete event simulation," *Operations Research* 23, 33–45.

18. H. Cramér, Sur un nouveaux theorème-limite de la théorie des probabilités. Actualités Scientifiques et Industrielles 736, pp. 5–23, Colloque consacré à la théorie des probabilités, Vol. 3. Hermann, Paris, October 1938.
19. H. Chernoff, "A measure of asymptotic efficiency for tests of a hypothesis based on the sum of observations," *Annals of Mathematical Statistics,* Vol. 23, pp. 493–507, 1952.
20. K.L. Chung, *A Course in Probability Theory,* 2nd ed. Academic, New York, 1974.
21. A. Dembo and O. Zeitouni, *Large Deviations Techniques and Applications*, 2nd ed. Springer-Verlag, New York, 1998.
22. L. Devroye, *Non-Uniform Random Variate Generation.* Springer-Verlag, New York, 1986.
23. M.D. Donsker and S.R.S. Varadhan, "Asymptotic evaluation of certain Markov process expectations for large time, II," *Communications in Pure and Applied Mathematics*, Vol. 29, pp. 279–301, 1976.
24. R. Ellis, "Large deviations for a general class of random vectors," *Annals of Probability,* Vol. 12, No. 1, pp. 1–12, 1984.
25. R. Ellis, *Entropy, Large Deviations, and Statistical Mechanics*, Springer-Verlag, New York, 1985.
26. A. Erdelyi (ed.), *Tables of Integral Transforms*, McGraw-Hill, New York, 1954.
27. K.-T. Fang, S. Kotz, and K.-W. Ng, *Symmetric Multivariate and Related Distributions*, Monographs on Statistics and Applied Probability No. 36, Chapman and Hall, New York, 1990.
28. W. Feller, *An Introduction to Probability Theory and its Applications, Vol. I*, 3rd ed. Wiley, New York, 1968.
29. W. Feller, *An Introduction to Probability Theory and Its Applications, Vol. II*, 2nd ed. Wiley, New York, 1971.
30. R. Gallagher, *Information Theory and Reliable Communication*, Wiley, New York, 1968.
31. J. Gärtner, "On large deviations from the invariant measure," *Theory of Probability and its Application,* Vol. 22, pp. 24 –39, 1977.
32. S. Geman and D. Geman, "Stochastic relaxation, Gibbs distributions and the Bayesian restoration of images. *IEEE Transactions on Pattern Analysis and Machine Intelligence*, No. 6, pp. 721–741.
33. P. Glasserman and Y. Wang, "Counter examples in importance sampling for large deviation probabilities, *Annals of Applied Probability*, Vol. 7, pp. 731-746.
34. I. Gradshteyn and I. Ryzhik, *Table of Integrals, Series, and Products*, 4th ed. Academic, New York, 1965.
35. R.M. Gray, "On the asymptotic eigenvalue distribution of Toeplitz matrices," *IEEE Transactions on Information Theory,* Vol. IT-18, pp. 767–800, 1972.
36. G. Grimmett and D. Stirzaker, *Probability and Random Processes*, 3rd ed. Oxford University Press, Oxford, England, 2001.
37. P. Hahn and M. Jeruchim, "Developments in the theory and application of importance sampling," *IEEE Transactions on Communications*, Vol. COM-35, No. 7, pp. 706–714, July 1987.
38. P. Heidelberger, "Fast simulation of rare events in queueing and reliability models," *ACM Transactions on Modeling and Computer Simulation*, Vol. 5, No. 1, pp. 43–85, 1995.
39. P. Hellekalek, "Good random number generators are (not so) easy to find," *Mathematics and Computers in Simulation*, Vol. 46, pp. 485–505, 1998.
40. V. Hunkel and J. A. Bucklew, "Fast simulation for functionals of Markov chains," *Proceedings of the 22nd Annual Conference on Information Sciences and Systems*, Princeton University, 1988.

41. M.C. Jeruchim, P. Balaban, and K.S. Shanmugan, *Simulation of Communications Systems: Modeling, Methodology, and Techniques*, 2nd ed. Kluwer Academic/Plenum, New York, 2000.
42. N.L. Johnson and S. Kotz, *Continuous Univariate Distributions–I*, Wiley, New York, 1970.
43. D. L. Iglehart and G. S. Shedler, *Regenerative Simulation of Response Times in Networks of Queues*, Springer-Verlag, New York, 1980.
44. M. Iltis, "Sharp asymptotics of large deviations in $\mathcal{R}^d$". *Journal of Theoretical Probability*, Vol. 8, No. 3, pp. 501–522, July 1995.
45. M. Iltis, "Sharp large deviations in $\mathcal{R}^d$ for general state space Markov-additive chains," *Statistics and Probability Letters*, Vol. 47, No. 4, 365–380, May 2000.
46. I. Iscoe, P. Ney, and E. Nummelin, "Large deviations of uniformly recurrent Markov additive processes. *Advances in Applied Mathematics* No. 6, pp. 373–412.
47. N.L. Johnson, S. Kotz, and A.W. Kemp, *Univariate Discrete Distributions*, 2nd ed. Wiley, New York, 1992.
48. J. Kiefer and J. Wolfowitz, "Stochastic estimation of a regression function," *Annals of Mathematical Statistics*, Vol. 23, pp. 462–466, 1952.
49. T. Kuczek and K. N. Crank, "A large-deviation result for regenerative processes," *Journal of Theoretical Probability*, Vol. 4, No. 3, pp. 551–561, 1991.
50. K.B. Letaief and J.S. Sadowsky, "Computing bit-error probabilities for avalanche photodiode receivers by large deviations theory," *IEEE Transactions on Information Theory*, Vol. 38, No. 3, pp. 1162–1169, 1992.
51. K.B. Letaief and J.S. Sadowsky, "New importance sampling methods for simulating sequential decoders," *IEEE Transactions on Information Theory*, Vol. 39, No. 5, pp. 1716–1723, 1993.
52. T. Lehtonen and H. Nyrhinen, "Simulating level crossing probabilities by importance sampling," *Advances in Applied Probability*, Vol. 24, pp. 858–874, 1992.
53. T. Lehtonen and H. Nyrhinen, "On asymptotically efficient simulation of ruin probabilities in a Markovian environment," *Scandinavian Actuarial Journal*, pp. 60–75, 1992.
54. D. Lu and K. Yao, "Improved importance sampling techniques for efficient simulation of digital communication systems," *IEEE Journal on Selected Areas in Communication*, Vol. 6, No. 1, pp. 67–75, January 1988.
55. G. Marsaglia, "Keynote address: A current view of random number generators," *Proceedings, Computer Science and Statistics: 16th Symposium on the Interface*, Elsevier, New York, 1985.
56. G. Marsaglia, "Monkey tests for random number generators," extracted from an article in *Computers and Mathematics with Applications*, 9, 1–10, 1993.
57. G. Marsaglia, "The Marsaglia Random Number CDROM," available from Dept. of Statistics, Florida State University.
58. G. Marsaglia and A. Zaman, "The KISS generator," Tech. Report, Dept. of Statistics, Florida State University.
59. A. Marshall, "The use of multi-stage sampling schemes in Monte Carlo computations," in M. Meyer (ed.), *Symposium on Monte Carlo Methods*, Wiley, New York, pp. 123–140.
60. N. Metropolis, A.W. Rosenbluth, M.N. Rosenbluth, A.H. Teller, and E. Teller, "Equations of state calculations by fast computing machines," *Journal Chemical Physics*, Vol. 21, pp. 1087–1092, 1953.
61. H. Miller, "A convexity property in the theory of random variables on a finite Markov chain," *Annals Mathematical of Statistics,* Vol. 32, pp. 1260–1270, 1961.

62. A. Nádas, "An extension of a theorem of Chow and Robbins on sequential confidence intervals for the mean," *Annals of Mathematical Statistics*, Vol. 40, No. 2, pp. 667–671, 1969.
63. P. Ney, "Dominating points and the asymptotics of large deviations for random walk on $\mathcal{R}^d$," *Annals of Probability*, Vol. 11, No. 1, pp. 158–167.
64. P. Ney and E. Nummelin, "Markov additive processes:I. Eigenvalue properties and limit theorems, II. Large deviations," *Annals of Probability*, Vol. 15, No. 2, pp. 561–609, 1987.
65. S. Parekh and J. Walrand, "A quick simulation method of excessive backlogs in networks of queues," *IEEE Transactions on Automatic Control*, Vol. 34, pp. 54-66.
66. D. Remondo, R. Srinivasan, V.F. Nicola, W.C. van Etten, and H.E.P. Tattje, "Adaptive importance sampling for performance evaluation and parameter optimization of communication systems," *IEEE Transactions on Communications*, Vol. 48, No. 4, pp. 557–565, April 2000.
67. B.D. Ripley, *Stochastic Simulation*, Wiley, New York, 1987.
68. R. T. Rockafellar, *Convex Analysis*, Princeton University Press, Princeton, NJ, 1970.
69. J.S. Sadowsky, "Large deviations and efficient simulation of excessive backlogs in a GI/G/m queue," *IEEE Transactions on Automatic Control*, Vol. AC-36, pp. 1383-1394, 1991.
70. J. S. Sadowsky and R. K. Bahr, "Direct sequence spread spectrum multiple access communications with random signature sequences: A large deviations analysis," *IEEE Transactions on Information Theory*, Vol. IT-37, pp. 514–527, 1991.
71. J.S. Sadowsky and J.A. Bucklew, "On large deviations theory and asymptotically efficient Monte Carlo simulation," *IEEE Transactions on Information Theory*, Vol. 36, No. 3, pp. 579–588, May 1990.
72. H.-J. Schlebusch, "On the asymptotic efficiency of importance sampling techniques," *IEEE Transactions on Information Theory*, Vol. 39, No. 2, pp. 710–715, March 1993.
73. G. S. Shedler, *Regeneration and Networks of Queues*, Springer-Verlag, New York, 1987.
74. D. Siegmund, "Importance sampling in the Monte Carlo study of sequential tests," *Annals of Statistics*, No. 4, pp. 673–684, 1976.
75. P.J. Smith, M. Shafi, and H. Gao, "Quick simulation: A review of importance sampling techniques in communication systems," *IEEE Journal on Selected Areas in Communications*, Vol. 15, No. 4, pp. 597–613, May 1997.
76. P.J. Smith, "Underestimation of rare event probabilities in importance sampling simulations," *Simulation*, Vol. 76, No. 3, pp. 140–150.
77. J.C. Spall, "Multivariate stochastic approximation using a simultaneous perturbation gradient approximation," *IEEE Transactions on Automatic Control*, Vol. 37, pp. 332–341, 1992.
78. R. Srinivasan, "Some results in importance sampling and an application to detection," *Signal Processing*, pt. 1, Vol. 65, pp. 73–88, February 1998.
79. R. Srinivasan, *Importance Sampling–Applications in Communications and Detection*, Springer-Verlag, Berlin, 2002.
80. R.J. Wolfe, M.C. Jeruchim, and P. Hahn, "On optimum and suboptimum biasing procedures for importance sampling in communication simulation," *IEEE Transactions on Communications*, Vol. 38, No. 5, pp. 639–646, May 1990.

Index

Springer Series in Statistics

(continued from p. ii)

Huet/Bouvier/Poursat/Jolivet: Statistical Tools for Nonlinear Regression: A Practical Guide with S-PLUS and R Examples, 2nd edition.
Ibrahim/Chen/Sinha: Bayesian Survival Analysis.
Jolliffe: Principal Component Analysis, 2nd edition.
Knottnerus: Sample Survey Theory: Some Pythagorean Perspectives.
Kolen/Brennan: Test Equating: Methods and Practices.
Kotz/Johnson (Eds.): Breakthroughs in Statistics Volume I.
Kotz/Johnson (Eds.): Breakthroughs in Statistics Volume II.
Kotz/Johnson (Eds.): Breakthroughs in Statistics Volume III.
Küchler/Sørensen: Exponential Families of Stochastic Processes.
Kutoyants: Statistical Influence for Ergodic Diffusion Processes.
Lahiri: Resampling Methods for Dependent Data.
Le Cam: Asymptotic Methods in Statistical Decision Theory.
Le Cam/Yang: Asymptotics in Statistics: Some Basic Concepts, 2nd edition.
Liu: Monte Carlo Strategies in Scientific Computing.
Longford: Models for Uncertainty in Educational Testing.
Manski: Partial Identification of Probability Distributions.
Mielke/Berry: Permutation Methods: A Distance Function Approach.
Pan/Fang: Growth Curve Models and Statistical Diagnostics.
Parzen/Tanabe/Kitagawa: Selected Papers of Hirotugu Akaike.
Politis/Romano/Wolf: Subsampling.
Ramsay/Silverman: Applied Functional Data Analysis: Methods and Case Studies.
Ramsay/Silverman: Functional Data Analysis.
Rao/Toutenburg: Linear Models: Least Squares and Alternatives.
Reinsel: Elements of Multivariate Time Series Analysis, 2nd edition.
Rosenbaum: Observational Studies, 2nd edition.
Rosenblatt: Gaussian and Non-Gaussian Linear Time Series and Random Fields.
Särndal/Swensson/Wretman: Model Assisted Survey Sampling.
Santner/Williams/Notz: The Design and Analysis of Computer Experiments.
Schervish: Theory of Statistics.
Shao/Tu: The Jackknife and Bootstrap.
Simonoff: Smoothing Methods in Statistics.
Singpurwalla and Wilson: Statistical Methods in Software Engineering: Reliability and Risk.
Small: The Statistical Theory of Shape.
Sprott: Statistical Inference in Science.
Stein: Interpolation of Spatial Data: Some Theory for Kriging.
Taniguchi/Kakizawa: Asymptotic Theory of Statistical Inference for Time Series.
Tanner: Tools for Statistical Inference: Methods for the Exploration of Posterior Distributions and Likelihood Functions, 3rd edition.
van der Laan: Unified Methods for Censored Longitudinal Data and Causality.
van der Vaart/Wellner: Weak Convergence and Empirical Processes: With Applications to Statistics.
Verbeke/Molenberghs: Linear Mixed Models for Longitudinal Data.
Weerahandi: Exact Statistical Methods for Data Analysis.
West/Harrison: Bayesian Forecasting and Dynamic Models, 2nd edition.